W9-ACU-258

HAZARDOUS WASTE SITE REMEDIATION

The Engineer's Perspective

HAZARDOUS WASTE SITE REMEDIATION

The Engineer's Perspective

O'Brien & Gere Engineers, Inc.
Syracuse, New York

Robert Bellandi
Technical Editor

VNR VAN NOSTRAND REINHOLD
New York

Library of Congress Catalog Card Number 88-5698
ISBN 0-442-27210-3

Printed in the United States of America

Van Nostrand Reinhold
115 Fifth Avenue
New York, New York 10003

Chapman & Hall
2-6 Boundary Row
London SE1 8HN, England

Thomas Nelson Australia
102 Dodds Street
South Melbourne, Victoria 3205, Australia

Nelson Canada
1120 Birchmount Road
Scarborough, Ontario M1K 5G4, Canada

16 15 14 13 12 11 10 9 8 7 6 5

Library of Congress Cataloging-in-Publication Data

Hazardous waste site remediation.

Includes bibliographies and index.
1. Hazardous waste sites—United States—Management.
2. Hazardous wastes—United States—Safety measures.
3. Hazardous wastes—United States—Management.
I. O'Brien and Gere Engineers, Inc. II. Title:
Remediation.
TD811.5.H437 1988 628.4′4 88-5698
ISBN 0-442-27210-3

Preface

This book addresses the immediate issue of investigating, designing, and implementing technology to remedy the most challenging hazardous waste sites. The perspective presented is that of the practicing engineer, the one who must see that the solutions proposed actually will reduce, to acceptable levels, the risks posed by the site. This book is written specifically for the upper-level college student in a technical program. However, it also is intended to furnish timely and insightful information to the corporate engineer planning or reviewing remediation work, the plant manager concerned with remediation of a hazardous waste site, and the state or federal government official regulating remediation.

The key to successful remediation is education: education of the parties potentially responsible for the wastes in disposal sites, education of the government regulators, and education of the engineers and scientists who must implement the remediation technology. The objective of this book is to further that educational process.

O'Brien & Gere Engineers, Inc. and its affiliated companies OBG Technical Services, Inc. and OBG Laboratories, Inc. have participated in the remediation of hazardous waste sites for nearly a dozen years. The experiences gained in working on over 100 disposal sites—fifteen of which are on the National Priority List—are the basis for this book. Each chapter was written by an officer, manager, or senior staff person who is directly involved in the conduct of hazardous waste site remediation. In most chapters, the implementation of engineering solutions is illustrated with case studies drawn from the authors' field experiences in order to make the information practical.

Two hundred forty million tons of hazardous waste are generated each year by industry in the United States—one ton for each person each year. That figure is startling. Even more startling is the fact that this massive volume of hazardous waste has been generated for years, yet most of it has been neither appropriately treated nor correctly disposed of. The landscape in industrial regions of the country is veritably littered with hazardous waste disposal sites—twenty thousand separate sites according to one reliable count. These disposal sites are not benign. Many of them detrimentally affect natural resources, notably ground water, and many of the sites put receptors, including some human populations, at risk.

The problem of hazardous waste disposal is multifarious: it affects conservation of scarce resources; it can present a danger to public health; it increases corporate liability; and, it presents a major financial cost to the nation. Remediation of an uncontrolled hazardous waste site—correcting the damage manifest by improper hazardous waste disposal—is a challenge. Remediation is an environmental challenge because at least one environmental medium has been contaminated by human activities and must be corrected. It is a technical challenge because every site is unique, and every site harbors many uncertainties; the technology is doubly challenged because the remediation must be completed economically. Ultimately, however, remediation is a social challenge because the process of remediation tests the ability of the institutions of society to reduce the risk of danger to the members of society. Society must respond to minimize the jeopardy of improper hazardous waste disposal, and two measures are mandatory:

- The volume of hazardous waste generated today must be minimized.
- The monstrous legacy of yesterday must be rectified.

To implement the first measure, industry requires incentives to reduce the rate at which it generates hazardous waste. Several approaches rather than a single technique are necessary: changes in production, improved efficiency, recycling, and efficient treatment of waste at the point of generation. It is justifiable for society to demand a reduction by half of hazardous waste generation within ten years. This is an objective that can be realized.

With regard to the second measure, society, functioning through the political process, emphatically has declared that existing uncontrolled hazardous waste sites will be corrected. The Comprehensive Environmental Response, Compensation and Liability Act of 1980 (CERCLA) articulated society's intention to reduce the present danger of improperly disposed hazardous waste, and that resolve was underscored when the Act was reauthorized on October 17, 1986. In economic terms alone, the magnitude of the national effort is noteworthy. A determined Congress and a reluctant president authorized new taxes and spending of $9 billion, a significant portion of the nation's treasury. Private interests will spend several times that amount in remediation efforts. The reauthorized CERCLA also put social muscle behind its rhetoric by enacting stronger civil and criminal sanctions than the original Act carried.

Today, the amendments to CERCLA are beginning to have a profound effect on the remediation of uncontrolled hazardous waste sites. Regrettably, the technology employed to date has focused on stabilizing wastes on the site of original disposal, disposing of wastes with high energy content via off-site incineration, and disposing of wastes low in organic content at secure landfills. Under Section 121 of the 1986 amendments, the preferred remedy will be that "which permanently and significantly reduces the volume, toxicity, or mobility of the hazardous substances, pollutants, and contaminants . . . ." Furthermore, the Act encourages innovative and alternative treatment of hazardous wastes on uncontrolled hazardous waste disposal sites.

Both the urgency of the problem of uncontrolled hazardous waste sites and the force of the 1986 amendments to CERCLA for improved remediation will cause the technology to advance. In situ bioreclamation utilizing organisms with engineered genes, in situ air stripping, catalytic oxidation, continuous soil contaminant extraction, and on-site microwave-, ultraviolet-, and infrared-based soil treatment will serve as tools of the hazardous waste specialist of the 1990s. The field is advancing rapidly, and it is an exciting time to be active professionally in hazardous waste site remediation. The authors look forward to joining with the new generation of engineers and scientists, those who will build upon the content of this book, in cleansing the environment and developing new technologies.

Cornelius B. Murphy, Jr., Ph.D.
Senior Vice President
O'Brien & Gere Engineers, Inc.
Syracuse, New York

Acknowledgments

Any book is the product of diverse minds and hands, but this one is particularly so. The experiences of the O'Brien & Gere organization in solving various site remediation problems for many clients have served as the basis for this text.

This book is really the product of the staff of the O'Brien & Gere family of firms. It benefited from the talents and contributions of many of the officers, managers, and senior staff people of O'Brien & Gere Engineers, Inc., OBG Technical Services, Inc., and OBG Laboratories, Inc.

Several staff members assisted with the research and the development of the text and figures. O'Brien & Gere acknowledges their assistance with gratitude and with pride in their dedication and efforts.

Certain individuals were especially important in writing this book. John R. Loveland, president of O'Brien & Gere, is acknowledged for his aggressive support. Robert Bellandi, of O'Brien & Gere Engineers, served tirelessly in the role of internal technical writer and editor. His organization and encouragement assisted all the contributors in completing their assignments. The book's editors at the Van Nostrand Reinhold Company were persistent and helpful in their advocacy for this book. Without their efforts, it would not have been written. O'Brien & Gere thanks them for their resolute support.

All the contributors to this book extend their thanks to the firm's clients. It was they who provided us the opportunity to refine and develop our skills in the area of remedial technology.

Cornelius B. Murphy, Jr., Ph.D.
O'Brien & Gere Engineers, Inc.

Acknowledgments

Contents

HAZARDOUS WASTE SITE REMEDIATION

The Engineer's Perspective

PART I
ASSESSMENT

Chapter 1
Introduction to Remediation of Hazardous Waste Sites

It is difficult to pinpoint an exact time when the environmental conscience of American society reached the threshold that triggered the environmental movement, a movement that has marked federal legislation for the last two decades. Since the manufacture of synthetic chemicals began, well-intentioned individuals had cautioned against the indiscriminate application and disposal of potentially harmful compounds. However, prior to the enactment of the Comprehensive Environmental Response, Compensation and Liability Act in 1980,[1] commonly known as CERCLA or Superfund,[2] there was no straightforward regulatory or legal mechanism to protect the public interest from the risks created by past chemical disposal practices. CERCLA was the finale of an enlightened environmental policy that sought to protect the water, air, and land resources of the United States.

Perhaps the kindling of the environmental movement, the harbinger of CERCLA, was the publication of Rachel Carson's *Silent Spring* in 1962.[3] This book described the connection between exposure to various chemicals and the onset of life-threatening diseases such as leukemia. When it was printed, most scientists and professionals considered the book an oversimplification of the situation, possibly more fiction than fact. Only in a few knowledgeable quarters was the book given credence. However, the public embraced it, and it was widely read. *Silent Spring* was timely, also, because it was published when many of society's traditional values were being challenged, questioned, and occasionally dismissed. Viewed in the historical perspective, *Silent Spring* has proved uncannily accurate. It should be read by anyone concerned with the environment.

ENVIRONMENTAL PROTECTION: THE NEW WAVE

Whether Carson actually triggered the environmental movement or not is a moot point. The fact is that environmental awareness was one of the many aspects of society irrevocably changed in the 1960s. The awakening social

---

This chapter was developed by Edwin C. Tifft, Jr., Ph.D. of O'Brien & Gere Engineers, Inc.

demands, keenly evidenced by such demonstrations as Earth Day in 1970, showed that society was ready for, and even insisted on, an end to the unregulated and uncontrolled pollution of the environment. This position was espoused not only by conservationists and student groups, but by municipalities and the chemical industry. As a result of widespread support, the U.S. Congress passed a series of environmental laws. Congress began with the National Environmental Policy Act in 1969 (NEPA),[4] a law that requires environmental impact statements for federal actions. Another of the first steps was the formation of the U.S. Environmental Protection Agency (U.S. EPA),[5] which consolidated the functions of several predecessor agencies such as the Federal Water Quality Administration Agency and the U.S. Public Health Service. Through the U.S. EPA, the role of the federal government in environmental management was significantly expanded. The U.S. EPA today is one of the most visible and powerful federal agencies; its 1987 budget was $4.15 billion, and it has a staff of several thousand stationed in its headquarters and throughout its ten administrative regions.

NEPA was quickly followed by stringent measures to protect environmental resources. The initial focus of environmental legislation was the control of ongoing processes such as air emissions, wastewater discharges, and hazardous waste disposal. The harm precipitated by past disposal practices was not fully appreciated until the regulatory programs for ongoing processes were implemented. A brief review of these programs will place CERCLA into perspective.

The first major piece of environmental legislation was the Clean Air Act of 1970.[6] The Act and its subsequent amendments firmly established a program to regulate and control the emission of airborne pollutants. It affected common compounds such as sulfur dioxide as well as a growing list of what are commonly considered hazardous chemicals such as benzene and asbestos.

The next significant piece of legislation was the Water Pollution Control Act of 1972,[7] now commonly called the Clean Water Act. This Act and its subsequent amendments, especially the 1976 amendments, gave the U.S. EPA authority to regulate industrial and municipal discharges into public sewer systems and surface waters. Although the development and enforcement of specific regulations was slow, a controversial issue itself, significant progress has been made to keep pollutants from the water. It appears inevitable that, within a few years, all such discharges will be controlled and reduced to acceptable levels. It is worth noting that the Clean Water Act also provided massive amounts of federal aid, approximately $40 billion, to municipalities for the construction of wastewater conveyance and treatment systems. The public and legislative commitment to a clean water program and a clean environment was demonstrated when Congress overrode President Reagan's veto of the 1986 reauthorization of the Act's Construction Grants Program.[8]

In 1974, Congress passed and the president signed the Safe Drinking Water Act.[9] Under this law, the U.S. EPA is required to establish and to update continuously drinking water standards for public consumption. Although not directly related to pollution control, these standards have gained widespread use as the indicators of acceptable levels of contamination. This Act also contains provisions for the regulation of certain types of underground injection systems. The 1986 amendments to the Safe Drinking Water Act[10] require that the U.S. EPA publish a list of contaminants that occur in public water supplies and that may merit federal regulation; this list is required to mesh with the hazardous substances specified by CERCLA.

By the mid-1970s, it was apparent that solid wastes were being disposed of in a largely unregulated manner. With great fanfare, Congress and the president enacted two major pieces of legislation in 1976 that were meant to close the loop on pollution control and provide "cradle-to-grave" control of all solid wastes, including chemical wastes:

- The Toxic Substances Control Act (TSCA)[11] mandated testing prior to the commercial manufacture of any new chemical and the disclosure of information about its toxicity. It was intended that TSCA prevent the inadvertent or unintentioned distribution of highly toxic materials. Polychlorinated biphenyls (PCBs) as a class were singled out for specific regulation under TSCA because of early knowledge of their toxicity and their nearly ubiquitous distribution. To this day, a distinction exists between the regulation of PCBs and all other hazardous wastes, at least on the federal level.
- The Resource Conservation and Recovery Act (RCRA)[12] was drafted to regulate the generation, transportation, storage, treatment, and disposal of waste materials that met the definition of *hazardous waste*. This law and its regulations and subsequent amendments have been difficult to implement. The first attempt at rule making in 1978 prompted comments from all sectors of society, and the published record stacked to 7 ft (2 m). By 1980, however, a system was essentially in place to regulate the hazardous wastes that were being generated. In 1984, many of the deficiencies of RCRA were remedied by the Hazardous and Solid Waste Amendments (HSWA).[13] These amendments virtually rewrote RCRA by adding many new provisions. Congress showed its will for implementation by asserting that certain measures would become regulations automatically unless the U.S. EPA acted by specified dates. In Subtitle I of HSWA, a new set of regulations was established to deal with underground storage tanks.[14] Federal environmental officials today suspect that a significant amount of hazardous waste is illegally and intentionally disposed of in a manner inconsistent with RCRA regulations. Compliance with RCRA is an area targeted for increasing enforcement.

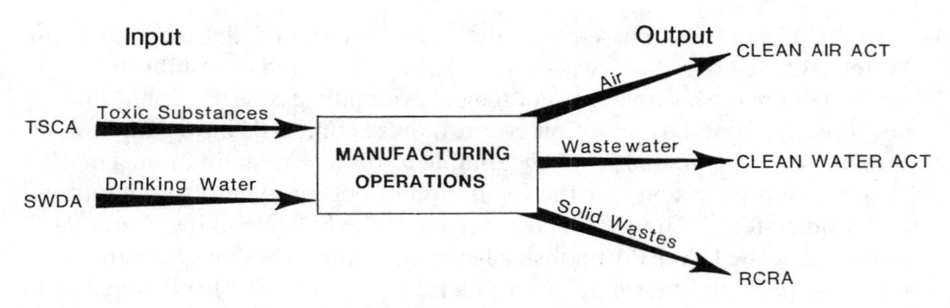

Figure 1-1. Closing the loop of environmental pollution (1970-1976).

Figure 1-1 diagrams the scope of these environmental laws.

In addition, several other significant environmental control programs were enacted:

- The Federal Insecticide, Fungicide, and Rodenticide Act (FIFRA).
- The Occupational Safety and Health Act (OSHA).
- Acts relating to the control of radioactive wastes.
- Various regulations of the U.S. Department of Transportation governing the movement of hazardous wastes.
- Various right-to-know laws, such as those requiring employers to disclose to workers or surrounding communities the hazardous substances used in the workplace.

Federal environmental policy of the 1970s and early 1980s was designed to regulate all facets of pollution, and, notably, marked the first time the federal government had seriously attempted to control pollution on a national, uniform basis. Many of these laws are not directly relevant to CERCLA, yet the intent of all the regulations adopted pursuant to the laws passed by 1976 was to prevent the uncontrolled release of wastes.

Almost all the states have followed the pattern of the federal government; most have enacted a legislative agenda for comprehensive environmental protection and have combined preexisting agencies or established new agencies responsible for pollution control. The state agencies are known by a variety of names such as Department of Environmental Conservation (New York), Department of Natural Resources (Michigan), and Environmental Protection Agency (Illinois). Whatever their names, their functions are similar: they share the responsibility for pollution control with the U.S. EPA in their respective states.

There was, by 1976, a perceptible sigh of relief; finally society had the problem of environmental contamination under control. However, this relief quickly became dismay with the discovery in the late 1970s of numerous abandoned, uncontrolled sites where hazardous wastes had previously been deposited in an unsound manner, and where the wastes were polluting the water, air, earth, and ground water. Perhaps the most notorious of these sites was the Love Canal in Niagara Falls, New York.

Love Canal was a shock to the environmental consciousness of Americans. In the summer of 1978, the public learned that hundreds of tons of chemical waste, waste later determined to contain a variety of highly toxic substances including dioxins, had been deposited in an abandoned canal in the 1940s and early 1950s. Although this location was not an unreasonable site, at least by the standards of practice that prevailed at the time, subsequent land development in the area for housing and an elementary school led to the release of the chemicals from the original disposal site and exposure of the residents in the area to the chemical wastes. Love Canal was the first site to gain extensive publicity, partially because of its proximity to the large, urban population of Niagara Falls and Buffalo, New York. At the same time, comparable sites were discovered and reported by the national media: the Valley of the Drums in Kentucky, the Stringfellow Acid Pits in California, the Seymour Recycling Facility in Indiana, the Pollution Abatement Services Site in New York, and many others.

The initial perception, subsequently verified, was that there might be hundreds if not thousands of such sites throughout the United States, posing an unacceptable risk to society. The regulatory agencies responsible for pollution control quickly found that none of the existing laws provided a framework or mechanism for regulating abandoned hazardous waste disposal sites. In fact, unless a site could be shown to pose a direct, immediate threat to the public health, there was little that anyone could do other than commence a civil lawsuit against the alleged responsible parties, if those parties even were known. In cases of true public health emergencies, the authorities did indeed act to avoid immediate problems. Nonetheless, hundreds of sites, which posed future dangers to the environment and to human populations because of long-term gradual releases of hazardous substances, could not be remedied. It became clear that for many of the sites, there was no identifiable responsible party who could be pursued, even if a means of pursuit existed. The parties were unknown or perhaps bankrupt; they were sometimes corporations no longer in existence or parties otherwise unavailable to perform remediation. Intense public frustration developed when the public and environmental regulatory agencies became aware that the pollution loop had not been closed.

Congress responded to this alarming situation by enacting the Comprehensive Environmental Response, Compensation and Liability Act in December 1980. CERCLA was designed to resolve all the issues associated with abandoned, uncontrolled, inactive hazardous waste disposal sites. The program is still in its infancy and is beset with controversy—primarily over the mode of funding—but it is evident that its original objective is being satisfied: several thousand sites have been identified for remediation, a mechanism is in place to address the risks of these sites, studies have commenced at several hundred sites, and remedial action has been completed at some of these sites.

CERCLA was initially authorized for a five-year period. There is considerable disagreement as to the effectiveness of the CERCLA program during its first years, an issue that has been addressed by several commentators.[15] In 1985 and 1986, the exchange between Congress and the administration over reauthorization was turbulent and raged for months. The debate culminated in the enactment of the Superfund Amendments and Reauthorization Act of 1986 (SARA).[16] Today, there truly are emerging mechanisms to regulate and control all aspects of environmental pollution. (See how Figure 1-1 has been modified in Figure 1-2.)

Before we proceed, it is worthwhile to review briefly the usage of the word "hazardous" as it applies to different environmental programs. A considerable amount of confusion and misunderstanding is apparent in the popular use of the term, but a clear understanding is necessary for the purpose of this book.

- *Hazardous waste:* This term applies strictly to those substances that exhibit the characteristics defined in RCRA and its regulations.[17] However, even among professionals who work with RCRA, "hazardous waste" often is given the definition of "hazardous substance" (see below). In this book, "hazardous waste" is used with its broader, popular connotation.
- *Hazardous substance:* Those substances specifically classified as "hazardous" in CERCLA are included here. The definition in CERCLA's regulations

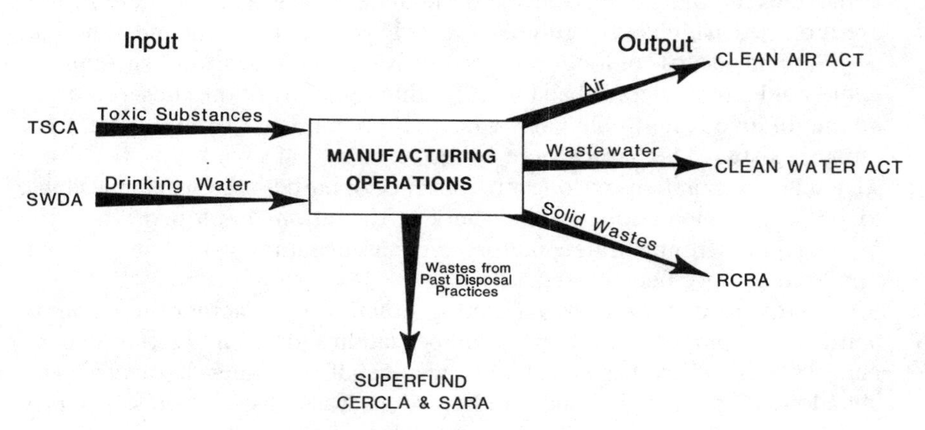

Figure 1-2. The pollution loop closed (1980).

borrows from other federal environmental laws: The Clean Water Act, the Clean Air Act, the Solid Waste Disposal Act, and the Toxic Substances Control Act.[18]

- *Hazardous material:* Regulations and programs of the U.S. Department of Transportation use this term. It applies to a designated substance or material determined to be capable of posing an unreasonable risk to health, safety, and property when transported in commerce.[19]

Other terms widely used and widely misunderstood are "toxic waste" and "toxic substance." The former has no regulatory significance and no precise definition. The term "toxic substance" has relevance only with respect to the Toxic Substances Control Act. Moreover, the word "toxic" was originally used in RCRA to characterize hazardous wastes. "Hazardous" and "toxic" are mistakenly used interchangeably. They are not synonyms and should not be substituted for each other. In the field of hazardous waste site remediation, their correct meanings are derived from specific federal legislation. The pertinent terms are defined in summary form in Table 1-1.[20]

TABLE 1-1. Definitions of "hazardous."

HAZARDOUS WASTE

U.S. EPA definition: (40 CFR 260.10) A solid waste that may cause or significantly contribute to an increase in mortality or an increase in serious irreversible, or incapacitating reversible illness; or pose a substantial present or potential hazard to human health or the environment when it is improperly treated, stored, transported, disposed of or otherwise managed; and, the characteristic can be measured by a standardized test or reasonably detected by generators of solid waste through their knowledge of their waste. The characteristics of a hazardous waste are: ignitability, corrosivity, reactivity, EP toxicity. (40 CFR 261.20-261.24) Hazardous wastes are listed in 40 CFR Subpart D (Parts 261.30-261.33).
The *U.S. DOT* also subscribes to this definition.

HAZARDOUS SUBSTANCE

Definition of CERCLA: Any substance designated pursuant to Section 102, Section 307(a), and Section 311(b)(2)(A) of the Federal Water Pollution Control Act. Any hazardous waste having the characteristics identified under or listed pursuant to Section 3001 of the Solid Waste Disposal Act. Any hazardous air pollutants listed under Section 112 of the Clean Air Act. Any imminently hazardous chemical substance or mixture with respect to which the Administrator of the U.S. EPA has taken action pursuant to Section 7 of the Toxic Substances Control Act.
Definition of the U.S. DOT: A material and its mixtures or solutions identified in 49 CFR 172.101 when offered for transportation under specific conditions of packaging and when the quantity of the material equals or exceeds the reportable quantity.

HAZARDOUS MATERIAL

U.S. DOT definition: A designated substance or material that has been determined by the Secretary of Transportation to be capable of posing an unreasonable risk to health, safety, and property when transported in commerce.

COMPONENTS OF CERCLA

CERCLA has four major components, whereby it:

1. Establishes a fund to pay for investigations and remedial actions at sites where the responsible parties cannot be found or will not voluntarily contribute (Superfund).
2. Establishes a priority list of abandoned or inactive hazardous waste disposal sites for remediation (the National Priority List [NPL]).
3. Establishes a mechanism to determine the appropriate action to take at abandoned or inactive hazardous waste disposal sites (the National Contingency Plan [NCP]).
4. Establishes a system of liability for potentially responsible parties (PRPs) to clean up, or pay to clean up, sites.

Following next is, first, a discussion of the general requirements of CERCLA. Subsequent to that, each of CERCLA's three components is explained to place the subject of site remediation—the technical aspects of which constitute the remaining chapters of this book—into its legal and social perspective. It should be noted that the discussion here is limited to the major provisions of CERCLA. The law itself and the attendant regulations should be consulted for details.

The breadth of the CERCLA liability scheme deserves examination. Four categories of persons are strictly liable for the release or threatened release of hazardous substances:[21]

- Owners or operators of facilities from which a release occurs.
- Owners or operators of facilities at the time of disposal.
- Persons who arranged for disposal, treatment, or transport of wastes, including waste generators.
- Persons who accept the material for transport to disposal or treatment facilities.

According to SARA, liability is defined thus:

> 107(c)(1)(D)(2) . . . the liability of an owner or operator or other responsible person . . . shall be the full and total costs of response and damages . . . . (3) If any person who is liable for a release or threat of release of a hazardous substance fails without sufficient cause to properly provide removal or remedial action upon order of the President pursuant to section 104 or 106 of this Act, such person may be liable . . . for punitive damages in an amount at least equal to and not more than three times the amount of any costs incurred by the Fund . . . .[22]

When CERCLA was first drafted,[23] the measure included language that made the liability strict, joint, and several. *Strict liability* is liability without fault; that is, "neither care nor negligence, neither good nor bad faith, neither knowledge nor ignorance"[24] can be claimed as a defense. Congress correctly predicted that there would be a significant number of sites for which the responsible parties would contest their contribution of hazardous wastes. These parties would, then, be unwilling or unable to agree on their proportional share of the costs, or whether they had any responsibility at all. Therefore, Congress drafted the provision ordering that a PRP is liable even if the method of disposal was completely in accordance with the prevailing standards, laws, and practices at the time of disposal. Under the original language of CERCLA, strict liability would have been imposed for all costs of removal of the wastes and for damages to natural resources.[25] Today, if a site poses a threat to public health, welfare, or the environment, the responsible parties are liable for the cost, regardless of any other argument. In other words, CERCLA is a "pay now-argue later" statute, and strict liability applies.

The "joint and several" provision, also removed from the original language of the House and Senate bills, was meant to avoid delays while the PRPs debated their relative responsibilities and liabilities. Under *joint and several liability,* each PRP is liable to the government for the entire cost of clean-up.[26] Although the language specific to "joint and several liability" was removed from the version of CERCLA ultimately enacted, the courts have interpreted the law as though the language were included. With regard to CERCLA, joint and several liability means that if a party contributed any wastes to a site, that party can be held accountable for all costs associated with the remedial program. If a party believes that it has been forced to pay more than its fair share, it may attempt to recover costs from other parties through civil actions, but it cannot contest its initial payment for all costs. These concepts of strict, joint, and several liability were strongly reaffirmed in SARA. If a party refuses to pay for a remedial program, the federal government through the Justice Department and the U.S. EPA can act to recover costs, and actions have successfully been pressed to recoup costs. In certain cases, CERCLA allows for the recovery of three times the government's cost: the "treble damages" provision.[27]

CERCLA's liability scheme demonstrates that Congress intended that hazardous waste sites be corrected in a timely manner. The provisions have caused great concern on the part of those industries that have abundant financial resources, the "deep pockets." Consequently, there is an increasing tendency for the PRPs to take direct control of remedial programs.

CERCLA applies to publicly owned sites to the same degree that it applies to private sites. For sites owned by the federal government itself, the original statute was unclear; however, SARA and interagency agreements or memoranda of understanding have elucidated that CERCLA applies to all parties without distinction. The U.S. Department of Defense (DOD) has adopted its own procedures for site investigations and remedial programs. As a result, DOD uses a different terminology for its hazardous waste sites, though the nature and intent of its actions are consonant with those of CERCLA. For private sites where the federal government undertakes the design and implementation of a remedial program, the U.S. Army Corps of Engineers, through an interagency agreement with the U.S. EPA, administers the design and implementation phases of remedial programs.

Superfund. The Superfund of CERCLA itself was designed to be a revolving fund used to finance the diagnostic investigations, the engineering studies, and the design and implementation of remedial programs. Superfund was initially underwritten at $1.6 billion under the premise that it could be used to pay for the timely performance of remediation work if the responsible parties could not be found or if they contested their responsibility. Superfund first was raised from general revenue and from a tax on the production of the petrochemicals that have the greatest likelihood of leading to hazardous wastes if improperly managed. It was the intention of those who drafted CERCLA regulations that the fund be reimbursed if costs were successfully recovered from the responsible parties. Thus, CERCLA provides the funding mechanism for sites where the PRPs cannot be identified or for bankrupt (orphan) sites. The fund is, too, the source of money for the oversight and administration of the program. Many states have adopted their own "state superfunds" to pay state costs. The regulations also allow the responsible parties to pay directly for the remedial program if they wish and if they provide timely notification of their willingness to manage the work.

The source of the financial resources of Superfund has been one of CERCLA's most controversial provisions. After its initial five-year authorization, reauthorization of the law was stalled for months while Congress debated various funding mechanisms. The resulting legislation was threatened with a presidential veto. SARA ultimately was passed and signed into law with provisions for a five-year $8.6 billion fund to be raised from taxes on petroleum products ($2.75 billion), business income ($2.5 billion), chemical feedstocks ($1.4 billion), and general revenues.

The National Priority List. The National Priority List (NPL) identifies the sites that pose the greatest health risk and are eligible for CERCLA reimbursement.[28] First published in 1982, it is continuously updated. The NPL currently lists over 700 sites, although it is suspected that potentially there may be in excess of twenty thousand sites in the United States.

The NPL was originally formulated from notification procedures and existing information sources. One of the original requirements of CERCLA was that all parties who knew or suspected the existence of a former waste disposal site for which they had responsibility were required to notify U.S. EPA of it before 1981. The notification was to contain a brief description of the nature, quantity, and location of the waste materials. In response to that stricture, corporations and individuals submitted tens of thousands of notices. Many parties took the conservative position that even a potential, unverified release was grounds for notification, a decision presumably based on the desire to avoid any accusations of failure to notify. At least one major petroleum company reported each of its gasoline stations. There was one unconfirmed report that even the driveway of a private residence was placed on the list. Since 1981, whenever a party has become aware of an abandoned or inactive hazardous waste disposal site for which it may have had responsibility, it has been required to notify the U.S. EPA. Failure to make the report can result in civil and criminal penalties and fines.

The U.S. EPA or the appropriate state agency follows a defined series of steps to determine whether or not a site on the NPL merits further action. Although these steps have different names, depending upon which agency takes the lead, they have the same objective: to determine if there is any immediate or long-term risk to human health or the environment associated with the site. A numeric ranking system, known as the MITRE Model or Hazard Ranking System, was developed for this purpose. If immediate risk is determined, the U.S. EPA is empowered to take whatever action is essential to reduce the risk to a level where it can be addressed in an orderly and rational manner. These first actions, known sometimes as initial response measures (IRMs), may range from fencing and securing the site to the removal of hazardous liquids or hazardous solids. Various criteria are used to evaluate sites where either immediate or long-term risks are present. Those sites with the most acute risks are remedied first.[29] It is the position of the U.S. EPA that when a hazardous waste disposal site poses a potential risk to the environment and human health, remedial action must be taken.

The National Contingency Plan. First published in 1982 and modified in 1985, 1987, and scheduled for revision in 1988, the National Contingency Plan (NCP) provides detailed direction on the action to be taken at a hazardous waste site,[30] including initial assessment to determine if an emergency or imminent threat exists, emergency response actions, and a method to rank sites and establish priority for future action. Most ranking and evaluation are undertaken by the U.S. EPA or its contractors; the Hazard Ranking System, or MITRE Model, is often used. When there is sufficient indication that a site poses a potential risk to the environment, a detailed study is required.

The NCP describes the steps to be taken for the detailed evaluation of the risks associated with a site. Such an evaluation is termed a remedial investigation (RI). The process of selecting an appropriate remedy is termed the feasibility study (FS). The remedial investigation and the feasibility study are often combined into a single measure, known popularly as a remedial investigation/feasibility study (RI/FS). The requirements of the RI/FS are usually outlined in a written work plan, which must be approved by the relevant federal and state agencies before it may be implemented.

A remedial investigation includes the development of detailed plans that address the following items:

- *Site characterization:* A description of the hydrogeological and geophysical sampling and analytical procedures to be applied in order to discover: (a) the nature and extent of the waste materials; (b) the physical characteristics of the site; (c) any receptors which may be affected by the wastes at the site.
- *Quality control:* The guidelines to be enforced to insure that all the data collected from the characterization program are valid and satisfactorily accurate.
- *Health and safety:* The procedures to be employed to protect the safety of the individuals who will work at the site and perform the site characterization.

When the investigation has been completed, the RI activities and a subsequent evaluation of the data gathered are presented—an activity often termed a risk assessment or an endangerment assessment—in a document often known as the remedial investigation report. The precise name of the report is dictated by the lead regulatory agency.

The remedial investigation report serves as a basis for the feasibility study, which evaluates various remedial alternatives. The review criteria include effectiveness in reducing risk, technical feasibility, public acceptance, and cost. The process of selecting a remedial alternative is rapidly becoming one of the most controversial aspects of the CERCLA program. All the remedies selected must be capable of reducing the risk at the hazardous waste site to an acceptable level. And, in general, the lowest-cost alternative that achieves this objective is chosen as the course of action. The results of the feasibility study are presented in a written report, a document that serves as a preliminary basis for the design of the selected alternative.

One of the keys to the National Contingency Plan is that it specifies that the degree of cleanup be selected in accordance with several criteria including the degree of hazard to the "public health, welfare and the environment."[31] Therefore, there is no predetermined level of remediation that can be required or must be achieved at any site. Rather, the degree of correction is established on a site-by-site basis. What is acceptable in one location is not necessarily acceptable in another.

A remedial investigation/feasibility study may be undertaken by the U.S. EPA, a state agency, or the potentially responsible parties, but, in any case, it must conform to the process set forth in the National Contingency Plan. In some cases, the U.S. EPA undertakes the remedial investigation/feasibility study, and the PRPs take responsibility for the design and implementation of the selected remedy. It is difficult to determine when it is in the best interests of both sides—U.S. EPA and PRPs—for one side or the other to take control of the RI/FS. This decision should be carefully considered by PRPs when the offer to assume responsibility for the process is made. The ability of PRPs to participate in the selection of remedies may be severely hindered if the U.S. EPA administers the project. (This issue and the related issue of recovering costs are complex legal subjects and are beyond the scope of this book. Anyone confronted with these issues should seek the advice of experienced counsel.)

Upon completion and approval of the RI/FS, the next step is the preparation of plans and specifications for the selected remedy—the remedial design. This design can be performed in a number of ways, depending on whether the actual construction is to undergo public bidding, or the PRPs will implement the design. To complete the process, the actual construction and other activities are undertaken in accord with the plans and specifications.

Superfund Amendments and Reauthorization Act (SARA). SARA reaffirmed and strengthened many of the provisions and concepts of CERCLA. Comments here on how SARA will affect the CERCLA program and the remediation process are limited to the broad concepts set forth in the law itself. Remedial programs probably will become more complex and expensive.

SARA clearly expresses a preference, but not a requirement, for remedies such as incineration or chemical treatment that render a waste nonhazardous.[32] Although it is an oversimplification to generalize, permanent remedies are significantly more costly, often as much as an order of magnitude more expensive, than alternatives such as on-site containment. It remains to be seen exactly how this preference of SARA will be applied in practice.

Another aspect of SARA is the requirement that the level of cleanup achieve compliance with state and federal standards.[33] It also remains to be seen how this requirement will be reconciled with the concepts of risk assessment and risk reduction.

SARA significantly strengthens the requirement to consider damages to natural resources, especially off-site. CERCLA also required such observance, but few sites included this factor in practice. SARA provides a mechanism to include the issue in future investigations and remedies.

In response to concerns over lack of progress in the completion of remedial design and implementation, SARA establishes strict schedules for action.[34] It is now required that not fewer than 275 programs begin by October 1989. Remedial activities—on-site implementation—must begin on 175 sites by that time, and on 200 additional sites—375 in all—by October 1991. Such a stringent schedule will require manifest effort on the part of the U.S. EPA and responsible parties.

SARA also required that the U.S. EPA revise the obligatory method of remediation. At the time this book went to press, a draft of the revised criteria of the NCP was available. Although those criteria may change in the U.S. EPA's final draft, it is important to include them here:

1. Overall protection of human health and the environment
2. Compliance with Applicable or Relevant and Appropriate Requirements (ARARs)
3. Long-term effectiveness and permanence
4. Reduction of toxicity, mobility, or volume
5. Short-term effectiveness
6. Implementability
7. Cost
8. State acceptance
9. Community acceptance.

(For the current criteria, the reader is advised to consult 40 CFR 300.)

At this early date, it is impossible to know the effect of the new criteria. Even after the U.S. EPA promulgates them, regulators and other concerned parties will require 12 to 24 months or more to understand how they will be implemented.

SARA contains many other provisions that address issues such as liability, negotiation, settlements, and judicial review. Basically, it strengthens and substantiates the concepts of CERCLA.

SYNOPSIS

This book addresses the approach the practicing engineer uses to implement the National Contingency Plan at sites governed by CERCLA regulations. The various techniques used to develop components of the remedial investigation are set forth in the remainder of Part 1, Chapters 2 through 7. Chapter 2, "Purpose and Execution of Field Investigations," and Chapter 3, "Analyzing Hydrogeological Conditions," address how valid data are developed. The practices presented in these chapters are the fundamental building blocks for the entire remediation effort because they gather the data that define the hazardous waste site.

Those initial data are reviewed to appraise the potential jeopardy the hazardous waste site presents to human and wildlife populations and to the environment. The process used is a risk assessment, and the qualitative and quantitative forms of that evaluation are explained in Chapter 4, "Assessing Risk."

All field programs work hand in glove with the laboratory. The sampling protocols must be orthodox, and sample handling must be correct to permit the laboratory to perform the appropriate, required analyses. Data from the laboratory, together with information about the laboratory's quality control and quality assurance practices, reveal to investigators the dimension of contamination of the site's soil, surface water, ground water, and air. The operation of the laboratory is described in Chapter 5, "The Role of the Laboratory in Remediation Work."

A topic that is of paramount importance whenever workers collect data in the field or when they implement the remedial measures at hazardous waste sites is explained in Chapter 6, "Health and Safety at Hazardous Waste Sites." During each phase of the work, the jeopardy that workers are exposed to must be managed.

Chapter 7, "Ground Water Models: Tracking Contaminant Migration," addresses a special topic related to field investigations. When data sufficient to describe the migration of the contaminated ground water plume cannot be gathered in the field, modeling techniques—largely mathematical—are used to describe the movement, to interpret the data in hand, or to predict future movement. The elucidation of these techniques shows how the uncertainty in data can be reduced.

Part 2, on remediation, begins with Chapter 8, "Developing the Feasibility Study," a pivotal chapter. It explains how a remedial alternative, composed of one or more remedial technologies, is selected. The remedial alternative must fulfill demanding criteria in order to assure government regulators and designers that it will be effective in reducing the site's risk to an acceptable level.

The most common approaches to remediation of hazardous wastes on-site are presented in Chapter 9, "The Recourse of Closure On-Site." Techniques used when a leak or spill of a hazardous substance has occurred are presented in Chapter 10, "The Disposition of Ground Water." Chapter 11, "In Situ Biological Treatment of Ground Water," introduces a technique that may hold promise for the future. The manner in which leaks occur and the way they are remedied under RCRA are described in Chapter 12, "Correcting Leaking Underground Storage Systems."

A remediation measure that government regulatory agencies, now prodded by SARA, are encouraging in place of land burial is presented in Chapter 13, "Incineration as a Disposal Alternative." The technology for incineration will likely be improved in the future with its increasing use as the ultimate solution to hazardous waste disposal.

When hazardous wastes must be moved from the site in which they were originally interred, many problems surface. The off-site disposal of wastes is always an option worth considering, but it is a formidable one to implement. Factors to weigh before wastes are moved and during the move are related in Chapter 14, "Closure Through Off-site Remedies."

As many of the chapters suggest, extraordinary care must be exercised in all phases of the remedial investigation/feasibility study. The field work and laboratory work are performed meticulously; the alternative is chosen painstakingly. Finally, the remedial alternative must be executed with the same measure of attention. Chapter 15, "Implementing the Remedial Measures," shows how the remediation project is organized and executed to insure that the alternative is put in place correctly.

The method of remedying hazardous waste sites, although barely formed, is already undergoing a revolution. In 1986 when Congress enacted the Superfund Amendments and Reauthorization Act, the emphasis clearly shifted from the burial of hazardous wastes to their destruction, from possibly stopgap expedients to lasting remediation. The way that SARA's preference is actuated by regulatory agencies and responsible parties will define the effectiveness of hazardous waste remediation. The practicing engineers of the 1990s will participate in that revolution.

NOTES

1. Enacted by Public Law 96-510, 11 Dec. 1980, 94 Stat. 2767; 42 U.S.C. 9601.
2. The term "Superfund" is actually a misnomer because it applies to only one of the three major provisions of the Act. However, this term has achieved such high visibility and common use that it is now accepted. Therefore, for most purposes, the terms "Superfund" and "CERCLA" are interchangeable.
3. Rachel Carson. *Silent spring.* Boston: Houghton Mifflin Company, 1962.
4. PL 91-190 of 1 Jan. 1970, 42 U.S.C.A. 4321 *et seq.*
5. Reorganization Plan No. 3 of 1970 by President Richard M. Nixon, effective 2 Dec. 1970.
6. PL 91-604, Amendments to PL 90-148; 42 U.S.C. 7401 *et seq.*
7. PL 92-500, as amended. 33 U.S.C. 1251 *et seq.*
8. PL 100-4.
9. PL 94-523, as amended. 42 U.S.C. 300f *et seq.*
10. PL 99-339.
11. PL 94-469, as amended. 15 U.S.C. 2601 *et seq.*
12. PL 94-580, as amended. 42 U.S.C. 6901 *et seq.*
13. PL 98-616.
14. See Richard C. Fortuna and David J. Lennett. *Hazardous waste regulation: The new era.* New York: McGraw-Hill Book Company, 1987.

15. See, for example, (1) Bureau of National Affairs, Inc. Superfund II: A new mandate. Washington, D.C., 1987. (2) Fred Smith. The flawed logic of Superfund. *Cato Policy Report.* 7(Nov./Dec. 1985):6-9. (3) U.S. Office of Technology assessment. Superfund strategy. Washington, D.C., Apr. 1985. (4) Theodore G. Brown III. Superfund and the National Contingency Plan: How dirty is "dirty"? How clean is "clean"? *Ecology Law Quarterly.* 12(1984):89ff. (5) U.S. General Accounting Office, Community and Economic Development Division. *Hazardous waste sites pose investigation, evaluation, scientific, and legal problems.* Washington, D.C., 24 Apr. 81.

16. PL 99-499.

17. 40 CFR 261.3.

18. 40 CFR 300.6.

19. 49 CFR 171.8.

20. References are as indicated in Table 1-1 and were also taken from: (1) G. William Frick, Ed. *Environmental glossary.* Rockville, Md.: Government Institutes, Inc., 1984. (2) U.S. EPA (Office of Waste Programs Enforcement; Office of Solid Waste and Emergency Response). RCRA Ground-water monitoring: Technical enforcement guidance document. Sept. 1986. OSWER-9950.1.

21. CERCLA Section 107(a)(1). See also (1) Superfund liability. *Virginia Law Review.* 68(1982):1157. (2) Anita M. D'Arcy. Joint and several liability under Superfund. *Loyola University Law Journal.* 13(Spring 1982):489-522.

22. CERCLA, Section 107(c)(1)(D), subsections (2) and (3).

23. See H.R. 7020 and S. 1480 of the Ninety-sixth Congress.

24. *Black's law dictionary.* St. Paul: West Publishing Co., 1968.

25. Superfund liability. *Virginia Law Review.* 68(1982):1160-1161.

26. D'Arcy. Joint and several liability under Superfund. *Loyola University Law Journal.* 13(Spring 1982):492.

27. Section 107(c)(1)(D)(3)

28. 40 CFR 300, Appendix B.

29. See 40 CFR 300 and Guidance on remedial investigations under CERCLA (prepared for Hazardous Waste Engineering Research Laboratory Office of Research and Development and Office of Emergency and Remedial Response, Office of Waste Programs Enforcement, and Office of Solid Waste and Emergency Response, U.S. EPA. May 1985).

30. 40 CFR 300.

31. 40 CFR 300.65(f).

32. CERCLA, Section 121(b).

33. CERCLA, Section 121(d).

34. CERCLA, Section 116.

Chapter 2
Purpose and Execution of Field Investigations

The object of the field investigation is to collect representative samples from a hazardous waste site in order that the contaminants there can be analyzed and the risks defined. This chapter discusses the many aspects of acquiring representative samples. (Detailed information on the hydrogeological investigation is presented in Chapter 3.) The subject matter of this chapter stops at the laboratory door; analytical methods are presented in Chapter 5. The dimension of the field investigation is determined by the importance of the potential hazard at the site and by the resources and time available to the investigators. The goal of the field investigation is to characterize accurately and sufficiently the physical system at hand; therefore, the objectives of the field program are defined before the specific field activities are determined. The program is then structured to provide the information prerequisite to the rational progression of the entire remediation scheme. An ever-present limitation is the cost of obtaining the data; the scarcity of resources is a decided constraint on any field investigation program.

When engineers, scientists, and government regulators are first presented with a locale where the soil and water are known or suspected to be laden with chemical substances, the presence and concentration of the contaminants are determined through site exploration. That exploration furnishes two essential sets of data:

- The physical characteristics of the site are determined. Therefore, its geology, hydrology, and topography, and the interrelations among these parameters, are described.
- Identification of the contaminants is established, and their penetration into the environment is estimated; samples are taken of all potentially contaminated matrices. From the analytical data of those samples, the polluted area is delimited, and the contaminants on the site are specified.

This chapter was developed by Frank D. Hale and Douglas M. Crawford of O'Brien & Gere Engineers, Inc.

The information developed during field investigations is the cornerstone upon which the remediation of hazardous waste sites is based. It is used to identify past, present, and potential dangers to the environment and to the health of any people inhabiting the site's environs. If the site requires remediation, the data acquired through the field investigation are the basis for development of alternative remedies. In sum, the field investigation is critical to the success and effectiveness of the overall site remediation program.

HISTORY OF USE AT A CONTAMINATED SITE

The past use of a contaminated site is the seed of its current crisis. All relevant facts about historical uses are marshaled as the field investigation is planned. Germane items are the operation conducted on the site and the materials used, treated, stored, and disposed of; the disposal practices are especially important. All past land uses are reviewed because not only are changes in land use generally accompanied by changes in the types of waste at a site, but such changes also may alter the characteristics of the wastes already present. Further, the wastes may react and create other compounds, and the means by which those wastes could migrate from the site into the environs can be modified by alteration of the land's surface or stream channels. During the land use study, the following questions require explanation:

- What disposal practices were employed throughout the history of the site?
- What raw materials were on the site, and where were they used or stored?
- Where did the disposal of materials take place?
- In what condition were the wastes when discarded?
- How could the wastes be expected to change as a result of the method of disposal employed?

Corporate and municipal records register changes in ownership and land use. These records also list corporations and people within those organizations who could be contacted for details on past land uses and on the history of the site. Maps and historical aerial photographs also provide important data, and federal and state regulatory agencies may supply valuable information. Regulatory agencies often will have conducted an investigation to generate preliminary data regarding environmental contamination at the site, and the sources of their information can lead to additional facts.

If specific data describing the waste products discarded on the site are unavailable, the information can be inferred from records about the manufacturing or other industrial operations practiced there. If the operations that took place through the history of the site can be discovered, the way that various wastes may have interacted can be deduced with some assurance.

FACTORS OF THE FIELD INVESTIGATION

The field survey at a hazardous waste site attempts to define two essential factors:

- Waste materials potentially present.
- Mechanisms that can transport those materials.

The first task of the investigation is to develop a program of sampling and field analyses in order to estimate the extent and concentration of the contamination in all affected media and in all three spatial dimensions. The history of the site's use is the fundamental data base; it often tells not only how, when, and where raw materials were used and stored, but also how, when, and where wastes were treated, stored, or disposed of. Depending upon the adequacy of the historical data and the sufficiency of details about the potential contaminants on the site, the sampling program can be designed to search for specific waste compounds or for indicator parameters to be used in the laboratory analysis. For example, if only capacitors and transformers were discarded on a site, the sampling and analytical programs would be designed to determine the extent of contamination by polychlorinated biphenyls (PCBs) specifically. However, if the only useful information about a site were that at one time it was used as a landfill for industrial wastes, comprehensive sampling and analytical programs would be mandatory and could include indicator parameters such as pH, conductivity, heavy metals, and total organic carbon. In many instances where the data are equivocal, the sampling and analytical programs are phased. The initial survey is broad in scope, and, as information is developed, additional sampling and analyses are conducted to furnish specificity.

The second task of the field investigation is to distinguish how the topography, geology, hydrology, and climate of the site affect the movement of the contaminants present. Wastes migrate from the source of contamination through a variety of routes: mobilization with ground water, the flushing of sediment with surface water or overland runoff, the ferrying of soil particles in the wind, and ingestion and conveyance by biota.

Ground water wells are installed to provide information on ground water hydraulics and quality. Contaminants often are transported with ground water flow, and the area of such contamination can be widespread. Ground water models and solute transport models are used regularly to determine the areal extent of pollutant transport by ground water. If surface water is the dominant mode of transport, regional rather than just local pollution problems could ensue; wet weather surveys and surface water modeling are used to determine the significance of surface water in the transport of contaminants.

The sampling plan also is designed to determine whether contaminants are being transported in the water phase or being transported when adsorbed to particles. Where the wind is expected to play a role in the transport of contaminated materials, air quality monitoring is conducted. In some instances, air quality models are developed.

Vast quantities of data can be generated at a hazardous waste site, yet little useful information can result unless specific measures are taken to minimize sources of unexplainable and unverifiable errors. Therefore, quality control procedures are essential as samples are collected. These procedures address the specific sampling techniques, handling of the sample, calibration of the field instruments, decontamination, documentation, and recording of data. Guidelines for each aspect of the program are developed and implemented, together with appropriate training of the field crew, well before the beginning of the sampling effort. With the pertinent guidelines in practice, the field investigator can be assured of collecting representative samples, and cross-contamination will be obviated.

The breadth of the site investigation and the number of samples generated necessitate stringent management of data. A plan must be developed to control the data before any are collected; failure to develop such a plan could result in significant expense later in the program.

Before sampling activities are initiated, a document is prepared to summarize all pertinent site information and all planned sampling and sample-handling activities. This is the *sampling plan.*[1] The U.S. EPA specifies the desirable contents of a sampling plan; they are listed in Table 2-1. In addition, the sampling plan takes into account any requirements of the particular state where the field work will be conducted in order to incorporate specific state stipulations.

TABLE 2-1. Necessary components of sampling plans according to CERCLA.

Investigative objectives	
	Map of locations to be sampled
Site background	
	Sample locations and frequency
Analyses of existing data	
	Analytical procedures
Analytes of interest	
	Operational plan and schedule
Types of samplers	

The collection and handling of samples is performed carefully. In order to demonstrate that a sample was collected using the appropriate practices, the entire procedure is recorded. A standard form, similar to that illustrated in Figure 2-1, is begun when the sample is collected. As possession of the sample is relinquished by one technician and transferred to another, the chain of custody document is so revised. The chain of custody form also contains pertinent information:

- Required analyses.
- Type of sample: water, soil, air.
- Sampling technique: grab sample or composite sample.
- Location from which sample was taken.
- Date and time of sample collection.

Each field investigator also summarizes the events and conditions of the work in a log book. Workers and other staff, weather conditions, samples taken, and other pertinent data—notably abnormal occurrences—are recorded. The log book is usually maintained by the field coordinator.

A photographic record is an ideal complement to written records for a field investigation. Photographs are useful for both technical and legal purposes, as they furnish an unequivocal and permanent account of each sampling location. The photos of the sampling sites are arranged to include reference objects in order that the precise point can be relocated.

SAMPLING VARIOUS MATRICES

Samples are collected from all the physical features on the site that are known or surmised to be contaminated; the matrices that might be contaminated are water, soil, air, biota, and artifacts. Specific protocols must be followed in collecting a sample for each different matrix. (Sample techniques for collecting ground water are not discussed in this chapter; see instead Chapters 3 and 10.)

Whenever samples are taken, the objectives of the sampling program dictate the methods employed. Parameters that require definition before sampling is initiated are:[2]

- Site of interest.
- Compounds of interest.
- Qualitative specificity (class or compound specificity); that is, determination of whether it is adequate to establish, for example, just total organics, or whether it is necessary to identify each specific organic compound.
- Mandatory performance parameters of the sampling methods employed.

O'BRIEN & GERE

CHAIN OF CUSTODY RECORD FOR SAMPLES

SURVEY	SAMPLER: (Signature)
ABC LANDFILL	John C Sampler

STATION NUMBER	STATION LOCATION	DATE	TIME	SAMPLE TYPE WATER Comp	Grab	AIR	SEQ. NO.	NUMBER OF CONTAINERS	ANALYSIS REQUIRED
1	GROUND WATER WELL 1	6/30/87	0800		✓		1	3	Sb, As, Cd, Cu, Pb, TOC, SULFATE, CHLORIDES
2	GROUND WATER WELL 2	6/30/87	0900		✓		2	3	Sb, As, Cd, Cu, Pb, TOC, SULFATE, CHLORIDES
3	GROUND WATER WELL 3	6/30/87	1000		✓		3	3	Sb, As, Cd, Cu, Pb, TOC, SULFATE, CHLORIDES
4	GROUND WATER WELL 4	6/30/87	1100		✓		4	3	Sb, As, Cd, Cu, Pb, TOC, SULFATE, CHLORIDES
5	SURFACE WATER - STATION 1	6/30/87	1400	✓			5	2	Sb, As, Cd, Cu, Pb, TDS
6	SURFACE WATER - STATION 2	6/30/87	1430	✓			6	2	Sb, As, Cd, Cu, Pb, TDS

RELINQUISHED BY: (Signature)	RECEIVED BY: (Signature)	Date/Time
RELINQUISHED BY: (Signature)	RECEIVED BY: (Signature)	Date/Time
RELINQUISHED BY: (Signature)	RECEIVED BY: (Signature)	Date/Time
RELINQUISHED BY: (Signature)	RECEIVED BY MOBILE LABORATORY FOR FIELD ANALYSIS: (Signature)	Date/Time
DISPATCHED BY: (Signature) Date/Time	RECEIVED FOR LABORATORY BY:	Date/Time
METHOD OF SHIPMENT:		

Figure 2-1. Sample chain of custody document.

- Physical state of interest (gas, particle, or total).
- Cost restraints.
- Required degree of accuracy of data.

In sum, sampling methods are predicated upon the type of data required to characterize the physical conditions and extent of contamination at the site.

The qualities of the contaminants tested affect the sampling plan; the sampling equipment, procedures, containers, and preservation techniques all are selected to relate to the tested parameters. For example, certain compounds that associate with oils tend to sorb to the surfaces of a sampling device, and if the device is not submitted for analysis, false low readings can be obtained. In one case in the authors' experiences, over 90 percent of the PCBs in a water sample remained on the surfaces of the glass jar used to hold the sample.

Surface Water

Flow Measurement. When collecting samples from the surface water coursing through or near a hazardous waste site, the investigator is interested in the concentration of specific substances and in the fractions of the substance that are soluble and that are particulate. In addition to collecting samples, the flow rate of the stream is estimated in order to assess adequately the transport mechanism. The laboratory determines the concentration of a contaminant in the water, but the field investigator measures the stream flow. It is often useful to combine flow monitoring with a rain gauge to facilitate modeling conditions of high flow, as atmospheric precipitation can significantly affect contaminant transport.

Several methods are available to measure the flow of surface water in open channels. Most of the methods are based upon the continuity equation:

$$Q = AV$$

This equation states that the flow rate (Q) is proportional to the product of cross-sectional area of flow (A) and the velocity of flow (V). To simplify data gathering, average values of the flow rate, cross section, and velocity of flow are usually acceptable. The average velocity of an open channel is measured with a velocity meter, such as a gurley meter or current meter, placed at various locations in the cross section of flow. The average depth of the channel is normally determined by using a ruler or weighted tape measure, depending on the size of stream. If the stream is quite large, the U.S. Geological Survey may maintain a gauging station which could provide useful data.

When measuring flow, weirs or flumes are usually employed in conjunction with head-measuring devices such as bubbler systems, pressure transducers, and dippers. The weirs or flumes are calibrated so that if the depth of water flowing over a certain section is known, the flow is proportional to the depth of flow. Calibration curves, known as *head-discharge curves,* relate head (depth of flow over the appropriate section of the weir or flume) to the flow rate. These curves are based upon the premise that, under normal flow conditions, there is only one flow rate for a given depth of flow.

The flow rate in an open channel can be determined in other ways. If certain hydraulic parameters are known, others can be determined. Knowing the slope of the channel (an approximation of the slope of the hydraulic grade line), the hydraulic radius (the cross section of flow divided by the wetted perimeter), and depth of flow, one can calculate the flow rate using an empirical formula.[3] Flow rates also can be calculated by using the time-volume method of flow determination. A known volume of water is collected over a predetermined length of time; the volume : time ratio (volume divided by the time required to collect the volume) is the volumetric flow rate of the stream.

In another technique used to determine the flow rate, a conservative chemical, one that does not react significantly such as lithium chloride, is introduced at a known concentration and rate into the stream to be measured; this is the dilution method of flow estimation. Water samples are collected at a downstream location and analyzed for the introduced chemical. The flow rate is calculated using the known stock, background, and downstream concentration of the chemical and the rate at which it was injected.

Sample Collection. In water sampling, the distribution of contaminants between the particulate fraction and the soluble fraction often is relevant to the selection of a remedial approach appropriate to the hazardous waste site. If the sample is to be filtered, the accepted way to differentiate between the filtered sample and the nonfiltered sample is to use a 0.45 μ membrane filter.[4] Relatively clean surface water can be filtered directly using a syringe with a Swinnex™ filter arrangement or a Millipore™[5] type apparatus. Where solvents are a concern, pressure filtration is preferred over vacuum filtration. Certain substances adsorb to the filtered glass support during filtration, and such a contingency should be addressed by the sampling plan. If the water bears a heavy sediment load, it is accepted practice to allow the sediment to settle for a few minutes and then to prefilter the water with a coarse medium prior to final filtration with the 0.45 μ filter.

Sampling Techniques. Sampling water at or near the surface is usually accomplished with grab sampling techniques. Typically, the sampling device

TABLE 2-2. Typical field instruments for monitoring surface water

Thermometer	Depth transducer
pH meter	Current meter
Salinity meter	Secchi disk (turbidity)
Dissolved oxygen meter	

is simply lowered into the water just below the surface, and the water flows into the sampler. A number of conventional instruments are utilized to describe the characteristics of the water; see Table 2-2.

Some contaminants, such as oil, spread slowly despite the wave actions that tend to homogenize surface conditions on large bodies of water; so it is important to take samples at a number of locations in order to define fully the horizontal extent of contamination. In cases of such sheen-producing substances, it may be necessary to induce artificial turbulence to acquire a uniform sample. For instance, oil containing PCBs often remains on the surface of water, and a sample taken from beneath the surface may not accurately quantify the transport of PCBs from the source of contamination.

Samples of surface water bodies at various depths are taken with an instrument such as a VanDorn sampler, illustrated in Figure 2-2. This sampler, on a rope, is lowered into the water with the plungers at each end in the open position. When the sampler reaches the desired depth, a messenger—a weight that slides freely along the rope—is dropped to the sampler. Upon impact, the messenger trips a mechanism that releases the plungers, which close the ends and impound the water sample. The sampler is then raised, and the sample is withdrawn.

Contaminant concentrations will vary with depth in surface water for several reasons—discharge configurations, currents, source-sink relationships, and diurnal effects, for example. In one waste pond the authors examined, the conductivity of the surface 12 inches (0.30 m) was approximately 200 μmhos/cm, while that of the next 3 ft (1 m) measured 15,000 μmhos/cm. This considerable difference in conductivity foretold additional environmental contaminants of concern. The sources of the differences were the geometry of the influent and effluent pipes and the characteristics of the wastewater. Variations over time or flow rate can be accounted for by collecting composite samples. *Time-proportional composites* are comprised of subsamples obtained over discrete time periods. *Flow-proportional composites* incorporate equal volumes of subsamples collected after a discrete volume of flow has passed a reference point. By installing a continuous-reading flow meter in a culvert with a flume, it is possible to signal an automatic sampler to take a sample from the turbulent portion of the flume after a discrete volume has passed.

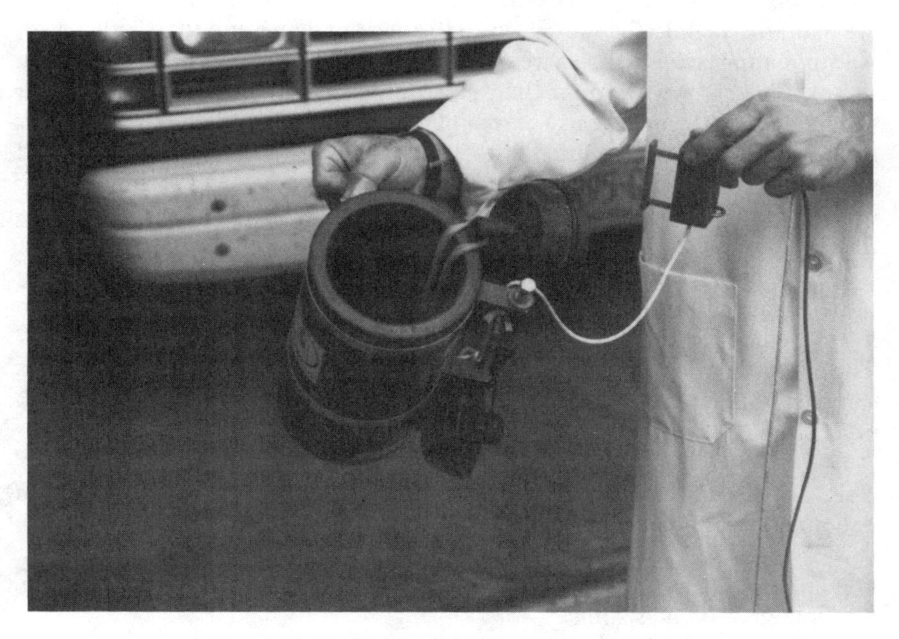

Figure 2-2. VanDorn sampler for the water column.

Sediment. A considerable exchange of contaminants occurs between the sediments of a surface water body and the water column, as sediments may act as both a source and a sink for contaminants. Therefore, sediments are frequently sampled when surface waters are contained within or adjacent to a hazardous waste site. The transport of contaminants from sediment rarely exceeds 1 ft (0.3 m) in depth, but major storms may expose contaminated sediments and thus alter transport mechanisms; consequently, cores of 0.3 to 1 m (1 to 3 ft) are regularly collected and submitted for analysis. The samples of sediment are segregated prior to analysis, and the rationale for the separation can be their observed physical differences—differences such as color and texture—which can be useful to aid characterization during implementation of the remedial procedure. Alternatively, the samples can be separated for analysis by using specific depths from grade or from elevations.

Sediment samples can be taken in either disturbed or undisturbed form. If satisfactory characterization can be achieved with the mean level of contamination, a disturbed sample is acceptable and can be obtained by using grab samplers such as the Ponar grab sampler or Eckman dredge; see Figure 2-3. These samplers are attached to a rope and dropped into the water column at the sampling locus. Their weight enables them to penetrate the sediment. When retrieved, the jaws close and contain the sample.

Undisturbed samples can be obtained in shallow water, under 10 ft (3 m), using core samplers such as Lexan corers or stainless steel corers. The corers are hand-driven into the sediments to the desired depth and then withdrawn. Another approach is to use a weighted corer dropped from a boat. Obtaining samples in this way permits characterization of parameters related to the sediment's depth, and the use of retainers and Lexan liners enables relatively undisturbed cores of sediments to be collected. To procure undisturbed samples from deeper surface water bodies, hollow stem auger techniques are used in conjunction with floating platforms or barges. Chapter 3 explains these techniques.

When cores are collected, the recovered depth is often less than the depth of penetration. Both values are logged to assist investigators when the data are evaluated for the quantity of contamination and the volume of sediment. The cores are frozen and split longitudinally with a saw. One section furnishes samples of the strata; the other is used in the analysis, sometimes to form a composite of the entire core.

Soil. The surface of soil, that layer typically within 3 ft (1 m) of grade, can become contaminated in a number of ways:

- Inappropriate waste management practices, such as indiscriminate dumping.
- Accidents, such as spills.
- Deposition of airborne waste particles.

Figure 2-3. Ponar™ grab sampler for sediments. Side view, opened and closed.

Contaminated soil poses a health risk, because of the potential for direct contact, because the wastes could volatilize, and because the waste could be broadcast by the wind, surface water, or ground water. If the historical use of the site is well known, sampling sites can be selected based upon specific information about "hot spots"; however, this method of sampling, often referred to as "judgmental sampling," should be used cautiously, as false conclusions can be drawn from it.

A grid is employed when few facts about the site are available and when contamination is expected over a large area. To fix sampling locations in surface soil, the grid is superimposed on the site. The grid is based either on general information about the site or on a statistical method. It is possible for some hot spots to be left unsampled by this approach, but the grid can be combined with a phased sampling program. During the first phase of such a program, the general contamination levels are determined, and a more dense grid is superimposed in areas with the higher concentrations of contamination; then additional sampling is conducted. If nothing is known about the site, statistical methods using random number theory can be used as an objective way to fabricate the sampling points. At the outset, the number of samples required for a given area and the statistical confidence level are balanced against the financial resources available to the investigator. A grid is drawn on the site map, and the coordinates of the initial sampling location are established randomly by selecting a pair of numbers from a random number table. Subsequent locations proceeding from the antecedent point are selected similarly.

Various techniques are used to collect undisturbed surface soil samples in cores. A sampler—usually a tube of Lexan, stainless steel, or another suitable material—is pushed or driven into the soil to a depth of 1 to 3 ft (0.3 to 1 m); the sampler with soil is then twisted to fracture the soil at the end of the corer. If the soil is incoherent—for example, sand—an "eggshell sampler" is installed at the bottom to prevent the sample from slipping out as the corer is withdrawn. When it is necessary to obtain samples from depths below 3 ft (1 m), a posthole digger or power auger is used first to reach the desired depth, and then the corer is used to collect the sample. The strata of the sample are identified and segregated in the manner used to classify stream sediments.

Air. An uncontrolled hazardous waste site can release into the air compounds such as halogenated solvents and fugitive dusts containing metals or organic substances. These contaminants are dispersed during work at the site such as excavation, capping, or land-spreading of wastes. To protect the health and safety of those working on the site and in the site's environs, the ambient concentrations of these emissions are monitored. If the concentrations approach a hazardous level, measures are implemented to abate the release, or protective safety equipment is employed.

Occasional Field Measurements. Various field instruments are available to sample air quality. Their purpose is to monitor the air for public health and safety and to identify compounds for specific laboratory analysis.

Combustible gas indicators determine the combustible gas levels in potentially hazardous environments, primarily enclosed spaces where the concentrations of solvents may reach hundreds of thousands of parts per million. The instrument is calibrated with hexane or methane; if these gases are unsuitable, calibration is performed with the gas of interest. When high levels are measured, the investigator in charge of the sampling may recommend protective respiratory equipment. When the lower explosion limit is reached—a mixture of the volatile compound and air that could support an explosion—additional precautions are mandatory.

Oxygen meters often are used intermittently in conjunction with combustible gas indicators in potentially hazardous environments. It is imperative to be aware of depressed oxygen levels, to interpret the readings of combustible gas indicators, and to establish respiratory protection. The oxygen meter allows oxygen molecules to diffuse through a selective membrane into its sensor assembly. After passing through the membrane, the molecules react electrochemically at two electrodes and produce a flow of electrons. The current produced is proportional to the oxygen level and is displayed by the instrument.

The *organic vapor analyzer* (OVA) furnishes a measure of the organic vapors in the air by utilizing hydrogen flame ionization. The instrument may be used intermittently or continually. The OVA is significantly more sensitive than the combustible gas indicator, reaching detection limits on the order of 1 ppm, and is often used to develop data in order to select the appropriate respiratory protection equipment. When continuous samples are taken, air is drawn into a probe and moved by a pump with a known flow rate through particle filters to the detection chamber, where the sample is exposed to a hydrogen flame. The flame ionizes organic vapors; as they burn, the vapors leave positively charged carbon-containing ions that are collected by a negatively charged electrode in the chamber. As the positive ions are collected, a current is generated, the strength of the current corresponding to the rate of collection. This current is displayed by the organic vapor analyzer.

When applied to monitoring, the organic vapor analyzer gives an estimate of the organic concentration of the atmosphere expressed as the gas used for calibration, and the true contaminant is not reported. The analyzer detects methane, a gas released during anaerobic degradation such as occurs at municipal landfills; consequently, at sites where hazardous solvents may be mixed with municipal garbage, it must be used carefully. The instrument can be modified to work as a gas chromatograph in order to identify contaminants, but maintaining quality control in the field is challenging.

Photoionization detectors (the abbreviation is HNU) utilize the principle of photoionization to detect trace gases in the air. These instruments can be used intermittently or continuously. The photoionization detector reports the concentration of organic substances in the air expressed as the calibration gas. Its detection limit is on the order of 1 ppm, and although it provides limited information, it is often used to determine whether or not respiratory protection is needed. This detector's sensor is an ultraviolet light source. The photons emitted act to ionize organic compounds, yet do not affect the common gases in the air. Two electrodes are adjacent to the light source in the chamber. When a positive potential is applied to one electrode, ions formed by the absorption of ultraviolet light are driven to the collector electrode. The current produced is proportional to the concentration of organic compounds and thus provides the measurement.

It is prudent to utilize a *radiation detector* when the information about a site is limited in order to determine at the outset whether or not a radiation hazard exists. The safety officer usually samples the area for radiation during the preliminary site inspection, and a Geiger counter is employed. Radiation ionizes the gas contained in a tube in the instrument; an electric current is produced when the gas is ionized, and the instrument measures the current. The Geiger counter detects beta and gamma radiation.

If the vapor of interest has been identified, it is often more economical to use *colorimetric indicators* than combustible gas indicators, organic vapor analyzers, or photoionization detectors. A tube in this instrument contains chemically treated solids that react with the calibration gas to produce a color change. By controlling the volume of air that enters the instrument and by noting the color change, it is possible to quantify the concentration of the substance contained in the air. The nature of the chemical reaction is broad enough that positive interference is possible; for example, if trichloroethylene is present, it will also react with the tetrachloroethylene monitoring tube to display a reading higher than the actual concentration of tetrachloroethylene in the air.

Continuous Air Sampling. Air at a hazardous waste site is sometimes sampled continuously to describe its quality. An air sampling program provides more comprehensive and specific data than that attained with occasional field measurements. The goal of an air sampling program is to specify which compounds are present and their concentrations. It is used, for example, to discover whether fugitive emissions are the by-product of remedial activities. An important purpose of the sampling is to monitor conditions for worker safety during remediation.

In devising a sampling plan for the air, several factors unique to the locale are considered: wind velocity, precipitation, range of air temperatures, and potential confounding influences such as automobiles or nearby industrial emissions. The requirements of a sampling plan often dictate that samples be collected during a variety of meteorological conditions to permit a thorough assessment of transport mechanisms.

Air samples are, as a rule, collected in sorbent tubes. Sampling is described as short-term if the time period is less than 8 hours. High volume ("hi-vol") samplers draw air at a known flow rate into a chamber for a specified time. The collected air is analyzed by a laboratory, and the resulting data are expressed as the mass of the compound per volume of air sampled per unit of time; thus the data are normalized. Long-term sampling is conducted if the values must be averaged over time.

Biota. Biological samples provide valuable information about the health effects of pollutants at a hazardous waste site. By evaluating the micro- and macro-organisms exposed to contaminants, the investigator can project the gross effects—for example, changes in growth and reproduction—and specific effects—such as bio-accumulation in the food chain—of long-term exposure. Sampling methods are specific to the organism under study and also to the nature of the data required.

Fish provide information on both gross and specific effects of exposure. The sampling area is determined by the species and quantity of fish required for the sample, and electroshocking equipment or nets are used to take fish. The current from an electroshock enters the water and temporarily stuns the fish, which rise to the surface and are collected alive. Gill nets, suspended vertically in the area to be sampled, capture fish, which are killed by the nets. Fish are also caught by using hooks and lines.

There is prodigious variability among the plants that might be sampled at a hazardous waste site. This chapter cannot begin to describe all the sampling techniques appropriate to different plants, but methods are described in several reference sources.[6] Plants absorb contaminants through their roots and sustain direct contact with airborne, waterborne, or soil-resident contaminants.

In general and for purposes of sample collection at hazardous waste sites, there are two broad categories of plants:

- Phytoplankton and other microscopic aquatic plants.
- Macrophytes.

Phytoplankton and other microscopic aquatic plants and floating aquatic macrophytes are sampled with a fine-mesh tube-shaped net, towed behind a boat at the desired sampling depth. The net is then retrieved, and is rinsed over a collection bucket, transferring the plants to the bucket. The plants are taken to a laboratory for analysis. Rooted macrophytes, both aquatic and terrestrial, are cut as close to the root as possible or uprooted when it is necessary to collect the plant with roots intact. They, too, are taken to a laboratory for analysis.

Animals, whether they are aquatic, amphibious, or terrestrial, are mobile and may become directly exposed to contaminants if they encounter a hazardous waste site. In addition, if the contaminants are transported beyond the waste location, animals in the site's environs may be affected. Traps are set at feeding areas or on pathways to and from feeding and nesting areas to collect animals as samples. Both live and dead specimens are utilized.

Artifacts, Including Containers. Buildings and other structures, including tanks, contaminated by hazardous substances present special problems to the field investigator. The development of methods appropriate to the study of contaminated artifacts is being pioneered, and few guidelines state unequivocally which levels of contamination are acceptable. Depending upon the contaminants, structures or tanks may be decontaminated and reused, or they may be demolished and scrapped. Samples are collected from artifacts by wiping the contaminated surface or by taking cores. Drums, buckets, and other closed containers holding unknown materials are the most common artifacts discovered at hazardous waste sites.

Containers. The fact that materials are in containment reflects that they are concentrated and, therefore, are more likely to pose a safety hazard than environmental samples. Consequently, the safety of the sampling crew is important, and strict protocols are established and obeyed. Contained wastes are usually combined and then stored in tanks prior to shipment. Bulk shipment is generally used for reasons of safety and economy. In combining wastes, a screening protocol is necessary to distinguish among:

- Organic and aqueous liquids.
- Acids and alkalies.
- Reactive substances such as cyanides and sulfides.
- Radioactive wastes.
- Explosives.

The tanks at a hazardous waste site may vary in size from those that are small and transportable by a fork lift to large underground concrete installations. Each container is assigned a number by the field investigator, and each sample is assigned a sample number by the laboratory that performs the analysis. In the field, the contents of the containers are screened to indicate how they should be categorized and segregated. Separate areas are often reserved for organic solvents, acids, alkalies, reactive solids, and other solids.

Solids, liquids, sludges, gases, or combinations of these substances can be present in containers. Solid materials can be sampled using sampling triers or scoops; liquids and sludges are generally best sampled with a glass tube sampler of 10 mm dia. See Figure 2-4 for an illustration of the equipment.

When large tanks are sampled, it is crucial to discern whether distinct strata of materials are present. A preliminary investigation of the tank is made by probing the stored material with a rod or pipe. By noting the depth of penetration, the presence of accumulated solids on the bottom can be ascertained; by examining the withdrawn probe, the viscosity of the materials present can be gauged. A probe is dedicated to each tank investigated because decontamination of probes between tests is impractical.

A conductivity meter may be attached to the probe and lowered into the tank to gather information about the properties of strata in the tank. Both the depth of the probe and the measured conductivity are monitored in order to delimit aqueous layers sandwiched between organic phases. The investigator should consider that the meter may have to be decontaminated or discarded if this procedure is employed.

Rigid tubing or pipe can be lowered into the tank to draw a representative column of the contents. If a vacuum is attached to the pipe, a composite sample of the contents can be withdrawn. Using the calculated volume of the pipe and taking a measured amount of sample yield a composite sample proportional to the quantity of the substance in the tank. The layers thus extracted can then be separated for analysis. If a larger volume of sample is necessary where comprehensive characterization or treatability testing will be performed, a similar method may be applied: the suction line is lowered to the desired stratum; the line is purged or the plug is removed; next, the required volume of sample is pumped to the sampling vessel. A stratum may also be sampled by attaching a container to the probe. The lid is removed when the stratum at the depth to be sampled is reached, and the container fills. As the open container is withdrawn through the upper strata of the tank, the sample normally is contaminated slightly, but this problem can be remedied in the field either by decanting the floating layer or by using a siphon to remove the lower layer.

Samples of similar wastes from multiple containers can be blended and submitted to the laboratory for detailed analysis. Organic solvents may be analyzed for energy content, ash content, sulfides, chlorides, PCBs, and viscosity to ascertain the feasibility of recycling or incineration. Acids might be analyzed for selected metals and anions to learn whether treatment using conventional hydroxide precipitation would be applicable and which chemicals would be required.

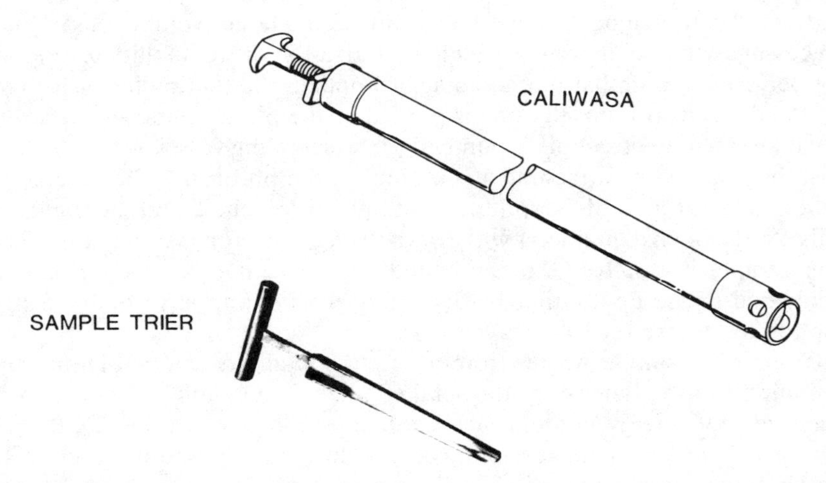

Figure 2-4. Sampling trier and composite liquid waste sampler (Coliwasa) for contained materials. Graphics supplied by Direct, Millville, N.J. Used with permission.

Structures. Organic substances may be adsorbed to surfaces of artifacts such as structures and mobilized by solvents or oils that contact the surfaces. For example, workers with oily hands could mobilize PCBs on surfaces they touch. Wet wipe samples are taken to detect organic substances. A cotton swab or filter paper is soaked in a solvent such as acetone or hexane, a measured surface area is carefully rubbed, and the swab or paper is stored in a sealed vial. The concentrations of particular substances are determined through analysis of the wipe, and the contamination is reported on an areal basis such as parts per million/cm^2.[7] An alternative method is to use the swab to collect the contaminants and then to rinse it in a defined volume of solvent; it is assumed that the contaminant is removed completely from the swab, and that only the solvent is analyzed by the laboratory. The mass of the contaminant in the solvent is used to provide information on the mass of contaminant per unit area of the surface.

Dry filter paper is used to collect dust from a surface. A specific area is wiped with the filter paper, and the paper is then analyzed to determine the level of contamination per unit of area.

Core samples are sometimes taken from, for example, wood block floors, ceiling tiles, or carpets, and the total contamination of the core is determined. However, such measures of contamination are difficult to interpret. Cores are taken by means of a hole saw, and the depth of the core, its diameter, and any sectioning of the core are dictated by the objectives of the sampling program. The holes left by the removed cores are plugged with a suitable replacement material.

EQUIPMENT AND SAMPLE CARE

To prevent cross-contamination of samples, all reusable sampling equipment is decontaminated after each sampling event. Decontamination procedures are specific to the different types of sampling equipment, the wastes, and the matrices.[8] For example, if the focus of the analytical plan is halogenated organic solvents, these solvents would be excluded from the decontamination routine. It is also important to perform the decontamination away from the work site. It is common, for example, to use acetone or methanol to clean soil-sampling equipment. These solvents should be contained and disposed of properly at a location other than the hazardous waste site, as the objective of succeeding studies of the same site may be to locate those very solvents. Were the solvents used in decontamination disposed of on the site, the future study would be needlessly confounded.

Most hazardous waste sites are remote from the laboratories that perform the analysis; so samples must be dispatched from the field site. The samples are packed to prevent breakage of the container and to maintain a cool temperature to preserve them. The U.S. Department of Transportation has promulgated regulations governing the shipment of chemical substances that are potentially hazardous wastes.[9] These guidelines define the type of containers that may be used and the labeling required of the package. Once properly packaged, samples may be shipped through a variety of channels: mail service, overnight or express couriers, air freight, buses, or personal vehicles.

LABORATORY COORDINATION AND QUALITY ASSURANCE

The sampling crew and field investigator work hand in glove with the laboratory. It is desirable to initiate coordination early in the planning phase of the sampling program so the analytical requirements will be understood by both parties. Early coordination allows the laboratory to prepare and test the equipment to be used in the field, and it minimizes potential difficulties at critical periods during implementation of the field investigation program. Additionally, it is critical that matters such as sample holding times, preservation techniques, and decontamination procedures be coordinated between the sampling crews and the laboratory.

The samples collected in the field are analyzed at a qualified environmental laboratory. The laboratory requires direction from the field investigator; it expects and requires guidelines about:

- The analyses required.
- The quality assurance and quality control procedures.

Many laboratories maintain their data and their analysis schedules with a computer. Samples are logged, analyses are recorded, and the schedule is accessed from a computer terminal. In many cases, it is possible to obtain preassigned sample numbers from the laboratory, and prepared sample containers can be supplied with preprinted labels. To mesh operations, the field crew often is directed to use computer codes compatible with the laboratory's operation when describing the samples. See Table 2-3, which gives the items of data used for such records.

This sample tagging method permits the field investigator to receive printed copies of the data generated periodically during a project. If an elaborate computer system is not available, the data can be transmitted to the laboratory as a data file and utilized by a microcomputer with data base or spreadsheet software. The data can then be reviewed and presented by the laboratory for the field investigator's examination.

SUMMARY

The field investigation is not designed until the objectives of the remedial investigation are defined. The data developed through the field investigation

TABLE 2-3. Items of data used by laboratory management information system.

TYPE OF SAMPLE:	Ground water, surface soil, surface water, air, biota, artifact.
LOCATION OF SAMPLE:	Numerical location; grid coordinates.
DEPTH OF SAMPLE:	Depth below grade, screen elevation, sample elevation.
DATE:	Date sample collected.
TIME:	Time sample collected.
SAMPLER:	Individual who took sample.

and subsequent laboratory analyses are the foundation for all remedial decisions and activities at a hazardous waste site. Therefore, the design and implementation of the field investigation must be proficient. Moreover, it is wise to delay development of the basis of design for any remedial work until the data from the field investigation have been evaluated to insure that the most comprehensive information available is applied to the site.

The design of the field sampling plan begins by incorporating all available information describing the previous uses of the site; it then develops information on all the physical conditions of the site. Quality control and quality assurance procedures are employed to manage the methods of sampling and to warrant that the samples sent to the laboratory represent bona fide field conditions. Records are maintained on all aspects of sample collection, including a chain of custody.

The field program can include the sampling of a variety of matrices with various techniques and diverse equipment. The goal of all field work is to acquire representative samples; so the equipment used is managed carefully. Calibration and decontamination of the equipment are critical to the validity of the field program.

Responsible engineers coordinate their activities with the laboratory analyzing the samples, both to foster agreement on sampling protocols and requisite sample care and to insure that the field data are accessible to all involved parties as the field program progresses and as remedial alternatives are formulated.

BIBLIOGRAPHY

Barcelona, M. J., J. A. Helfrich, and E. E. Garske. Sampling tubing effects on ground water samples. *Analytical Chemistry.* (57)1985: 460-464.

Bhatt, H. G., R. M. Sykes, and T. L. Sweeney. *Management of toxic and hazardous waste.* Chelsea, Mich: Lewis Publishers, Inc., 1985.

Bordner, R. and J. Winter (Eds.). *Microbiological methods for monitoring the environment—water and wastes.* U.S. EPA. Dec. 1978. EPA-600/8-78-017.

Brantner, K. A., R. B. Pojasek, and E. L. Stover. Priority pollutants sample collection and handling. *Pollution Engineering.* (March 1981): 34-38.

Chen, C. Y. and J. E. McCarthy. *RCRA inspection manual.* U.S. EPA Office of Enforcement. July 1980.

Cheremisinoff, P. N. and A. C. Moresi. *Air pollution sampling and analysis deskbook.* Ann Arbor, Mich: Ann Arbor Science Publishers, Inc., 1978.

deVera, E. R., E. P. Simmons, R. D. Stephens, and D. L. Storm. *Samplers and sampling procedures for hazardous waste streams.* U.S. EPA/MERL. Jan. 1980. EPA-600/2-80-018.

Furse, M. T., J. F. Wright, P. D. Armitage, and D. Moss. An appraisal of pond-net samples for biological monitoring of lotic macroinvertebrates. *Water Research.* (15)1981:679-689.

Gibb, J. P., R. M. Schuller, and R. A. Griffin. Procedures for the collection of representative water quality data from monitoring wells. Champaign, Ill: Ill. Dept of Energy and Natural Resources, et al. Cooperative Ground Water Report 7. 1981.

Greenberg, A. E., R. R. Trussell, and L. S. Clesceri (Eds.). *Standard methods for the examination of water and wastewater,* 16th Ed. Baltimore: Port City Press, 1985.

Greeson, P. E., T. A. Ehlke, et al. (Eds.). Methods for collection and analysis of aquatic biological and microbiological samples. *Techniques of water-resources investigations of the U.S. Geological Survey,* Book 5, Ch. A4. 1977.

Harris, D. J. and W. J. Keffer. *Wastewater sampling methodologies and flow measurement techniques.* U.S. EPA Region VII Surveillance and Analysis Division. June 1974. EPA-907/9-74-005.

Hazardous chemical spill cleanup. Pollution Technology Review No. 59. Park Ridge, N.J.: Noyes Data Corporation, n.d.

Mason, B. J. Preparation of soil sampling protocol: techniques and strategies. U.S. EPA Environmental Monitoring Systems Laboratory. May 1983. EPA-600/4-83-020.

New York State Department of Environmental Conservation, Division of Water. *Sample collection manual.* Dec. 1983.

Sweeney, T. L., Bhatt, H. G., R. M. Sykes, and O. J. Sproul. *Hazardous waste management for the 80s.* Ann Arbor, Mich.: Ann Arbor Science Publishers, Inc., 1982.

U.S. Environmental Protection Agency. *Guidance on remedial investigations under CERCLA.* May 1985.

———. Guide for decontaminating buildings, structures, and equipment at Superfund sites. Mar. 1985. EPA/600/2-85/028.

———. *Technical assistance document for sampling and analysis of toxic organic compounds in ambient air.* Environmental Monitoring Systems Laboratory. NTIS [PB83-239020]. June 1983. EPA-600/4-83-027.

———. *Addendum to handbook for sampling and sample preservation of water and wastewater.* Sept. 1982. Environmental Monitoring and Support Laboratory. Aug. 1983. EPA-600/4-83-039.

———. *Handbook for sampling and sample preservation of water and wastewater.* Sept. 1982. Environmental Monitoring and Support Laboratory. EPA-600/4-82-029.

———. *Safety manual for hazardous waste site investigations.* Sept. 1979.
———. *Handbook—industrial guide for air pollution control.* June 1978. EPA-625/6-78-004.
———. *NPDES compliance sampling manual.* Office of Water Enforcement. 1977.
———. *Sampling and analysis procedures for screening of industrial effluents for priority pollutants.* Environmental Monitoring and Support Laboratory. Apr. 1977.
———. *Handbook—continuous air pollution source monitoring systems.* June 1975. EPA-625-79-005.

NOTES

1. The requirements of a sampling plan are listed in the U.S. EPA document *Guidance on remedial investigations under CERCLA.*

2. U.S. EPA, EMSL. *Technical assistance document for sampling and analysis of toxic organic compounds in ambient air.* June 1983. EPA-600/4-83-027.

3. Formulas were developed by Manning, Chezy, Bazin, Kutter, Hazen-Williams, and Scobey. See H. M. Morris and J. M. Wiggert. *Applied hydraulics in engineering,* 2d Ed. New York: John Wiley & Sons, 1972.

4. A. E. Greenberg, R. R. Trussell, and L. S. Clesceri (Eds.), *Standard methods for the examination of water and wastewater,* 16th Ed. 1985.

5. Swinnex and Millipore are registered trademarks of Millipore Corporation.

6. See, for example, P. E. Greeson, T. A. Ehlke, et al. (Eds.). Methods for collection and analysis of aquatic biological and microbiological samples. *Techniques of water-resources investigations of the U.S. Geological Survey,* Book 5, Ch. A4. 1977. See also *Standard methods for the examination of water and wastewater,* 16th Ed. 1985.

7. M. P. Esposito, J. L. McArdle, A. H. Crone, and J. S. Greber (PEI Associates, Inc.) and R. Clark, S. Brown, J.B. Hallowell, A. Langham, and C. D. McCandlish (Battelle Columbus Laboratories) for U.S. EPA. *Guide for decontaminating buildings, structures, and equipment at Superfund sites.* Mar. 1985. EPA/600/2-85/028.

8. E. R. deVera, E. P. Simmons, R. D. Stephens, and D. L. Storm. *Samplers and sampling procedures for hazardous waste streams.* U.S. EPA/MERL. Jan 1980. EPA-600/2-80-018.

9. 40 CFR 262-263; 49 CFR 171-179.

Chapter 3
Analyzing Hydrogeological Conditions

When a chemical product or waste material, whether in liquid or solid form, is accidentally or purposely deposited on the land or into a stream, that material enters the hydrogeological system. On soil, components of the substance that are water-soluble may leach into the ground, or, if in liquid form, the substance may seep and migrate in a nonaqueous phase. In most cases, the quality of the ground water is affected. People are also affected because about 50 percent of all potable water used in the United States is derived from ground water resources, and many industrial processes rely on high-quality ground water. In an industrial society, undesirable substances inevitably will enter the ground water and affect its quality. Most often, the adverse environmental effects of human activity are accidental: a train derails and the drums of organic compounds it was carrying rupture and trickle into the soil. All too often, however, ground water quality has been degraded as a result of past ignorance: a landfill was located over a permeable sand unit, and the leachate now flows through the sand into the ground water system. Occasionally—though less often now than in the past—the contamination is willful and illegal: drums of wastes are dumped into a marsh on a farm instead of paying to put it into a secure landfill. No one can fully know the extent of harm to the ground water environment after such events. Information about the extent of the damage, if any, must be generated and that specialized information is developed through a hydrogeological investigation.

Reliable information about the extent of the adverse effects or the potential for adverse effects depends upon recognition of:

- The mechanisms by which ground water flows.
- The velocity of the flow.
- The quantity of the flow.
- The chemical quality of the ground water.

Those data are developed through a variety of field investigations designed and executed in a manner suited to the hydrogeological conditions of the

---

This chapter was developed by James T. Mickam, C.P.G., of O'Brien & Gere Engineers, Inc.

locale. Hydrogeologists determine the extent of contamination and develop data about the ground water system, the geology, and the contaminants. In addition to hydrogeologists, there are chemists, toxicologists, safety officers, engineers, and others who utilize those data. The hydrogeological data are also used to establish remedial design criteria, to set remedial priorities, and to rank remedial options. In this chapter, the techniques used to develop the data necessary to evaluate a site's hydrogeological characteristics are explained. The material in the first part of Chapter 10 also addresses this topic.

THE GROUND WATER SYSTEM

Ground water is one element of the hydrologic cycle. Other elements are precipitation, surface water runoff, infiltration, and evapotranspiration. By these means, the water resources of the world are continually recycled. Precipitation in the form of rain or snow falls to the earth, and it either evaporates, is transpired directly into the atmosphere, makes its way into surface waters via overland transport, or percolates into the ground water. Surface water and ground water are used for a variety of purposes and eventually return to the oceans. The cycle is continually repeated.[1] That pattern is illustrated in Figure 3-1.

Subsurface geological units that contain and convey water are defined on the basis of their relative ability to transmit ground water. *Aquifers* are subsurface geological formations that contain and transmit a sufficient volume of ground water such that water can be withdrawn for human uses. A formation with little or no capacity to transmit ground water is an *aquiclude*. *Aquitards* are intermediate between aquifers and aquicludes; they have a limited capacity to transmit ground water. Some authors prefer not to use the designation aquiclude because theoretically no, or very few, earth materials are unable to transmit ground water; instead, these authorities describe aquicludes of varying impermeability. All of these terms are relative, their use depending on the conditions at a specific locale. For example, a silty clay layer located between fine sand units may have a sufficiently low hydraulic conductivity or it may provide a marked contrast in hydraulic conductivity such that it retards the flow between the sand units. The silty clay unit would be an aquitard. At another site, a medium sand unit with moderate hydraulic conductivity may serve as an aquitard if it is located between two coarse gravel horizons.[2] An aquifer may be composed of unconsolidated sediments or lithified strata.

Two different types of aquifers behave differently hydrologically: the unconfined or water table aquifer and the confined or artesian aquifer. Figure 3-2 illustrates the two types. The *unconfined aquifer* is composed of permeable material that extends upward to the earth's surface; recharge into

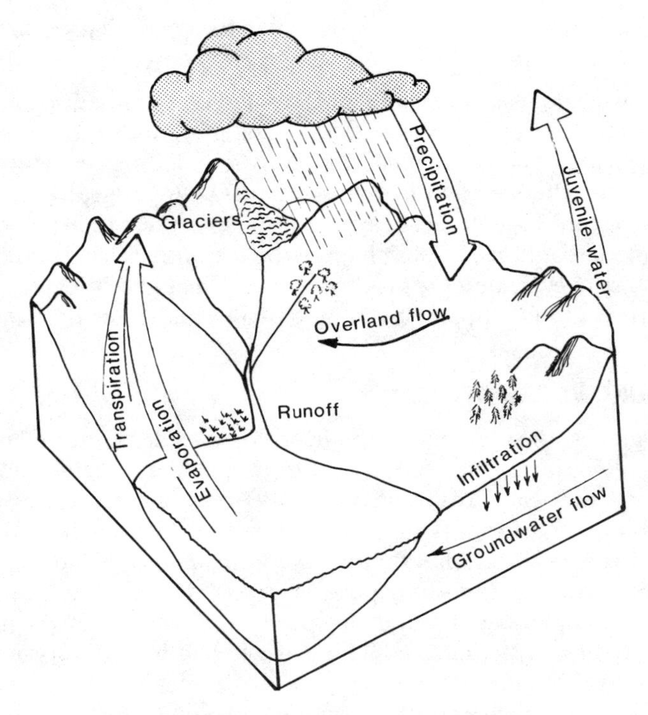

Figure 3-1. The hydrologic cycle.

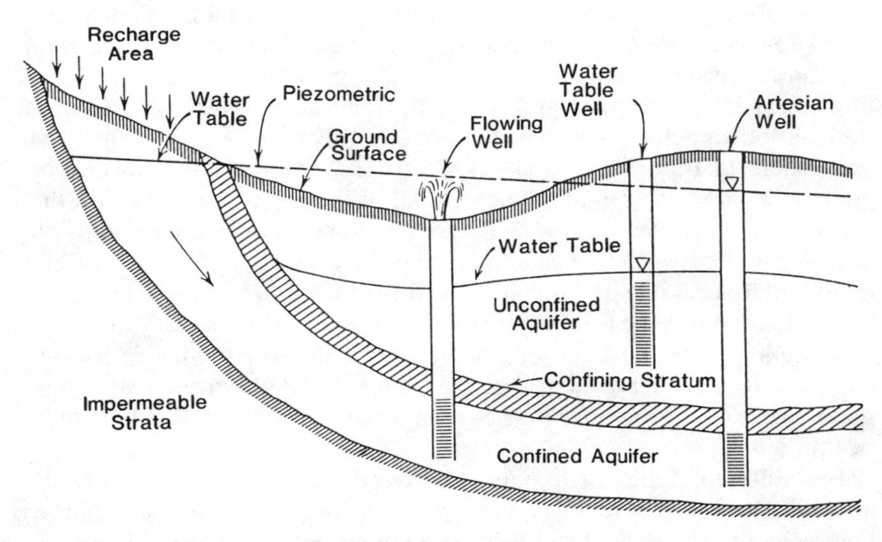

Figure 3-2. Schematic cross section of unconfined water table aquifer and confined artesian aquifer.

it is by way of direct percolation of precipitation. These aquifers are under atmospheric pressure, and the depth to ground water tends to be shallow. It is these aquifers that most often are affected by waste disposal practices, as they are directly exposed to activities at the surface. The *confined aquifer* generally occurs in the deeper subsurface and is most often overlain by an aquitard or aquiclude. It is under pressure greater than that exerted by the atmosphere. Water levels in wells completed in artesian aquifers are above the level of the upper aquifer boundary, as illustrated in Figure 3-2. The pressure in the artesian aquifer is sometimes, but not usually, sufficient to create a flowing well at the surface. Confined aquifers are not directly exposed at the surface; they receive recharge by slow percolation through the confining layer or from a remote location where the aquifer may outcrop and be exposed at the surface as an unconfined aquifer.[3]

Quantification of the elements that compose the hydrologic cycle is known as a *hydrologic budget* or *water budget.* The evaluation of an area's hydrologic budget is essential to an understanding of the ground water and surface water system of a specific site. It is particularly important in dealing with hazardous waste disposal sites because the budget defines the quantities of potentially affected ground water, the degree to which contaminants are diluted as they are transported through the ground water, the quantity of ground water that may be received by surface water, and, importantly, the quantity that may have to be collected during remedial efforts. The data collected during hydrogeological evaluations are used to formulate the hydrologic budget for a site.

Several factors are considered in developing a strategy to investigate ground water. First, the objectives of the investigation are articulated. At most hazardous waste sites, the initial goal is to define the extent and degree of contamination; a secondary goal is to examine remedial measures. From such goals, concise objectives are formulated, and a study is designed to develop the information needed to reach the objectives. The controlling regulations and the public's perception of the severity of the problem often influence the types of study elements of an investigation. Other guiding factors in the development of the study are the intended use of the data and the study's cost. A hydrogeological investigation generally costs a fraction of the total remediation program, but it can be expensive. A small study that evaluates available information and samples background conditions can cost less than $5000; an extensive study that covers a large area and requires detailed evaluations of the ground water hydrology can cost several hundred thousand dollars. If the approach to the study is planned and deliberate, the study can be both technically sound and economical.

Considering regulations, public interest, and cost, the following questions need to be addressed to identify the objectives of the study:

- What information is already known about the hydrogeology of the site and the region?
- Is there an immediate concern for negative effects on the environment, the wildlife, or the human population in the environs?
- Is the potential source of ground water contamination active, or has it terminated?
- Is the area rural or urban; is the terrain rugged or easily accessible?

Answers to these questions determine the type and level of detail of data required for the study, the time frame available, and any special logistical elements that require consideration.

In an extensive hydrogeological investigation, the study commonly is comprised of three phases:

1. Background information review.
2. Noninvasive field studies.
3. Invasive field studies.

The plan of study should provide flexibility to add, modify, or eliminate components because the study becomes redefined as it is executed. Built-in flexibility permits better technical management of the investigation and results in a more economical study because the later, more invasive and costly stages of the study are predicated on the results of the initial stages.

PHASE 1: REVIEW OF BACKGROUND INFORMATION

Before initiating a field investigation, it is advisable to perform a desktop study to assemble and review information. All available facts about regional conditions and conditions specific to the site under investigation are assembled and evaluated to determine the additional data necessary for an understanding of the effect of the waste site on the locale's ground water system. This is a logical and obvious first step, but it is often overlooked. It is economical because it insures that investigators will not waste money redeveloping existing data. The categories of information collected during this phase of the hydrogeological investigation are listed in Table 3-1.

A computer-assisted search of the literature is a rapid and economical method of discovering information, even information for a particular locale, for sites in North America and the Caribbean. The key words utilized in the search are the area under investigation and the type of waste disposal site.[4] Table 3-2 indicates where additional information can be found. It is also useful to contact the regulatory agency, if any, involved with the site to determine whether previous studies were conducted on the site or within the

TABLE 3-1. Data gathered during Phase 1 background investigation.

- Topography and drainage of study area.
- General characteristics of near-surface soils:
  - —Drainage.
  - —Ability to support vegetation.
- General subsurface stratigraphy:
  - —Consolidated or unconsolidated.
  - —Igneous, metamorphic, or sedimentary.
  - —Local and regional structural trend of fractures, bedding planes, joints, fissures.
- Structural features that influence ground water flow.
- Estimated depth to first-encountered ground water.
- Estimate of aquifers and ground water in use.
- Local climate.
- Characteristics of the disposal or spill site: incidence of solid waste, drummed liquid wastes, lagoons, underground storage tank or pipeline failures, and spills on the surface.
- Type and quantity of waste materials at the site.
- Accessibility of the site.
- Degree of exposure of wastes.
- Safety conditions.

TABLE 3-2. Sources of hydrogeological background data.

GOVERNMENT AGENCIES

U.S. Geological Survey
- Topographic maps
- Geological maps
- Satellite maps
- Water resource evaluations
- Stream flow data

U.S. Department of Agriculture
- Soil surveys
- Vegetation information
- Satellite and aerial photographs

U.S. and State Departments of Transportation
- Maps, aerial and satellite photos

U.S. Air Force and NASA
- Satellite photos

Local, Regional, and State Planning Agencies
- Master land use plans
- Water quality plans
- Plans for utilities (water, sewer, solid waste)
- Information on past land uses

PRIVATE SOURCES
- Historical site development
- Special geological reports
- Boring and well logs
- Hydrological information
- Ground water and surface water flow and chemistry

LOCAL COLLEGES AND UNIVERSITIES

JOURNALS OF TECHNICAL SOCIETIES

general area, as the regulatory agency would likely be aware of any existing studies. The studies themselves may or may not be available, depending upon the regulatory and legal status of the site.

With this background information, the level of effort required in the subsequent phases of the investigation can be gauged. Even if unambiguous answers to the questions are unavailable, attempts to develop the information are useful; the exercise defines the areas where information is deficient and will have to be developed. It is evident that without sufficient exploration of existing studies, the investigation would be crippled from the outset. In addition, without attention to information already accessible, the subsequent hydrogeological work would incur costs that might otherwise be avoided.

PHASE 2: NONINVASIVE METHODS OF INVESTIGATION

Whereas the Phase 1 work achieves a general understanding of the regional hydrogeological conditions and waste site characteristics, Phase 2 activities focus on evaluating the particular conditions of the site as they relate to the characteristics of the hydrogeology and waste deposits. It is the objective of the Phase 2 work to develop information specific to the site. Because the existing data have been scanned, information about the site can be expanded by noninvasive study methods in order to understand better the subsurface conditions to evaluate the effect of the waste site on the local ground water system. The Phase 2, noninvasive study is not always necessary. It is most often implemented for large study areas where little or no information is available on the geometry of the waste disposal site—the horizontal and vertical extent of contamination—or on the hydrogeological conditions. Phase 2 can also be a worthwhile preparation for developing a technically sound and economical invasive, or subsurface, study.

The term *noninvasive* refers to techniques that produce both specific and general information without penetrating or disturbing the waste site or the earth. These investigations are sometimes termed "surface investigations." Knowledge of the use of noninvasive techniques and the interpretation of data from those techniques is constantly changing, and methods that are applicable and successful at one site may not suit a second site even when the hydrogeological and waste characteristics appear parallel. The most commonly employed noninvasive techniques can be classified as:

- Remote sensing and aerial photography.
- Reconnaissance field mapping.
- Geophysical surveys.

Remote Sensing Imagery and Aerial Photography. The review of aerial photographs is commonly begun during Phase 1. Depending upon the time required to secure all the pertinent photos, however, these efforts are often completed during Phase 2. *Remote sensing* includes images created by enhancing nonvisible light waves, primarily those below infrared, to display various energy emissions from the earth. *Aerial photographs* are pictures of the earth in the visible wavelengths. Both techniques give the investigator a view of the hazardous waste site in the context of its surroundings.

Remote sensing imagery is developed from data gathered at high altitudes, often above the atmosphere by satellites, and the images are large-scale—an inch or centimeter on the photograph typically represents several miles or kilometers on the land surface. Because most hazardous waste sites comprise less than 100 acres (40 ha), these images generally are most useful in evaluating regional water resource conditions.

Aerial photographs, on the other hand, have a smaller scale and can be invaluable for assessing site characteristics. Air photos usually are viewed in pairs with a stereoscope in order to achieve an exaggerated vertical dimension. Some common features of interest discerned from aerial photos are listed in Table 3-3.

The direction of ground water flow in bedrock aquifers is controlled by the location and orientation of various structural features within the rock unit; for example, fractures, bedding planes, faults, and a variety of solution characteristics such as sinkholes and caverns. To evaluate potential routes of ground water and leachate migration, one must have a good understanding of the locations of the controlling influences. Therefore, structural evaluation of the bedrock is necessary at sites where a bedrock aquifer is the hydrogeological system of concern, and it should begin with an inspection of aerial photographs. The remaining steps of the investigation, particularly

Table 3-3. Data evident on aerial photographs.

Data evident on aerial photographs
Area geomorphology and surface drainage patterns.
Horizontal extent of waste materials.
Historical development of site.
Bedrock structure: location, orientation, and frequency of vertical fractures and karst features.
Areas of stressed vegetation adjacent to the site, possibly caused by waste leachate streams.
Ground water discharge points.
Waste leachate seeps.

Figure 3-3. Analysis of history of waste site with aerial photography. (a) 1980 photograph of waste site obscured by ground cover.

locating ground water monitoring wells, can then be concentrated in the identified routes of flow.

A comparison of historical with current air photos is useful to deduce and document the development of a waste site. Such a comparison is critical if no other records exist, and if no people familiar with the site are available for interviews. With this comparison, the locations of the primary material depositories can be determined (see Figure 3-3). In Figure 3-3(a), a photograph taken in 1980, the general location of an abandoned hazardous waste

(b) 1968 photograph of waste sites better illustrating boundaries of disposed activities.

disposal site is illustrated. The boundaries of the waste disposal activities are clearer in Figure 3-3(b), a photo taken in 1968. Such an evaluation can direct field activities by providing data unavailable on the ground. The technique requires little time and is economical.

Field Reconnaissance. The data base for the site is expanded through field reconnaissance activities, where the new data are most useful in fine-tuning the subsequent phases of the study. The objectives of the field reconnaissance are both to authenticate the information available from the background studies and to reevaluate the appropriateness of proposed study methods. Before field activities commence, however, serious reconsideration of the study is desirable and practical; both the field inspection goals and the logistics, which may affect the usefulness or ease of completion of certain work tasks, are reconsidered.

None of the investigatory techniques discussed to this point has required anyone to enter the site. When the investigation moves to the field, the background work proves its worth, as the investigators bring with them knowledge of the fundamental hydrogeology of the site. It is advisable to prepare a checklist of items to be inspected and thus utilize time in the field to best advantage. Significant observations of field reconnaissance mapping are listed in Table 3-4.

Geophysical Surveys. Earth and waste materials exhibit characteristic responses to physical stimuli. For example, different materials alter the earth's electromagnetic field uniquely or respond distinctly to stimuli such as electrical current or shock waves. The behavior of earth or waste materials under various physical conditions can be evaluated with geophysical techniques in the field. No penetration, or just limited penetration, of the earth is required, and these methods yield no samples; therefore, they are

TABLE 3-4. Important observations of field reconnaissance mapping.

TERRAIN OF SITE AND SURROUNDINGS
- Accessibility for geophysical surveys and drilling equipment
- Size of site compared with data from background information

UNIQUE SITE LOGISTICAL CONSIDERATIONS
- Presence of underground utilities
- Need for excavation/drilling clearance
- Access routes across private property
- Availability of clean water for decontamination

SITE GEOLOGICAL CONDITIONS
- Agreement of surface features with regional pattern of geology
- Locations of bedrock outcrops
- Consistency with background data

TOPOGRAPHY, DRAINAGE, AND VEGETATION
- Consistency with background data
- Locations of leachate or liquid waste discharges
- Location of stressed vegetation found during background study

CURRENT CONDITION OF WASTE MATERIALS
- Phase(s)
- Degree of exposure
- Activity of source(s)

MONITORING DEVICES IN STUDY AREA
- Confirmation of previously established devices
- Verification of depths of existing monitoring wells and ground water elevations

noninvasive. Geophysical surveys furnish data that can identify lithological and structural features below the surface.

The geophysical techniques discussed here are magnetics, electromagnetics, electrical resistivity, ground penetrating radar, and seismic refraction. These are the most common geophysical survey techniques used at hazardous waste sites although others, including seismic reflection and induced polarization, are available. The choice of the appropriate technique is a function of the purpose of the study, the hydrogeological conditions, the nature of the waste materials, and the accessibility of the site. (Borehole geophysical methods that collect data in a completed borehole are discussed later.) Important aspects of a number of geophysical survey methods are summarized in Table 3-5.

TABLE 3-5. Summary: geophysical survey methods.

SURVEY	APPLICABILITY	ADVANTAGES	DISADVANTAGES
SEISMIC REFRACTION/ REFLECTION Determines lithological changes in subsurface.	• Ground water resource evaluations • Geotechnical profiling • Subsurface stratigraphic profiling including top of bedrock	• Relatively easy accessibility • High depth of penetration dependent on source of vibration • Rapid areal coverage	• Resolution occasionally obscured in layered sequences • Susceptibility to noise from urban development • Difficult penetration in cold weather (depending on instrumentation) • Operation restricted during wet weather
ELECTRICAL RESISTIVITY Delineates subsurface resistivity contrasts due to lithology, presence of ground water, and changes in its quality.	• Depth to water table • Subsurface stratigraphic profiling • Ground water resource evaluations • Aquifer studies • Landfill evaluations	• Rapid areal coverage • High depth of penetration possible (400–800 ft [140–290 m]) • High mobility • Results that can be approximated in field	• Susceptibility to natural and artificial electrical interference • Limited use in wet weather • Limited utility in urban areas • Interpretation that assumes a layered subsurface • Lateral heterogeneity not easily accounted

(continued)

TABLE 3-5. *(continued)*

SURVEY	APPLICABILITY	ADVANTAGES	DISADVANTAGES
ELECTRO-MAGNETIC SURVEY Delineates subsurface conductivity contrasts due to changes in ground water quality and lithology.	• Subsurface stratigraphic profiling • Ground water contamination evaluations • Landfill evaluations • Ground water resource evaluations • Locating buried utilities	• High mobility • Rapid resolution and data interpretation • High accessibility • Effectiveness in analysis of very high resistivity • Equipment easily accessible	• Data reduction less refined than with resistivity • Use unsuitable in areas with surface or subsurface power sources, pipelines, utilities • Less vertical resolution than other methods • Limited use in wet weather
GROUND PENETRATING RADAR Provides continuous visual profile of shallow subsurface objects, structure, lithology.	• Locating buried objects • Delineation of bedrock subsurface and structure • Delineation of karst-type bedrock cavities • Delineation of the physical integrity of human-made earthen structures	• Great areal coverage • High vertical resolution in suitable terrain • Visual picture of data	• Depth of penetration limited (approx. 50-75 ft [15-23 m]) • Accessibility limited due to bulkiness of equipment and nature of survey • Interpretation of data not absolute • Limited use in wet weather
MAGNETICS Detects presence of buried metallic objects.	• Location of buried ferrous objects • Detection of boundaries of landfills containing ferrous objects • Location of iron-bearing rock strata and dikes	• High mobility • Data resolution possible in field • Rapid areal coverage	• Detection dependent on size and ferrous content of buried object • Difficult data resolution in urban areas • Limited use in wet weather • Data interpretation complicated in areas of natural magnetic drift

Magnetic Surveys. Magnetic surveys use an instrument called the magnetometer to delimit the location of buried ferrous objects—underground tanks, buried drums, and pipelines. Because the earth has a magnetic field, it acts as a large dipole magnet. Objects, either on the surface or below the ground, that themselves have a degree of magnetism disturb the earth's magnetic field. It is this distortion that the magnetic survey recognizes.

A variety of lightweight and portable magnetometers are available. (Geophysicists exploring for minerals utilize aerial instruments to survey large areas.) Magnetometers are capable of evaluating the horizontal extent of waste depositories. The mathematical calculations required to predict the vertical extent of the objects demand assumptions about the magnetic moment of the target body, so that the depth of objects can be predicted only with great uncertainty in a heterogeneous waste depository. The depth to which a magnetometer can detect objects is generally limited to between two and four times the largest representative dimension of the target. Artifacts on the site such as power lines, substations, and above- or below-ground structures and pipes can confound the magnetic survey.[5]

A magnetic survey of a hazardous waste site collects data at regularly spaced data points by means of a survey grid or traverse established on the ground. Initially, the grid is spaced widely, say every 50 or 100 ft (15 to 30 m); in a second pass, more readings are taken in locations where target objects were indicated. The collected data are plotted on the site map (plan view) or in profile, and the magnetic anomalies are identified. Figure 3-4 displays typical data. The earth's magnetic field fluctuates throughout the day; so it is necessary to return to an established base station every couple of hours in order to correct for variances. Depending upon the terrain of the site, between 5 and 10 acres (2 and 4 ha) can be surveyed within a day.

Electromagnetics. Electromagnetic surveys, or terrain conducting surveys, evaluate the variability of subsurface electrical conductivity. This technique can be used to investigate ionic and inorganic contaminants and is applicable if variable electrical conductance in the subsurface is present. The basis for electromagnetic surveys is Ohm's law, which declares that resistance, or the ratio of voltage to current, for all or part of an electrical current at a fixed temperature is generally constant. Electromagnetic surveys use an electromagnetic field induced by an electric current.[6]

Data from an electromagnetic survey can be presented in plan view or profile. Electromagnetic data are presented as conductance, the reciprocal of resistivity. The data can be used to evaluate the horizontal and vertical extent of waste materials, and they can also furnish information about the site's stratigraphy and provide information about the extent of a contaminant

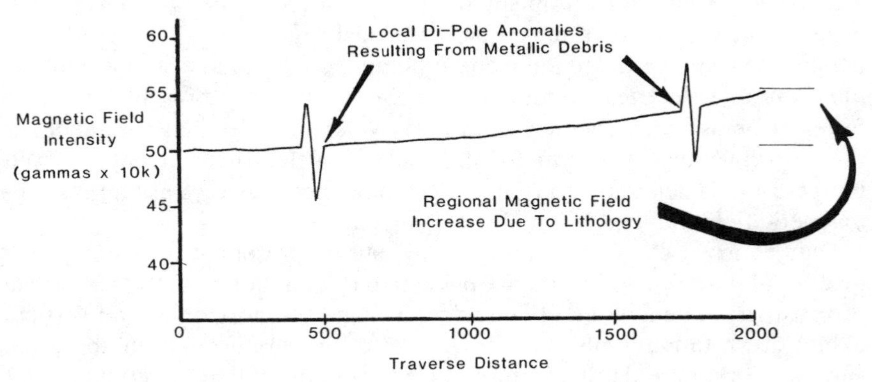

Figure 3-4. Results of magnetic survey.

plume—one with electrical conductance—in the ground water. As with other techniques, the data are commonly gathered from points on a grid.

Figure 3-5 illustrates data gathered on a site where a saturated brine solution had been released to an unconsolidated unconfined aquifer. As the figure shows, the electromagnetic survey effectively identified ground water with high electrical conductance, the areas contaminated with the brine. These data directed a drilling program and thus minimized the costs of exploration. Electromagnetic profiles can be combined with magnetic surveys to locate buried metallic objects.

Electromagnetic survey equipment is portable, and its use is rarely limited by the ruggedness of the terrain. The distance between the transmitting coil and the receiver coil can be varied to change the depth of penetration. The equipment may or may not contact the ground, and surveys, particularly profiles in which the coil spacing is fixed, are rapid. The equipment gives continuous readings of subsurface conductance, and data are available immediately in the field.[7]

Electrical Resistivity. Electrical resistivity surveys measure differences in subsurface electrical resistances. They can be used to correlate electromagnetic surveys, as resistivity is the reciprocal of conductance. These surveys are an application of Ohm's law. The earth's resistance is affected by the lithological composition and structure of rocks below grade and the presence and quality of ground water. Surveys of electrical resistance depend on differences in resistance between subsurface materials. If the horizontal and vertical resistances of surveyed strata are equivalent, the survey can detect variations in electrical conductance of the ground water. Thus the survey identifies the extent of contamination in ground water where the specific conductance of the fluid is affected by contaminants.

During a resistivity survey, an electrical current is induced into the subsurface through electrodes. The drop or change in electrical potential is measured at electrodes distant from the source electrodes. The change in electrical potential is a function of the length of the path of the current and the electrical resistance of the conducting media. A vertical resistivity profile is developed as the spacing of the electrodes is altered about a central point. From those readings, subsurface resistivities are calculated. Several different electrode configurations have been employed; the two most common are the Wenner and Schlumberger electrode configurations.[8]

The data from an electrical resistivity survey may be interpreted qualitatively and correlated with other known subsurface conditions. Alternatively, the data may be reduced mathematically to develop a model of the site below grade and predict the depths to the interfaces of the geological formation. The reduced data are presented both in plan view, to illustrate the variation

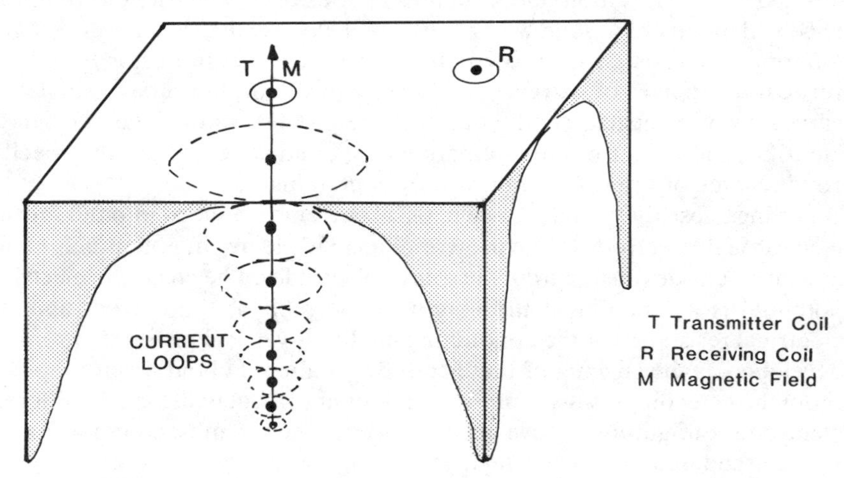

Figure 3-5. Electromagnetic survey.

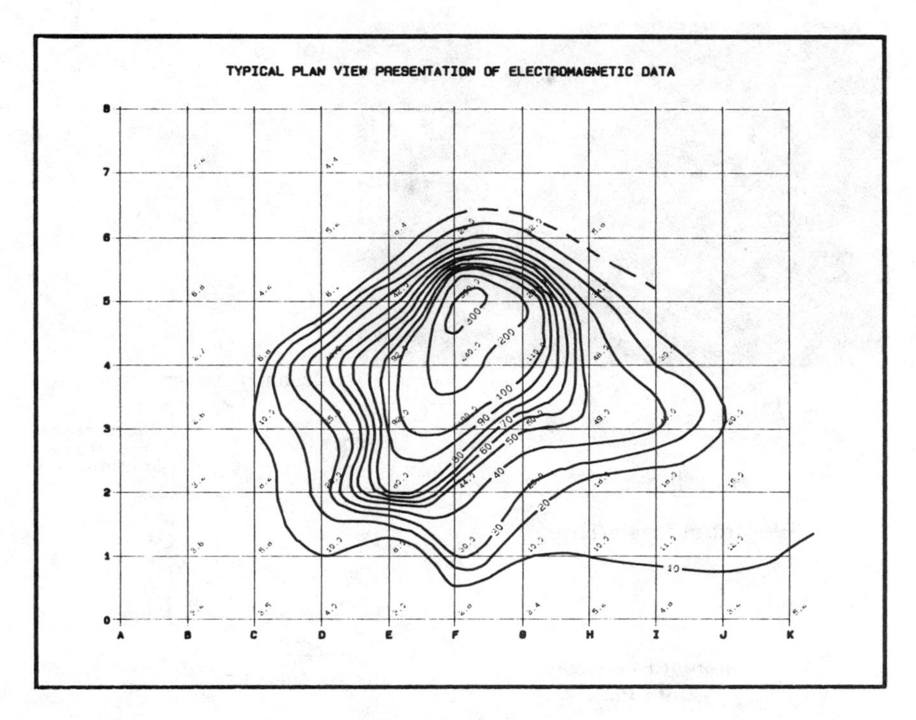

Figure 3-5. *(continued)*

in horizontal resistivity at specific depths, and in profile, to depict variations in vertical resistivity. Both arrangements are illustrated in Figure 3-6. Figure 3-6(a) portrays a survey completed in the midwestern part of the United States to evaluate the horizontal configuration of a contaminant plume originating from a surface impoundment containing metal pickling wastes. Figure 3-6(b) shows electrical resistivity data from a site in the Northern Plains and depicts the vertical extent of buried waste materials.

There are limitations to the use of electrical resistivity surveys. As with magnetic surveys, subsurface resistivity is affected by cultural features; therefore, such surveys are unsuitable for urban areas or where buried conductors are present. Where vertically extensive and highly resistive unsaturated material is present, it is difficult to achieve good contact between the electrodes and the ground. Further, because the interpretation of resistivity data assumes uniform strata, lateral subsurface changes confound readings.

The equipment used in electrical resistivity surveys is portable, and the work usually is not limited by rugged terrain, but it is easier to maintain the accuracy of the electrodes' spacing on gentle relief than on rough surfaces. On

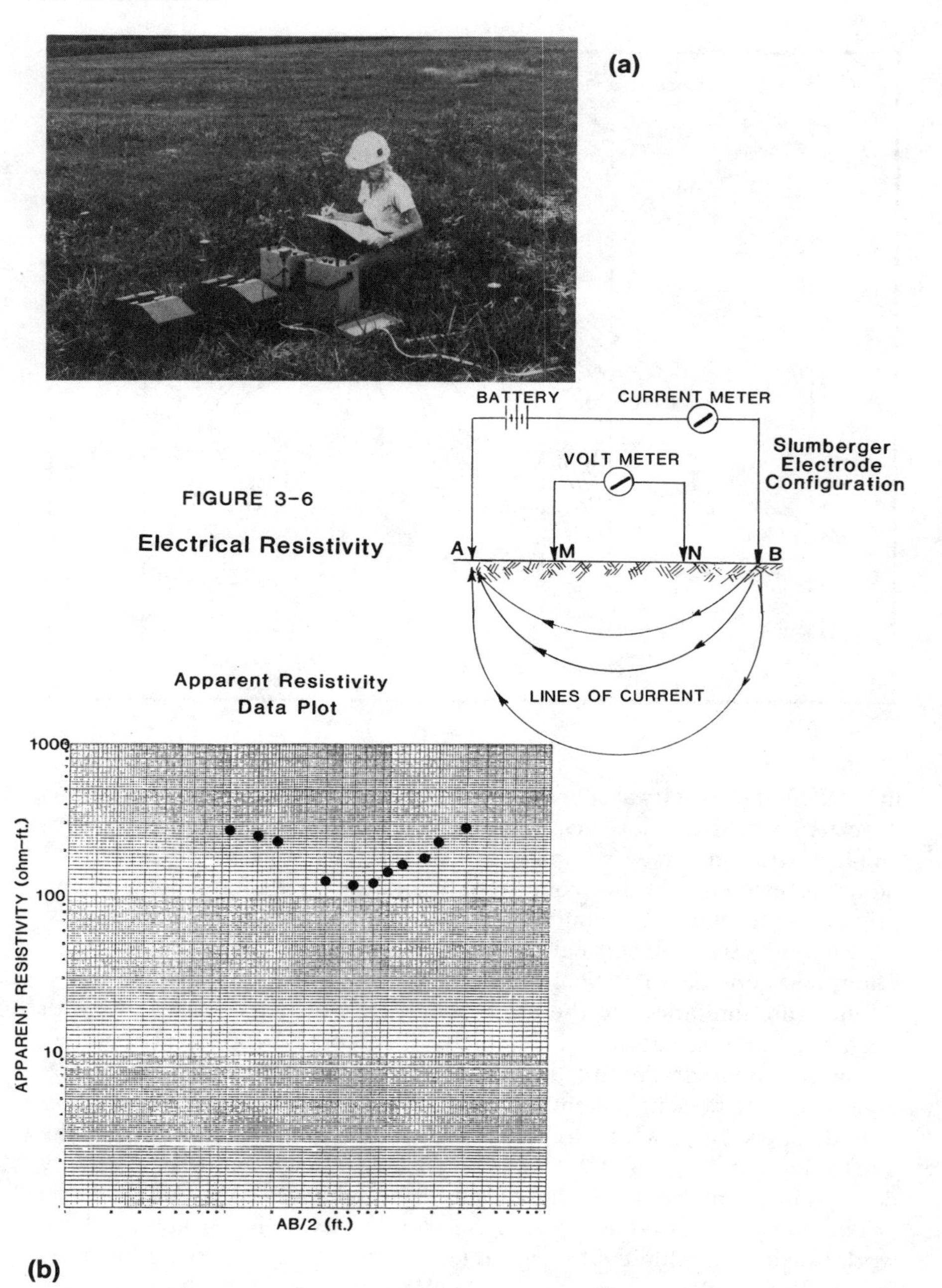

Figure 3-6. Electrical resistivity surveys.

sites with substantial relief, the electrodes are placed parallel to topographic contours. Depending on the depth of the survey, electrical resistivity surveys are relatively fast. Experienced survey crews can complete about 20 sounding stations daily in a survey with a maximum sounding depth of 150 ft (46 m) and readings taken every 20 ft (6 m).

Ground Penetrating Radar. Ground penetrating radar is applicable to locating buried objects and delimiting the extent of various subsurface features such as waste disposal trenches and pits. It has been employed successfully to locate bedrock surfaces, fracture zones, and karst features.

During the ground penetrating radar survey, a short repetitive electromagnetic pulse is radiated into the earth and reflects from subsurface features; the travel time of the pulses is recorded at a receiving antenna. If a source of electromagnetic pulses and a receiving antenna are towed across the landscape, a continuous profile can be generated. Although the depth of penetration is sometimes limited by the capability of the equipment available, it is primarily a function of the electrical conducting properties of the subsurface material. In general, greater penetration is achieved in rather resistive media, as materials with high electrical conductivity disperse the pulse. Penetration depths of 10 to 30 ft (3 to 9 m) are common, but penetrations deeper than 100 ft (30 m) have been reported. A major advantage of ground penetrating radar equipment is that it is generally unaffected by unrelated sources of electromagnetism.

Data from ground penetrating radar are usually presented as surface profiles. Figure 3-7 was generated at an industrial facility on the East Coast in which the task was to locate buried metallic and nonmetallic drums. The equipment necessary to conduct a ground penetrating radar survey is cumbersome. Specialized instruments are required for signal reception, filtering, and recording.[9]

Seismic Refraction. Changes in subsurface lithology can be detected with seismic refraction surveys. Such surveys are especially useful in delimiting the topography of bedrock overlain by unconsolidated sediments and the thickness of waste material at hazardous waste sites. The velocity with which seismic or shock waves travel through different media is related to density. To achieve usable data, a contrast in velocities is necessary; seismic velocity must increase with depth. In the ideal situation, low-density strata lie over high-density targets. If a dense layer is positioned between the surface and the target, data interpretation is difficult, and if layers of especially low velocity are at the surface, the depth of penetration is reduced. Extraneous sources of vibration, from construction activities or traffic, can confound the data.

Figure 3-7. Ground penetrating radar.

Velocity boundaries are the locations where media of contrasting densities meet. In a refraction survey, seismic waves are refracted through the medium at the boundaries.

Refraction surveys can establish the type of subsurface material. They can determine the bedrock topography and are useful at hazardous waste sites where the surface of the bedrock may have an influence on the flow of ground water in the unconsolidated zone or where detailed geological structural mapping is demanded. Data are gathered by generating a shock: the ground is struck with a weight, or an explosive charge is fired. The times required for the wave to penetrate the subsurface, to refract, to travel along the refraction boundary, and to return to the surface are measured, and these data are recorded at different distances along predetermined lines. An example of seismic refraction data is presented in Figure 3-8. The survey equipment is available in portable units, and terrain is not usually a limiting factor. The maximum practical depth of a seismic refraction survey with portable equipment and nonexplosive shock wave generating sources is generally less than 100 ft (30 m).

PHASE 3: INVASIVE FIELD METHODS

If the results of Phase 2 indicate that the lithology and the ground water hydrology require additional characterization, samples of the subsurface materials and of the ground water contained within them are necessary. Invasive methods are employed to collect lithological samples and to install ground water monitoring wells. Invasive investigative techniques encompass conventional excavation, soil profile probing, and borehole drilling. The equipment used in invasive investigation techniques can be as simple as a shovel or as complex as a drilling rig, depending on the subsurface geology, the proposed depth of penetration, the type of sample demanded, whether the sampling device is to be temporary or permanent, and whether the device is to collect fluids or gases. The invasive techniques presented here are those most commonly used for the investigation of hazardous waste sites.

Conventional Machine Excavation. Where detailed information from shallow depths is required, a trench is suitable. Trenches are a quick and economical means to assess the horizontal and vertical extent of buried wastes. They give the geologist the opportunity to inspect directly the details of shallow soil horizons. Conventional excavation equipment—usually small rubber-wheeled backhoes and sometimes mini-backhoes or large hydra-shovels—is used to open a trench in the earth. The investigator enters the trench, examines the exposed subsurface material, and collects samples of the desired subsurface unit. Trenches are not usually employed for explora-

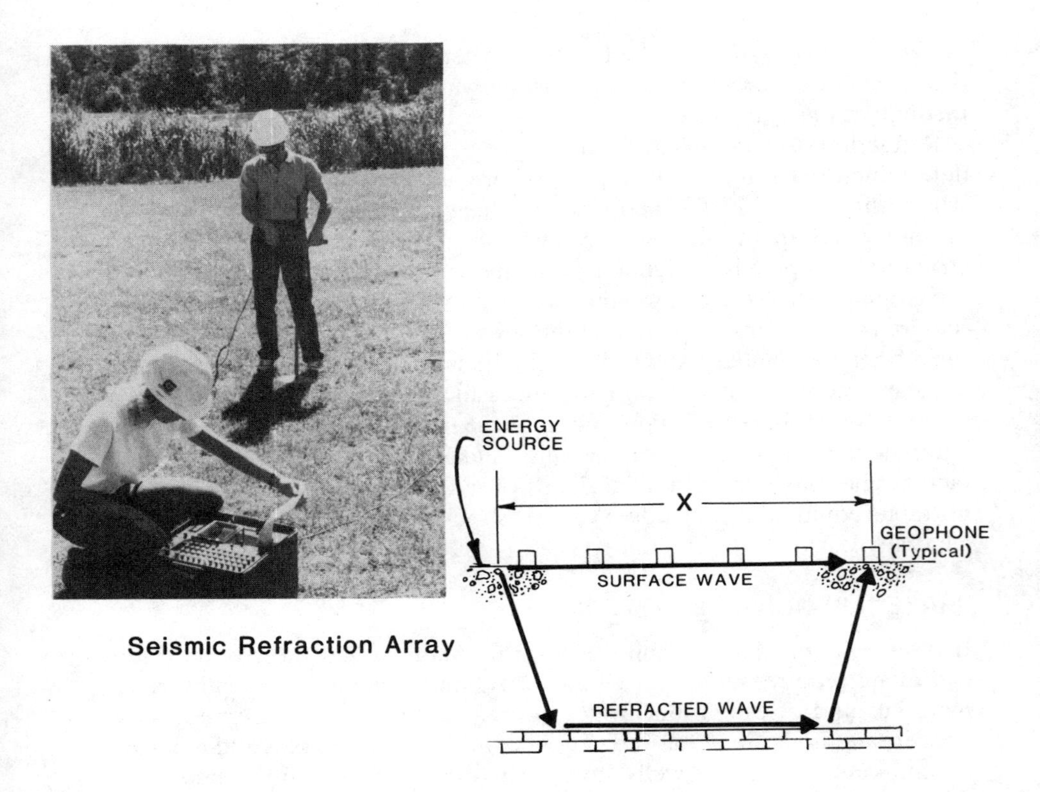

Seismic Refraction Output

TIME IN MILLISECONDS

5 10 15 20 25 30 35

GEOPHONE SPACING (Feet)

5 10 15 25 35 45 55 65 75 85 90 95

Figure 3-8. Seismic refraction survey.

tion at depths greater than 15 ft (4.5 m) for logistical and safety reasons, and they are not usually dug to depths below the local water table because such excavations are unstable, even in stiff and cohesive clays.

Excavation has been widely used at hazardous waste sites to collect large volumes of samples for physical and chemical analysis. Figure 3-9 illustrates trenching at a site where petroleum-based tars had been deposited. A majority of the tars were found to occur within 5 ft (1.5 m) of the surface. Trenching creates large volumes of spoil, and the technique may be uneconomical if soil material considered hazardous must be removed from the site and discarded according to pertinent regulations.

Developing Soil Profiles. A soil profile evaluates near-surface soil horizons and rock structure. In hazardous waste investigations, it can assess soil types and the extent of hazardous waste contamination. The profile is developed by completing a small excavation or borehole and by collecting, logging, and classifying a small disturbed sample of the desired subsurface or waste material.

Soil profiles, in the context of this book, usually are completed with manual equipment, shovels, and augers and limited to 10 ft (3 m) deep. A tripod drilling apparatus that relies on limited mechanical power also is used. The shovel or auger extracts a disturbed sample. With a tripod, the depth to which a relatively undisturbed soil core can be obtained is increased because a weight aids in advancing the core barrel. Terrain rarely limits the areas where soil profiles can be completed, as the tools are portable. It is not advisable to employ soil profiles to collect samples below the ground water table, particularly in incoherent soils, because the borehole probably will collapse before a specific sample can be collected.

Borehole Drilling. If the depth of excavation is greater than 10 to 15 ft (3 to 4.5 m), or if the excavation is below the local water table, borehole drilling is employed. Borehole drilling is the most widely used invasive investigatory technique at hazardous waste sites and for hydrogeological investigations in general. To drill a borehole, a mechanically powered drilling apparatus or drilling rig pierces the earth, and samples of the subsurface are removed. Boreholes are drilled at hazardous waste sites to evaluate subsurface lithologies and to install subsurface gas and ground water monitoring wells. During a remediation project, they are completed to install ground water collection wells and injection wells. Several different techniques are available for drilling boreholes. The physical principles have not changed in the past few decades, but the equipment is constantly improving.

Figure 3-9. Trenching method of hazardous waste site examination.

Auger Drilling. The most widely used type of borehole drilling equipment at hazardous waste sites is auger drilling, applicable only in unconsolidated material. The rotating screw of the drilling stem is able to penetrate the earth rapidly, and the equipment is relatively easy to operate and mobile. Two types of auger drills are used: solid stem and hollow stem. See example in Figure 3-10.

The auger used in solid stem auger drilling has a solid core attached to a continuous untapered screw making up a single piece for the drill tool. As the auger advances into the earth, subsurface material is displaced by the rotating auger flights. Penetration of the earth is rapid, 20 to 40 ft/hr (10 to 20 cm/min). Composite samples of the drill cuttings are deposited at the surface and collected at regular intervals, say every 5 ft (1.5 m). Discrete samples from specific depths cannot be obtained.

Solid stem auger drilling can be efficient and economical at depths less than 50 ft (15 m), but is difficult and unreliable below that, and the borehole loses stability and collapses below the ground water table. The interpretation of a composite sample taken from a range of depths cannot be precise. Materials from deep formations, particularly those below the water table, cannot readily reach the surface and are questionable. Therefore, many hydrogeologists find the method unreliable at hazardous waste sites and rely on solid stem auger drilling only when the nature of the geology is not at issue, and where only shallow borings are required, for example, to install equipment to a limited depth below the ground water table.

Hollow stem auger drilling also advances a screw-type drill stem into the earth and displaces the subsurface material, but the augers are hollow and permit discrete, relatively undisturbed samples to be collected from specific depths. Samples are usually collected every 5 ft (1.5 m) or at each change in the lithology. In unconsolidated material at hazardous waste sites, hollow stem auger drilling is the drilling method of choice. It enables the use of documented and defensible sampling methods: split spoon[10] and Shelby tube.[11] The sampling device, the split spoon or the Shelby tube, is attached to a section of special pipe called the drill rod and lowered inside the hollow stem auger. The split spoon is driven into the unpenetrated subsurface material below the auger stem; the Shelby tube is pushed. The drill rod and the sampling device are then removed from the borehole, and the sample is brought to the surface for recording and classification. The split spoon is driven with a standard weight of 140 lb (63.5 kg) dropped from a standard height of 30 inches (762 mm). The number of blows required to drive the sampler in each 6 inches (152 mm) is recorded. The number of blows is important information. It can be used both to evaluate the engineering bearing capacity of the subsurface material and to assess the relative density of the subsurface and, therefore, lithological differences.

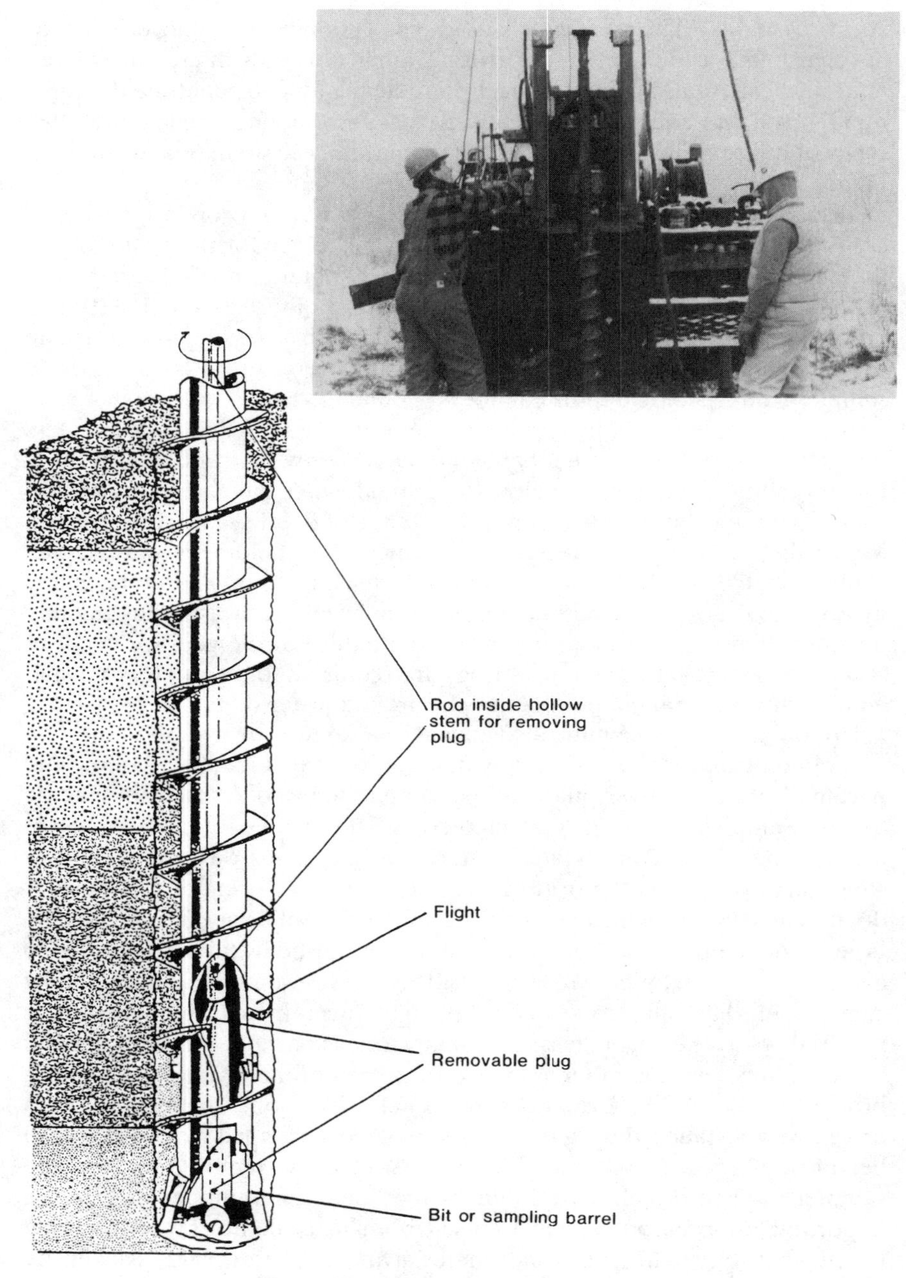

Figure 3-10. Auger drilling.

Although hollow stem auger drilling can collect accurate samples below the ground water table, saturated material, particularly sand, can heave into the auger under hydraulic pressure and create sampling difficulties. To compensate, a hydraulic head within the auger often is produced, at a force greater than or equal to that created by the aquifer by the addition of fluid; alternatively, heaved material is removed from the auger by injecting fluid down the drill rod stem and flushing or washing it to the surface.

Auger drilling is limited to unconsolidated materials, although very dense unconsolidated materials, such as some glacial tills, can resist the penetration of low-powered machines. The size of the borehole of auger drills can vary, but it is generally less than 8 inches (20 cm) for hazardous waste site investigations. Most manufacturers recommend against the use of auger equipment below 100 ft (30 m), but augers have been successfully operated at depths of 175 ft (53 m) in favorable geological conditions.

Conventional Rotary Drilling. Rotary drilling describes the creation of a borehole by means of the mechanical rotation of the drilling tool. Rotary drilling does not use a pitched screw to displace the earth; instead, fluid or air is forced down the drill stem to displace material and to remove it from the borehole via the annular space between the drill rod and the borehole wall. Figure 3-11 illustrates a fluid rotary drilling system.

Sodium bentonite or a synthetic material is most often added to water to form the drilling fluid, or mud, when rotary drilling is completed in unconsolidated material. The fluid both displaces the cuttings and stabilizes the borehole wall by forming a "mud cake" on the sides of the hole. In consolidated bedrock, water or air is commonly the displacing medium. Water or air rotary drilling also is used in unconsolidated materials, but it generally requires that a casing be carried down the advancing borehole to prevent collapse, particularly below the water table.

Rotary drilling equipment can be compact and mobile or large and cumbersome to transport, depending upon the depth it has the capacity to bore. The depth of penetration is limited by the rig's capacity to hoist the drill rod and to circulate fluid. Some small-sized rotary drilling equipment cannot penetrate deeper than 400 ft (120 m). Large rigs, used mostly in the petroleum industry, can reach thousands of feet. The diameter of the borehole varies typically between 4 and 8 inches (10 and 20 cm). Boreholes 24 inches (61 cm) or larger, a size necessary to install large-diameter recovery wells, have been completed using conventional rotary drilling methods. Samples may be collected by simply straining the returning drilling fluid and retrieving the cuttings it contains. The fine grains in the unconsolidated samples are lost, a condition acceptable in consolidated bedrock. Undisturbed samples in unconsolidated material can be collected by using the split spoon

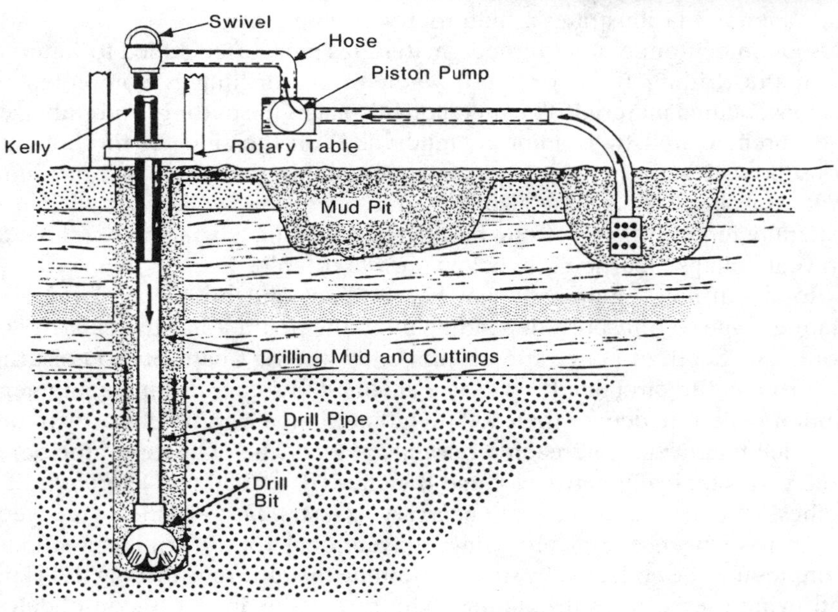

Figure 3-11. Operation of rotary drill.

or the Shelby tube; the drill stem is removed, and a drill rod with the sampling device is lowered into the borehole to collect the sample. Various coring tools are also available for use in both unconsolidated and consolidated formations. Efficiency is reduced when drilling is stopped, and fluid is circulated in order to collect undisturbed samples.

Penetration during rotary drilling is rapid. Advancement rates of 10 to 20 ft/hr (5 to 10 cm/min) in unconsolidated material and 5 to 15 ft/hr (2.5 to 7.5 cm/min) in consolidated material or greater can be achieved. At depths greater than 100 ft (30 m) and for boreholes greater than 8 inches (20 cm), penetration is slower.

Percussion Drilling. Percussion drilling is not commonly used for hydrogeological investigations but mostly is employed to install water supply or contaminant recovery wells. It is occasionally used for data gathering if auger or rotary equipment is unavailable, but, as a rule, it collects only disturbed samples. To develop the borehole, the earth is struck repeatedly with a cutting tool or drill bit. There are two types of percussion drilling procedures: conventional cable tool drilling and down-hole hammer drilling. Figure 3-12 illustrates these drilling systems.

Conventional cable tool drilling was one of the first modern drilling methods developed to complete deep boreholes. With this technique, a heavily weighted cutting tool or drill bit is attached to a cable string and repeatedly raised and lowered to strike the earth and dislodge the subsurface material. Cuttings are removed from the borehole by a hollow cylinder, called a bailer, with a check valve at the bottom. The bailer travels down the borehole and collects cuttings at the base of the hole. When cable tool drilling is completed in unconsolidated material, a casing is driven down the hole as the hole is advanced to prevent cave-in. In consolidated bedrock, short casing lengths are commonly used at the surface to keep the hole plumb when starting. See Figure 3-12(a).

Down-hole hammer drilling employs the same principle as cable tool drilling: the earth is repeatedly struck to dislocate materials at the base of the borehole. A flat-headed air-actuated tool similar to a jackhammer is attached to a conventional drill rod and advanced down the borehole. The cuttings are forced up the borehole by high pressure air from the same source used for operation of the hammer. See Figure 3-12(b).

The size of percussion drilling rigs varies with the rigs' capacities for depth and borehole diameter. Down-hole hammer rigs are often combination machines and have rotary capabilities as well as percussion applications. Rigs with down-hole hammer options may require support vehicles to carry the rods, bits, and air compressors if they cannot be mounted on the drill rig.

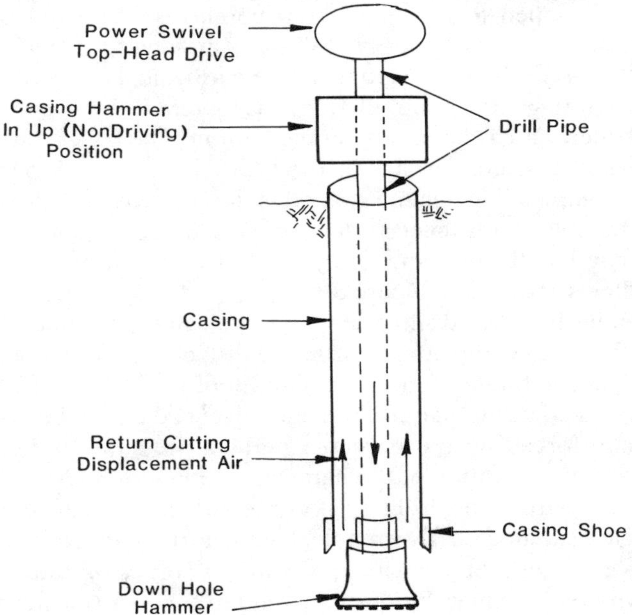

Figure 3-12. Down hole hammer drilling.

Conventional cable tool drilling penetrates rather slowly. In consolidated materials, the penetration rate may be only 1.25 to 5 ft/hr (0.6 to 2.5 cm/min) in bedrock; in unconsolidated material, the rate can be 5 to 20 ft/hr (5 to 25 cm/min). Down-hole hammer drilling has improved percussion drilling decidedly over the past two decades. In bedrock, the rate ranges between 10 and 50 ft/hr (0.4 and 2 cm/min); in unconsolidated material, the rate may be as rapid as 50 to 75 ft/hr (25 to 38 cm/min) and is often limited only by the rate at which the drill rod and casing connections can be made. The author observed rates between 20 and 30 ft/hr (10 and 15 cm/min) in an evaporite bedrock sequence—limestone, anhydrite, and gypsum—for a borehole 10 inches (25.4 cm) in diameter.

Borehole Geophysical Surveys. Having completed the borehole, the investigator has several methods at hand to evaluate further the subsurface lithology and the hydro-stratigraphy of the earth penetrated by the drilling rig. These means, known as borehole geophysical logging, are completed within open or cased boreholes and collect geophysical information from within the formation. These surveys can furnish information unavailable from inspection of the drill cuttings and are particularly applicable to boreholes completed without discrete sampling methods, such as those frequently employed with rotary or percussion drilling.

An extensive suite of borehole geophysical survey methods is available and usually used by the petroleum industry; however, the methods employed in water resources or hazardous waste investigations are generally limited to natural gamma ray logging, electrical logging, temperature logging, and borehole caliper logging. More specialized nuclear and sonic techniques are available, but they are uneconomical unless the circumstances of a specific site warrant them. Both portable and vehicle-mounted instruments are available to take measurements.

Natural Gamma Ray Logs. Each different subsurface material has a characteristic natural gamma ray decay rate. The natural gamma ray particle decay emitted by subsurface materials can be measured and recorded. The dominant isotope detected is potassium 40 (K^{40}), the isotope that accounts for approximately nine-tenths of all earth-material isotopes.[12] This isotope is produced when potassium silicates remineralize in clay sediments. Therefore, natural gamma ray logs are particularly useful in distinguishing clay layers and the variable clay content in coarser-grained sediments. Because formations have characteristic gamma ray counts, a continuous survey of gamma particle decay rates at different levels in the borehole can be used to distinguish lithologies. The casing on a well can attenuate up to one-third of the gamma ray particles, but gamma ray logs are completed in cased as well as uncased holes, and they are completed both with and without fluid in the borehole.

Natural gamma rays are detected and counted with either a Geiger-Mueller tube, a device that uses a gas chamber detection method (a Geiger counter), or a scintillation counter that employs a photoelectric crystal. A portable instrument generally is used to measure gamma ray intensity at a specific location in the borehole, and vehicle-mounted instruments can be lowered into a hole to measure gamma ray radiation continuously. Portable hand-cranked equipment can provide a continuous recording of gamma ray intensities throughout the length of a borehole. The natural gamma ray detector most often used for borehole geophysical logging does not distinguish between gamma ray particle energy intensities but records gross gamma activity. This method is acceptable, though, because the purpose of the logging is to provide information to assist in relative interpretation of the same site regarding the occurrence or absence of a particular lithology.[12]

Electrical Logs. Electrical logs measure the natural electrochemical spontaneous potential and the response to induced electrical stimuli in subsurface formations. The basis for these measurements is Ohm's law. Several types of electrical logs are available, including spontaneous-potential, single-point resistance, and multi-electrode resistivity logs. Various electrical borehole log configurations are shown in Figure 3-13. To employ any electrical log, the borehole must be uncased and filled with a conducting fluid such as drilling mud or water.

Spontaneous potential logs measure the natural electrical potentials that occur between the borehole fluid and the penetrated subsurface material. The electromotive force that creates spontaneous electrical potential can be an electrokinetic phenomenon caused by an electrolyte moving through a permeable medium or, more important, by electrochemical processes that occur at the boundaries of different materials. Differences in spontaneous potential due to different intensities of electrochemical electromotive forces obtained with similar fluids and dissimilar lithologies thus can be used to correlate lithologies. A lead electrode attached to an uninsulated wire is raised and lowered in the borehole; a stationary electrode is placed at the surface; an instrument such as a millivolt meter measures electrical potential. The traveling electrode in the borehole measures the potential recorded at different depths; the potential at the stationary, ground electrode is constant. Therefore, any measured potential is due to differences in potential obtained with the borehole fluid and the different lithologies.[12]

Single-point resistance logs measure the electrical resistance of the subsurface material between a movable borehole electrode and a stationary surface electrode. In practice, the potential difference, in volts, is measured

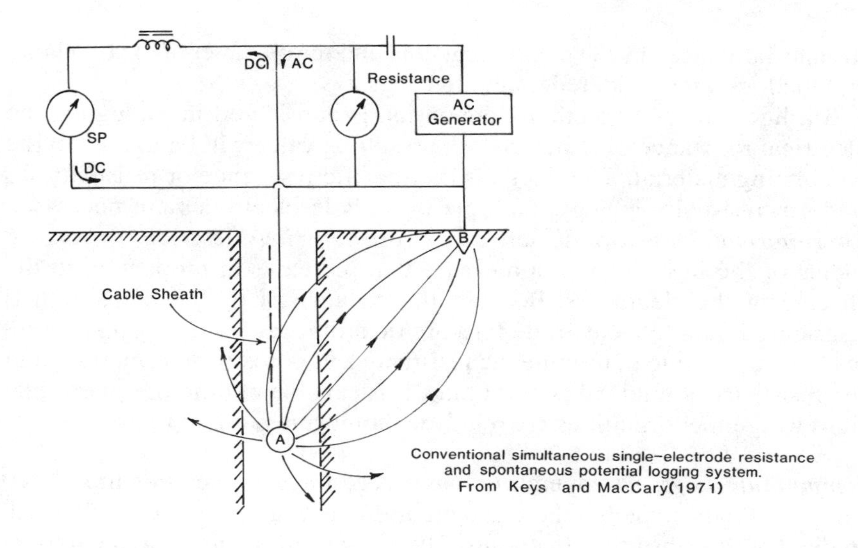

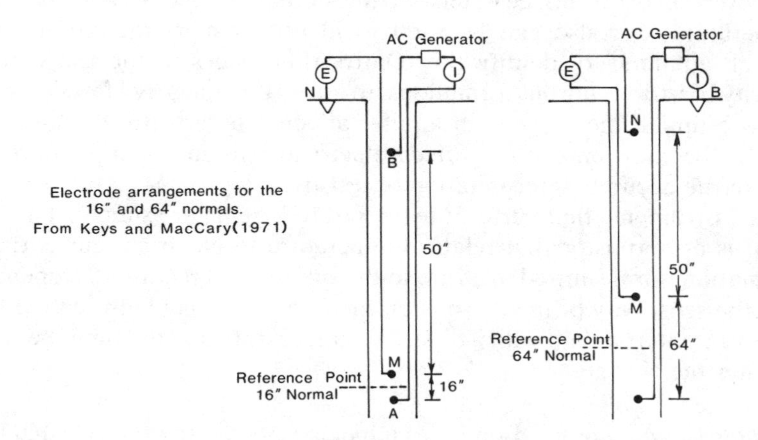

Figure 3-13. Electrical borehole log configurations. Source: *Occupational Safety and Health Guidance Manual for Hazardous Waste Site Activities.* NIOSH/OSHA/USCG/EPA, 1985.

and converted to resistance using Ohm's law. During the survey, a constant current is maintained. The equipment used is similar to that used for spontaneous potential logging and employs both alternating and direct current; thus, it can log both single-point and spontaneous potential resistance. With alternating current used in single-point resistance logs, the borehole and surface electrodes function as the current and potential electrodes. Therefore, the radius of investigation about the electrode is small and strongly influenced by the conducting fluid and the diameter of the borehole, and the logs are considered qualitative logs.[12]

If a linear or cross-sectional dimension is considered in evaluating the electrical resistance of a material, the resulting value will be unique to the conducting material. That value is the specific resistance or resistivity. To perform resistivity logging, multiple down-hole electrodes are necessary. *Short-normal, long-normal,* and *lateral resistivity logs* describe the relative radius of the investigation, a measure that increases in proportion to the spacing of the electrodes. Because the distance of the current's path is considered, multiple-electrode logs are quantitative. Such logging is especially competent to determine the resistivity and porosity of formations and the resistivity of mud cakes and fluids. It is capable of lithological correlations where the formations are relatively homogeneous and thick.[12]

Temperature Logs. Temperature logs record the variable amount of heat energy of fluids in the borehole and are useful in evaluating the geothermal gradient of the subsurface materials. Those data can be related to the source and movement of subsurface waters and the thermal conductivity of lithologies. Temperature logs also can be used to identify major fracture zones in bedrock aquifers, to identify the relative differences in the transmitting capacity of various porous formations, to assess the integrity of well casings, and to estimate the area of an aquifer affected by a waste discharge. In addition, because temperature affects electrical resistance, temperature logs assist in the accurate interpretation of resistivity logs.

The instruments that perform temperature logging are simple. The heat sensor is a thermistor that relates temperature to electrical current; that information is transmitted in a wire to the surface and recorded. Depending upon the sensitivity of the thermistor, there may be a lag between the time the sensor enters a fluid with a different temperature and the time the signal indicates the change.[12]

Borehole Caliper Logs. Borehole caliper logs record the variable diameter of a borehole. The diameter varies because of subsurface fractures and because of the drilling itself. Caliper logs are particularly useful in fractured bedrock or karst aquifers to chart the position and size of fractures and solution channels. In unconsolidated formations where fluid rotary drilling was completed, borehole caliper logs are useful for lithological correlations, as less cohesive formations tend to wash out during drilling. Borehole calipers are arm-type feelers or, less commonly, bowlike springs affixed to an induction coil. Changes in the diameter of the sensor induce proportional changes in the electrical current. Some systems record the current from each sensor separately and permit measurement of borehole asymmetry.

Ground Water Monitoring Wells. In most hydrogeological investigations at hazardous waste sites, invasive investigations culminate with the installation of ground water monitoring wells. These wells furnish an active window into the subsurface and permit measurement of the ground water level or hydraulic potential at a specific location within an aquifer. The wells also serve as collection points for ground water samples. By using a series of monitoring wells, investigators can determine the direction and rate of flow in an aquifer, the disposition and distribution of contaminants, and the hydraulic characteristics of the aquifer. Monitoring wells serve a number of purposes:[13]

- Determination of the hydrogeological properties of the contaminated formation.
- Determination of the water table, or potentiometric surfaces, of aquifers in the system possibly affected by contamination.
- Collection of water for analysis and determination of water quality.
- Monitoring movement of the contaminant plume.

In theory, monitoring wells are not true piezometers. A piezometer measures the hydraulic potential (or head) at a particular point in the water-bearing medium. The ideal piezometer is constructed of a casing capped at the end and placed at a discrete location in the subsurface. A small porthole is made at the lowest end of the assembly prior to installation; thus, the water level in the casing is dependent on the hydraulic head at the porthole. Monitoring wells are, as a rule, constructed of a length of well screen and do not measure discrete hydraulic potentials. The water level observed in the well is an average of the hydraulic potentials occurring along the length of the screen. Data from a monitoring well describing the ground water elevation or hydraulic head can, however, be used to determine horizontal ground water flow direction if the wells in the series are screened at a similar level in the aquifer. If the ratio of screen length to aquifer thickness is small, head data from a series of wells installed at the same location but at different depths also can be used to evaluate vertical ground water flow.

The design and installation methods of ground water monitoring wells depend upon the hydrogeological conditions of the site and the type of contaminants the well is intended to monitor. The primary variables are:

- Geology of the aquifer: unconsolidated, porous media, or bedrock.
- Location of the monitoring well: the upper, first-encountered ground water or a lower aquifer separated from the upper water table by an aquitard.
- Intended purpose of the well: temporary monitoring or long-term monitoring activities such that permanent pumping equipment may be required.

- Type of contaminant: organics, perhaps nonaqueous phase liquids, or inorganic compounds, perhaps in concentrations that are corrosive.

Guidelines for Well Installation. Most state environmental agencies offer guidelines on the installation of a monitoring well and, furthermore, require that a permit be secured before a monitoring well is installed. Certain states demand that specific criteria be incorporated into the proposed well design and installation protocol. Recently, the U.S. EPA developed comprehensive guidelines for well designs used in hydrogeological investigations proposed at sites regulated by RCRA.[14] Although these guidelines furnish review procedures for RCRA sites, they will probably also be used by government reviewers of non-RCRA sites such as sites governed by CERCLA. The American Society for Testing and Materials (ASTM) is also developing guidelines for monitoring wells and hydrogeological data collection methods. The remainder of this section presents typical well design and installation methods for unconsolidated and bedrock hydrogeological systems for single and multiple aquifers utilizing available guidelines. Geochemical conditions as they relate to well construction material and well placement issues are also discussed. Where the guidelines if used exclusively are not technically comprehensive or are inconsistent with current standard practice, options are offered. The information presented below is related to the subject of ground water collection, presented in Chapter 10.

At most hazardous waste sites, the primary target of the investigation is the aquifer in unconsolidated material that is first encountered, because the hydrogeological system nearest the land surface is first affected by waste components. The first aquifer is often a shallow unconfined (water table) aquifer, and the depth to ground water is often less than 50 ft (15 m). A typical water table monitoring well is depicted in Figure 3-14. In unconsolidated media, a well screen filters the grains of material and prevents them from entering the well. Well screens have specifically designed openings made by special cutting machines or by a continuously wound wire around a series of vertical support rods. The length of the screen is varied to accommodate specific applications. Use of a well screen can be illustrated through an example. A loss of petroleum products created a layer of free product, a nonaqueous layer, atop a water table aquifer. The elevation of the aquifer varied approximately 5 ft (1.5 m) annually. To insure that a segment of the well screen remained above the water table and able to detect the free product layer throughout the year, a 15 ft (4.5 m) length of well screen was installed 5 to 7 ft (1.5 to 2 m) below the seasonal low water table elevation. Had the contaminants of concern occurred in a thin and distinct subsurface horizon, a shorter length of well screen would have been appropriate.

A sealant for the annular space is required to block the migration of

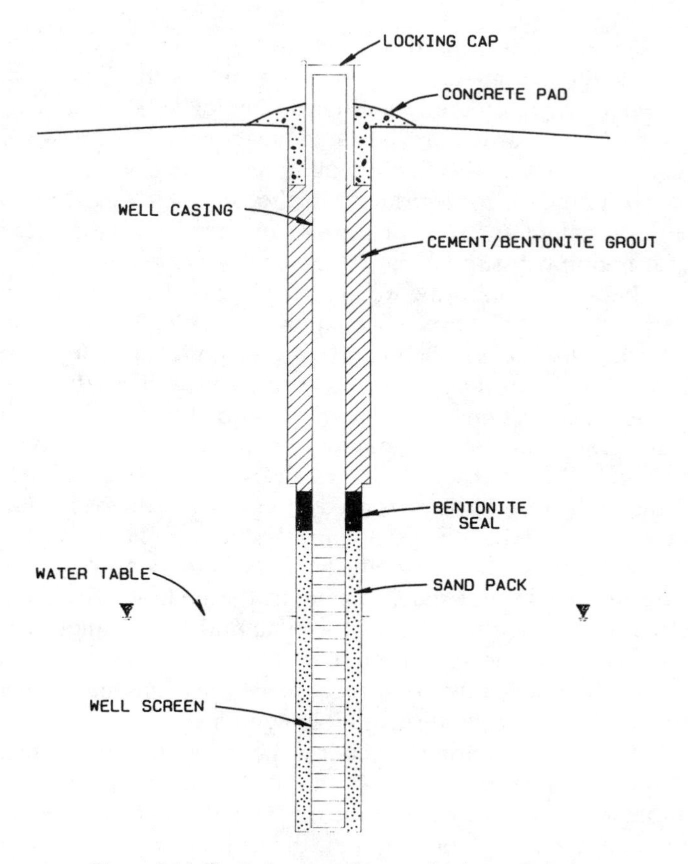

Figure 3-14. Typical water table monitoring well design.

contaminants into the sampling zone from the surface or intermediate zones; thus, cross-contamination between strata is forestalled. In certain native formations, a filter pack may not be required. A compatible well casing is attached to the well screen prior to installation and extended to the surface. The remaining annular space above the filter pack or collapsed native material is then filled with an annular sealant. The hydraulic conductivity of the sealant should be one or two orders of magnitude less than that of the surrounding formation. The standard sealants include sodium bentonite pellets and a mixture of bentonite and neat cement. Specialty grouts are available if hostile ground water chemistry is encountered. A layer of bentonite pellets, 1 to 2 ft thick (30 to 60 cm), is usually installed atop the filter pack before it is grouted to prevent the grout from invading the filter pack.

It is essentially true, as suggested by the U.S. EPA guidance document, that the design of monitoring wells installed in bedrock is not inherently different from the design of wells installed in unconsolidated material. However, depending upon the specific hydrogeological conditions, options are available to the investigator when studying consolidated flow systems in consolidated material. If the first encountered ground water at the site is below the level of the bedrock surface, and the local lithology is composed of a relatively competent and stable bedrock, and if the facts indicate that the bedrock formation can support an open borehole, then an open borehole design for a bedrock monitoring well may be justified to monitor the upper portion of the aquifer. Moreover, if a composite sample of a thicker sequence of a bedrock aquifer is desired and appropriate to meet the objectives of an investigation, then an open hole bedrock monitoring well may be satisfactory. Such a well is generally constructed with a length of surface casing installed to the level of more stable or competent bedrock above the ground water level. The borehole is then advanced into the water-bearing zone and left as an open hole in the formation. It is used as a screened unconsolidated well. The surface casing is typically grouted in a fashion similar to the casing of a screened monitoring well. If hydraulic head measurements or ground water samples are desired from a specific zone in the bedrock aquifer, an open hole well design is inapplicable. The well should be completed with well screen and casing similar to that required in an unconsolidated formation; thus, the zone from which the well receives its water can be isolated.

At many hazardous waste sites, multiple-aquifer systems operate, one over another. Often, a lower confined aquifer is the target formation because it is necessary to determine the vertical extent of affected ground water. In such cases, a monitoring well may have to be installed through an upper contaminated unconfined (water table) aquifer, through an aquitard, and into the lower confined (artesian) aquifer. Such wells require unique design considerations. Figure 3-15 depicts a typical well design arranged to reach a lower aquifer. The installation of such a well pierces the aquitard and could create a hydraulic connection between the two aquifers, but it is essential that the borehole not support communication between the aquifers either permanently after construction or temporarily during construction. To guarantee isolation of the aquifers, exacting care must be taken when the grout seal is installed in the aquitard. The double-wall well design, although costly, is probably the best design for monitoring wells installed through upper contaminated zones.

The U.S. EPA recommends that the diameter of the casing of monitoring wells be 4 inches (10 cm). The design criteria of a number of states also specify this dimension. However, most practicing hydrogeologists use a 2-inch (5-cm) diameter, a dimension preferred because it is less costly and easier to install. Furthermore, current sophisticated monitoring well sam-

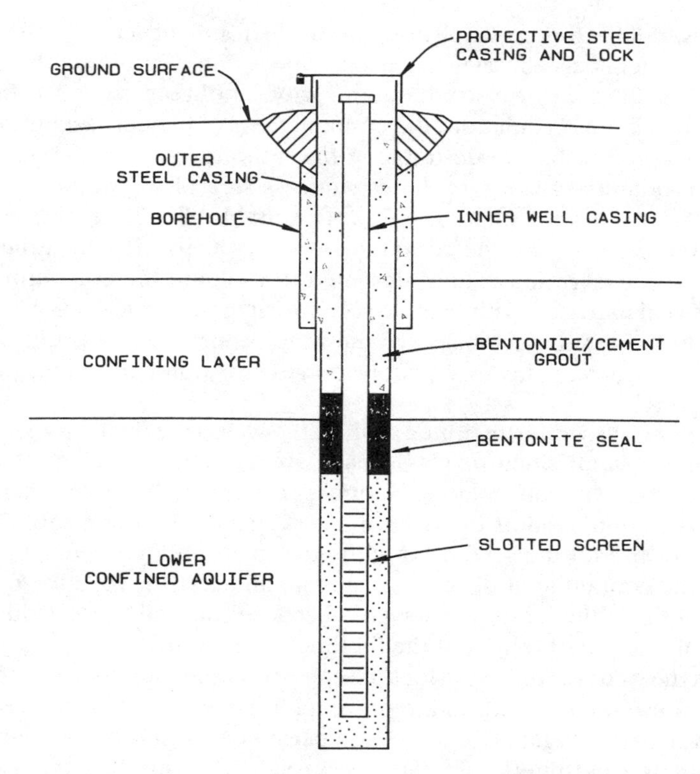

Figure 3-15. Multiple-aquifer well designs. Typical design of monitoring well for confined aquifers.

pling equipment is designed for the 2-inch (5-cm) well. The smaller well size is prudent, too, because at most hazardous waste sites the drill cuttings, well-development water, and sampling-purge water are considered contaminated and must be disposed of as hazardous wastes. The 2-inch (5-cm) hole will produce substantially less waste than the 4-inch (10-cm) hole.

The materials of construction for wells must be sufficiently durable to withstand the stresses of the specific geochemical environment, the expected lifetime of the monitoring program, the depth of the wells, and the expected contaminants. Yet, the materials should not, by their nature, affect the samples drawn. Various well casings and well screens are available: virgin fluorocarbon resins such as fluorinated ethylene propylene, polytetrafluoroethylene, Teflon™, stainless steel, polyvinyl chloride (PVC), polyethylene, epoxy biphenol, and polypropylene. Teflon™, stainless steel, and PVC are the most commonly used materials, and PVC, by far, is the least costly. None of these is suitable for every well installation environment, as each is degraded by certain chemicals and may leach substances that can modify the ground water or impair their own structural integrity. Until recently it was thought that when volatile organics are the contaminants of concern, stainless steel or fluorocarbon resins are desirable in the saturated zone, although a high corrosion potential further excludes stainless steel. PVC well casings and screens, according to the U.S. EPA, are suitable if only trace metals or nonvolatile organics are the contaminants of concern. Recently, however, sorption and desorption studies designed to evaluate the compatibility of different well materials with organic solvents suggested that PVC well materials do not affect the chemistry of the water sample if they are used with a sampling program that adequately purges the water in a well prior to the sampling.

The material used for the filter pack of the wells is typically rounded and clean quartz sand; silica or glass beads are sometimes, but rarely, used. Fabric filters are not acceptable. Natural gravel is suitable if it is chemically inert to the contaminants of concern and if a sieve analysis establishes the correct slot size for the screen. As with other materials, the sealant must be inert to the contaminant of concern, and no antagonistic reaction between the sealant and the waste is allowable because the seal must maintain its structural integrity throughout the lifetime of the well.

When the well is complete, a locking cap or compression seal is fastened in place to prevent tampering and to prevent foreign objects from entering the well. A vent permits gases to escape and permits the inside of the well casing to remain at atmospheric pressure. It is prudent to install bumper guards around the well, particularly in areas with heavy traffic, to inhibit accidental collisions between vehicles and the well and thus to obviate the costly replacement of the well.

Well Placement. Most organic and inorganic contaminants migrate as a function of the hydraulics of the ground water system, and ground water contamination, where it has been found, exists in three dimensions in the earth. An understanding of vertical as well as horizontal flow paths is indispensable; so monitoring wells are utilized to delimit both the horizontal and the vertical extent of contamination. Additionally, monitoring wells are placed both upgradient and downgradient of the suspected area of contamination, the number of wells necessary depending on the size and heterogeneity of the site. Upgradient wells are drilled beyond the upgradient extent of potential contamination to furnish samples of background water quality. Downgradient wells are placed just beyond and along the boundary of suspected contamination to furnish immediate notification of contaminant migration. If there is any possibility that volatile organics could escape to the

TABLE 3-6. Factors influencing the interval between downgradient monitoring wells.

INTERVAL CLOSER, IF:	INTERVAL WIDER, IF:
Aquifer manages or has managed liquid waste.	
Aquifer is very small in area.	
Fill material is near waste management units—where flow might occur.	
Buried pipes, utility trenches, etc., where point-source leak may occur.	
Complicated geology: • Closely spaced fractures. • Faults. • Tight folds. • Solution channels. • Discontinuous structures.	Simple geology: • No fractures. • No faults. • No folds. • No solution channels. • Continuous structures.
Heterogeneous conditions: • Variable hydraulic conductivity. • Variable lithology.	Homogeneous conditions: • Uniform hydraulic conductivity. • Uniform lithology.
Aquifer located on or near recharge zone.	
Steep or variable hydraulic gradient.	Low and constant hydraulic gradient.
Low dispersivity potential.	High dispersivity potential.
High seepage velocity.	Low seepage velocity.

unsaturated zone, a program of gas sampling and analysis may be instituted. The horizontal spacing between wells is based on the characteristics of the site. Several factors that affect the spacing interval of wells are summarized in Table 3-6.[15]

After the horizontal placement of wells is determined, the vertical placement is defined. The depth to which wells are installed depends upon the probable depth of the contamination and the pattern of ground water flow in the aquifer. Both the depth and thickness of the stratigraphic horizons that could serve as contaminant pathways must be known with acceptable accuracy in order to determine the depth of the wells. The chemical nature of the contaminants also affects the required depth of the wells. The mobility of the contaminant, its potential reaction products, its density, and its potential to degrade clays are considered. The contaminant may move not only with the advective flow of ground water but also by means of both chemical dispersion (moving faster than the flow of ground water in slowly flowing systems) and sorption (moving more slowly than the ground water).

Once the depth to which the wells will be placed is determined, the depth of sampling is fixed. The wells are screened in a manner that will detect migration. Again, the characteristics of the site are the deciding factor. Isopach maps illustrating the variable thicknesses of strata, coring data, sieve analysis, and fracture traces are used as sources of information to establish the sampling depth. The depth and length of well screens are adjusted to suit the nature of the site. Most sites with unconsolidated geology are heterogeneous and anisotropic, their permeability varying with depth because of interbedded sediments. The screen is situated so that it samples the discrete formation of interest. If the strata do not dip too markedly, both the upgradient and downgradient wells are screened at the same stratigraphic horizons to intercept the same uppermost aquifer, and they thus furnish comparable data on ground water quality. If the screen is too long, it could permit uncontaminated ground water to mix with the contaminated water from the affected formation. However, if hydraulic conductivity is low, longer well screens facilitate sampling by allowing water to enter the well in the volume necessary for sampling.

A single well cannot always adequately intercept and monitor the migration of a contaminant at a selected sampling site, and a well cluster or well nest—several wells in different boreholes installed to different stratigraphic horizons at the same location—is necessary. Table 3-7 lists instances where more than one well is required at a sampling site. In general, the number of required wells increases with the complexity of the site.[16]

TABLE 3-7. Conditions demanding well clusters.

ONE WELL PER LOCATION	MORE THAN ONE WELL
Absence of immiscible liquid phases.	Presence of immiscible liquid phases.
Low potential for vertical flow.	High potential for vertical flow.
Zone of thin flow relative to screen length.	
Homogeneous uppermost aquifer and simple geology.	Heterogeneous uppermost aquifer and complicated geology: • Multiple, interconnected aquifers. • Variable lithology. • Perched water zone. • Discontinuous structures.
	Discrete fracture zones.

Well Development. After a well is installed, it is developed to optimize efficiency and to remove fine materials that have entered. All well development equipment is decontaminated before use. Well development is a mechanical process: a flow of water is drawn into the well and flushed into the formation to remove the fine materials from the formation immediately surrounding the well screen. Development is continued until the movement of fines from the formation lessens and the well reaches its optimum yield. Development methods vary depending upon the aquifer materials, type of well construction, and well installation procedure.[17] Water from the formation usually is used in development, but if the yield of the formation is low, water from an external source, if the water is chemically compatible, may be used as a supplement. Unlike ground water collection wells (as discussed in Chapter 10), air should not be employed to develop monitoring wells if the proposed parameters in the ground water analysis include volatile compounds, and if sampling is to be completed within the short time frame of one to two weeks. Once the well is developed, it is maintained in order to insure that the ground water it furnishes for sampling is of suitably low turbidity and that it sustains an acceptable yield.

Many factors are important in the specification, construction, development, and use of a monitoring well, and all are recorded. Table 3-8 illustrates pertinent data about the well.[18]

TABLE 3-8. Information included in well documentation.

Date and time of construction.	Sealant materials.
Drilling method and drilling fluid used.	Sealant volume (weight/volume of cement).
Well location (±0.5 ft [15 cm]).	Sealant placement method.
Borehole diameter and well casing diameter.	Surface seal design and construction.
Well depth (±0.1 ft [3 cm]).	Well development procedure.
Drilling and lithologic logs.	Type of protective well cap.
Casing materials.	Ground surface elevation (±0.01 ft [30 mm]).
Screen materials and design.	Surveyor's pin elevation (±0.01 ft [30 mm]) on concrete apron.
Casing and screen joint type.	Top of monitoring well casing elevation (±0.01 ft [30 mm]).
Screen slot size and length.	Top of protective steel casing elevation (±0.01 ft [30 mm]).
Filter pack material and size; grain analysis.	Detailed drawing of well showing dimensions.
Filter pack volume calculations.	
Filter pack placement method.	

Evaluating Aquifer Characteristics. The purpose of constructing monitoring wells is to ascertain the characteristics of contaminated aquifers, particularly the direction and rate of contaminant movement. The most important characteristics are the flow, both horizontal and vertical, and hydraulic conductivity or permeability. Once a series of ground water monitoring wells is installed, the water levels are monitored and recorded, and tests are performed to determine the aquifer's hydraulic conductivity. This information is used to assess the direction and rate of contaminant migration.

The direction of flow of ground water is affected by a number of factors, including: well pumping, recharges, and discharges off-site; tidal processes or other intermittent natural variations such as the stages of a river; well pumping on-site; construction or changing land use patterns both off- and on-site; injection wells; waste disposal practices; and naturally occurring geological conditions. In order to gather data on the direction of ground water flow, observations are made through monitoring wells or boreholes. Data taken from piezometers or wells screened at equivalent stratigraphic horizons are developed into a potentiometric surface map. Piezometers installed in clusters provide the data that determine the vertical component of the flow. Generally, a cluster of piezometers or wells screened at different depths is used to measure the vertical variation in hydraulic head. The data are used to develop a flow net (a set of intersecting equipotential lines and flow lines representing a two-dimensional steady flow through the earth).

The ability of a water-bearing formation to conduct water is its *hydraulic conductivity (K),* which can also be defined as the capacity of a porous medium to transmit water. Ground water in an aquifer will move whenever there is a difference in head, moving from an area of high head to an area of low head. A relatively permeable formation is a potential pathway for contaminant migration. Therefore, it is important to know the hydraulic conductivity of each significant formation at a hazardous waste site.

According to the nineteenth-century French engineer Henri Darcy, the flow of water through the ground is similar to its flow through a pipe. Darcy's law describing the basic flow of ground water is:

$$V_{Df} = \frac{K(h_1 - h_2)}{L}$$

V_{Df} is the specific discharge and is also known as the Darcian velocity, $(H_1 - h_2)$ is the difference in hydraulic head, L is the length of the path between h_1 and h_2, and K is the hydraulic conductivity.[19] (Darcy's law, in modified form, will be utilized again in Chapter 7.) Additional important definitions are presented below and graphed in Figure 3-16.

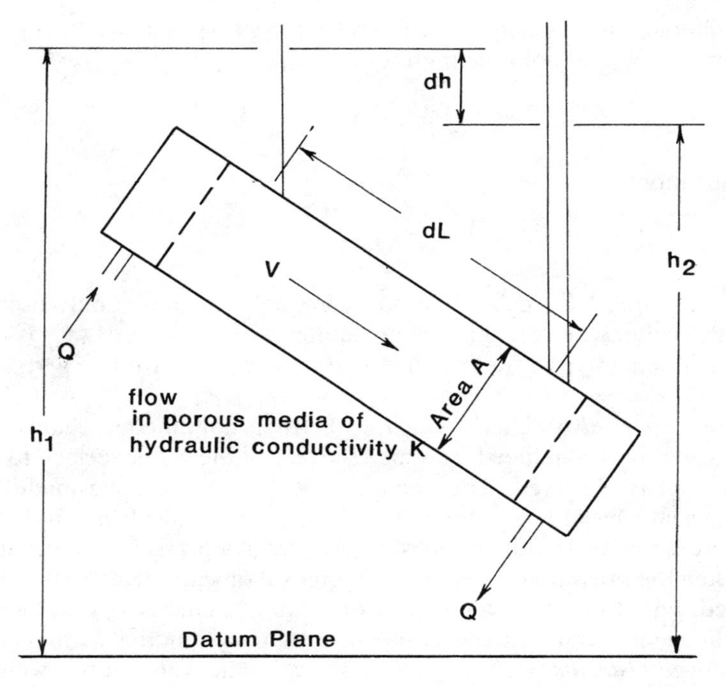

$$v = \frac{Q}{A}$$

and

$$Q = K\frac{dh}{dL}A$$

v - flow velocity
h_1 & h_2 - hydraulic head
dh - change in head
dL - change in length
Q - flow rate
A - flow through cross sectional area

Figure 3-16. Darcy's law.

The hydraulic gradient, or slope of water table *(I)*, is:

$$I = \frac{h_1 - h_2}{L}$$

Therefore, the specific discharge or the discharge per unit area becomes:

$$V_{Df} = KI$$

The discharge through a cross-section, Q, equals flow velocity V_{Df} multiplied by cross-sectional area, A; therefore:

$$Q = V_{Df}A$$

By substitution:

$$Q = KIA$$

This is the form of Darcy's equation hydrogeologists commonly use to calculate the volume of flow through an aquifer.

Hydraulic conductivity is determined in the field in two routine ways:

- *Single-well tests, also called "slug tests":* An instantaneous change in ground water level is induced, and the recovery of the water surface to its original level is observed. The change in water level can be made by removing water, by adding water, or by inserting a solid displacing rod. An alternative is pressurization of the well casing and depression of the water level; when the pressure is released, the removal of water from the well is simulated. Variations of Darcy's equation relate the change in water level to the change in duration to determine hydraulic conductivity.
- *Multiple-well tests, also called "pumping tests":* The water in one well is pumped, and the drawdown in nearby wells is observed. Data are taken from tests conducted with wells screened in the same water-bearing formation. To develop information about hydraulic communication, tests are taken in wells screened in different water-bearing strata. The greater the number of monitoring wells used, the greater is the amount of information gained on the spatial variation within the aquifer.

Slug test and pump test variables and configurations are depicted in Figure 3-17. A consideration whenever the pump tests are conducted is that potentially contaminated water may have to be stored or disposed of. In addition, the pumping may affect the waste plumes.[20]

The U.S. EPA provides general criteria for consideration during measurement of hydraulic conductivity:[21]

- Multiple-well tests are preferred to establish hydraulic conductivity. If single-well tests are used instead of multiple-well tests, more tests are required to characterize the site adequately.
- Data about hydraulic conductivity furnish average values for the entire area across a well screen. To secure data specific to limited depths, shorter well screens are necessary. If the average hydraulic conductivity for a

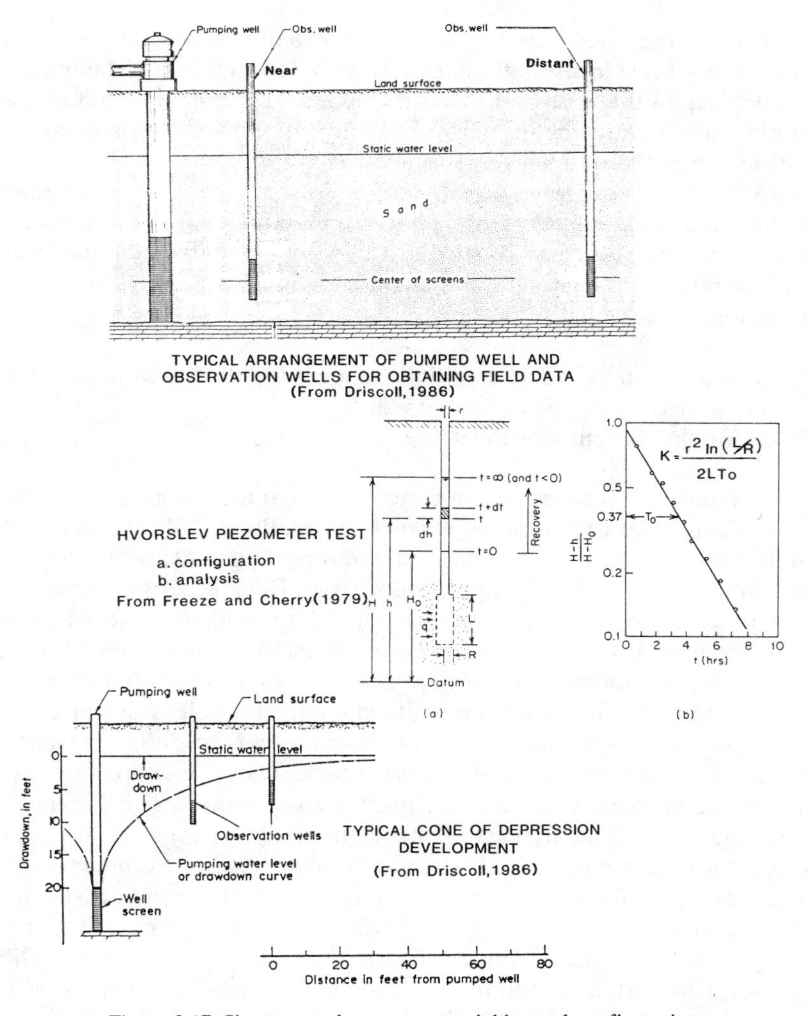

Figure 3-17. Slug test and pump test variables and configurations.

formation is required, entire formations may have to be screened, or data may have to be taken from overlapping clusters.

SUMMARY

In investigations of hazardous waste sites, a hydrogeological study is the essential first step that characterizes the environment and the jeopardy present. The investigation is planned carefully in order to achieve the stated goals, and it is executed meticulously in order to provide the necessary data economically. Most investigations are developed in phases, with later phases dependent upon the results of the earlier ones. Such scheduling maintains flexibility and permits the investigators to adjust the study at planned milestones to insure that it achieves its objectives.

Phase 1 of a hydrogeological investigation is the review of available background data. Existing reports, photographs, and records are scanned to collect pertinent data about waste disposal practices and locations of waste sites. It is important to garner as much information as possible from existing sources for two reasons:

- To describe with as much specificity as possible the environmental and safety hazards that must be addressed.
- To define the direction of the subsequent phases.

If the Phase 1 investigation indicates that continued work is warranted, the study continues with noninvasive techniques, Phase 2. These techniques include the use of remote sensing and aerial photographs, reconnaissance field mapping, and surface geophysical surveys. Remote sensing images are useful in gathering regional hydrogeological information. Aerial photographs are useful in identifying waste disposal site boundaries, drainage patterns, and locations of stressed vegetation, especially when historical photos are utilized. Reconnaissance site mapping through surface geophysical surveys provides a rapid and economical means to define subsurface lithology, bedrock structure, depth to ground water, locations of waste materials, and ground water contaminant plumes. Noninvasive geophysical surveys can be implemented with lightweight equipment suitable for use on hazardous waste sites, even those with rough terrain. The instruments develop records of electrical resistivity and conductivity and images of buried objects by use of sound and shock waves. Noninvasive techniques are effective tools in the preliminary assessment of subsurface conditions. They are particularly useful for extensive study areas where a general knowledge of the geology, depth to ground water, depth to bedrock, and thickness of waste materials is pertinent. They are more economical than invasive techniques, and they furnish direction for invasive work.

Geophysical surveys cannot substitute for invasive methods of investigation when detailed, specific information is demanded. Invasive techniques are used to collect lithological samples and to install ground water monitoring wells. A number of methods are commonly used in hazardous waste site investigations. Conventional machine excavation furnishes information about the soil horizon at depths usually less than 15 ft (4.5 m). Soil profiles catalog the subsurface waste materials and delimit the extent of contamination by means of a small excavation or borehole, usually penetrating less than 10 ft (3 m). Borehole drilling retrieves samples of the subsurface at greater depths than the other methods; it is also the technique used to install wells. The standard methods of borehole drilling are auger drilling, conventional rotary drilling, and percussion drilling.

Borehole geophysical surveys evaluate the lithology and hydro-stratigraphy of the earth through the window created by the borehole. The ordinary survey techniques used in the investigation of hazardous waste sites are natural gamma ray logging, electrical logging, temperature logging, and borehole caliper logging. Each of these conventional techniques is utilized to derive specific data for specific purposes. Additional techniques requiring more sophisticated instruments can be employed when specialized data are demanded.

The objective of the hydrogeological investigation is to understand the contaminated aquifer. The many techniques used to gather information about the site, both noninvasive and invasive, are designed to secure valid data about the mechanisms of ground water flows; the velocity of the flow, including the horizontal and vertical components of flow; the quantity of the flow, based upon the hydraulic conductivity and permeability of the aquifer; and the chemical quality of the ground water. When the aquifer has been satisfactorily characterized, the investigators can predict with reasonable accuracy the movement of the contaminants in the ground water. For example, it is important to know whether or not the contaminants could reach potable water wells. (Additional predictive techniques utilizing mathematical models are employed by the investigator when circumstances warrant them. These methods are introduced in Chapter 7.)

Data and evaluations from the hydrogeological investigation and from investigations that characterize other factors on the site, such as wastes in containers, structures, and gases, are applied in order to assess the risks presented by the site and to design remedial alternatives. The hydrogeological data are essential to establish remedial design criteria, to set remedial priorities, and to rank remedial efforts. With a well planned and phased hydrogeological study, the data pertinent to the investigation of a hazardous waste site can be collected. By maintaining a flexible approach to the study, the investigator gathers only the necessary information and emphasizes material data.

NOTES

1. *Journal of Freshwater.* 9(1985):2. D. K. Todd. *Groundwater hydrology.* New York: John Wiley & Sons, 1976.
2. *Journal of Freshwater.* 9(1985):2 and Todd.
3. *Journal of Freshwater.* 9(1985):2 and Todd, pp. 31-37.

4. An applicable computer-based bibliographic source is maintained by the National Water Well Association. Another is the Dialog™ Information Retrieval Service of Lockheed Corporation (3460 Hillview Ave., Palo Alto, Calif. 94304; telephone 800-334-2564).

5. W. M. Telford, L. P. Geldard, R. E. Sheriff, and D. A. Keys. *Applied geophysics.* Cambridge, U.K.: Cambridge University Press, 1976.

6. J. D. McNeill and Geonics Limited (1745 Meyerside Dr., Mississagua, Ontario, Canada L5T 1C5). Electrical conductivity of soils and rocks. Technical Note TN-5. Oct. 1980. Electromagnetic terrain conductivity measurement at low induction numbers. Technical Note TN-6. Oct. 1980.

7. J. D. McNeill and Geonics Limited. Electrical conductivity of soils and rocks. Technical Note TN-5. Electromagnetic terrain conductivity measurement at low induction numbers. Technical Note TN-6.

8. Telford, Geldard, Sheriff, and Keys.

9. J. J. Bowders, Jr., R. M. Koerner, and A. E. Lord, Jr. Potential use of GPR in assessing groundwater pollution in partially and fully saturated soils. Published by U.S. Environmental Protection Agency, Municipal/Environmental Research Laboratory, Edison, N. J. EPA Agreement #CR-804763.

10. American Society for Testing and Materials (ASTM). *1986 Annual book of ASTM standards,* Section 4, Construction. Vol 04.08, Soil and rock; building stones. pp. 298–303.

11. ASTM. *1986 Annual book of ASTM standards,* Section 4, Construction. Vol 04.08, Soil and rock; building stones. pp. 304–307.

12. W. Scott Keys and L. M. MacCary. *Techniques of water-resources investigations of the United States Geological Survey,* Chapter E1, Application of borehole geophysics to water-resources investigations. Book 2. U.S. GPO, 1971. This reference was utilized for background information on several of the techniques described below.

13. F. G. Driscoll. *Groundwater and wells.* St. Paul: Johnson Division, 1986. p. 715.

14. U.S. Environmental Protection Agency (Office of Waste Programs Enforcement; Office of Solid Waste and Emergency Response). *RCRA ground-water monitoring: Technical enforcement guidance document (TEGD).* Sept. 1986. OSWER-9950.1.

15. Adapted from Table 2-1, U.S. EPA *RCRA ground-water monitoring: TEGD.*

16 Adapted from Table 2-2, U.S. EPA *RCRA ground-water monitoring: TEGD.*

17. F. G. Driscoll and D. K. Todd. *Ground water hydrology.* New York: John Wiley & Sons, Inc., 1980.

18. U.S. EPA *RCRA ground-water monitoring: TEGD.* pp. 88–89.

19. H. Darcy. *Les fontaines publiques de la ville de Dijon.* Paris: V. Dalmont, 1856. Cited by Driscoll, p. 73.

20. *RCRA ground-water monitoring: TEGD.* pp. 32–33.

21. *RCRA ground-water monitoring: TEGD.* p. 34.

Chapter 4
Assessing Risk

A *risk assessment,* in the context of this book, is a measure of the potential for chemical wastes and physical agents to escape from a source and communicate with sensitive receptors in the environment. It is a crucial component of many remedial action programs because it is the construct utilized to determine the need for various remedial designs and to compare their effectiveness and applicability. The risk assessment interprets data about the source of contamination and articulates the effects the waste site exerts on the environment. The goal of a risk assessment is to identify the level at which the effects of a hazardous waste site become acceptable to current and future human and animal populations. Therefore, the risk assessment delimits both the aspects of the hazardous waste site that require remediation and the degree of remediation necessary. With that information, a remedial solution can be designed that will reduce unacceptable releases from the site to the targeted risk level. The method for the development of a risk assessment presented in this chapter is based upon the procedures of the U.S. EPA.[1]

It is a common misconception that chemicals are inherently dangerous at any level and that hazardous waste sites are a major health threat. This public sentiment has been institutionalized by many of the environmental policies recently enacted. The public perception of risk is sometimes at variance with the risk calculated at hazardous waste sites. One study reviewed the available information on the health status of residents near hazardous waste sites and concluded that there was little decisive evidence of serious health effects from chemical waste disposal sites. The report also emphasized that currently available data are insufficient to establish cause-and-effect relationships between chemicals at hazardous waste sites and human health, or to predict long-term effects accurately.[2]

Identifying chemical hazards and differentiating them from site risks is pivotal to an effective site evaluation. A *chemical hazard* is an inherent characteristic of a chemical compound such as toxicity, flammability, reactivity, or environmental mobility and persistence; the hazard is quantifiable, and it

---

This chapter was developed by Swiatoslov W. Kaczmar, Ph.D., C.I.H. of O'Brien & Gere Engineers, Inc.

measures the inherent property of a material to induce a particular adverse effect. Measures of hazards are standard parameters that are compiled in chemical and toxicological reference documents. *Risk,* on the other hand, is a site- or incident-specific probability that deleterious effects will occur. In order to calculate risk, the hazardous characteristics of a given chemical are applied to the specific circumstances in which the chemical is found. For example, the risk or likelihood of death following a given exposure to a particular chemical is dependent upon the dose administered, the weight and health of the exposed individual, the route of exposure, and other factors. There are, then, two activities that comprise the risk assessment:

- *Hazard characterization:* compilation of site information such as identification of wastes, definition of physicochemical and toxicological properties, and determination of the physical and demographic properties of the site under investigation.
- *Risk assessment:* consideration of all the hazard characteristics together and the potential adverse effects that they represent.

In its implementation, a risk assessment is an exercise that evaluates all the factors of a site that contribute to—in some cases, reduce—the likelihood of adverse effects to humans or the environment. In investigating a hazardous waste site, the object is to assess the actual effects of the site, not just to measure the waste residue. Simply finding a residue above background does not indicate that humans and animals are adversely affected. The completed assessment identifies and quantifies the effects of the existing and potential releases from the site under various scenarios. The investigators record the circumstances and facts supporting the conclusions of the risk assessment. Because this record will be scrutinized by regulatory agencies, attorneys, and the public, it must be competently done and defensible in court proceedings.

A risk assessment can be approached functionally in two tiers; the first, the qualitative assessment, is descriptive, whereas the second, the quantitative assessment, is measured or calculated. When the second tier is necessary, it builds upon the conclusions of the first.

- *Qualitative Assessment:* The basic aspects of the hazardous waste site are accounted in the qualitative risk assessment, and the analysis determines whether waste materials present at the site can be transported off-site. The analysis identifies the available pathways, determines whether any human or wildlife receptors can be reached by the wastes, and identifies a set of transport and exposure scenarios. Each scenario encompasses a complete exposure pathway. The potential consequence of each scenario is the transmission of risk to receptors in the environs. The probability, or risk, of any of these scenarios materializing is evaluated in the quantitative assessment.

The qualitative risk assessment produces a record stating the reasons why other waste components, transport routes, or receptors were not considered for further evaluation. It may also recommend that a particular source or transport route be remedied without recourse to the detailed analysis of the quantitative risk assessment. The qualitative assessment can also conclude that the ambiguity of data precludes the development of serviceable information, and thus it may recommend prompt remediation.

- *Quantitative assessment:* The scenarios devised in the qualitative risk assessment that display a potential of occurring are evaluated further in the quantitative risk assessment. These scenarios are analyzed to define the nature of any releases of wastes from the site, the rates at which the wastes are transported from the site, the specific populations of humans and animals exposed to the wastes, and the rates at which these compounds are entering the receptors' bodies. The exposure rates are juxtaposed with known toxicological effects and compared with acceptable exposure rates in order to determine the exigency for remediation.

With this two-tiered approach, the universe of possible scenarios is successively reduced, and only the most plausible scenarios are scrutinized. Such an approach conserves resources.

Performing a risk assessment at a hazardous waste site is a dynamic process. In most cases, the risk assessment furnishes definition and direction to the remedial investigation. Therefore, it is critical that the qualitative risk assessment be performed prior to or concurrently with the planning and implementation of the field investigation. When such an initial risk assessment is performed early in the site investigation, it forestalls the collection of irrelevant or superfluous field data and guarantees the collection of requisite data.

The design and execution of a risk assessment requires personnel with competent scientific judgment and objectivity. The investigators directing the risk assessment must understand the dynamics of chemicals in the environment, be able to assess dynamic physical forces, know the toxicological effects of chemical compounds on humans and animals, and know how to extrapolate toxicological data validly. No regulatory standards or generic remediation criteria can substitute for the methods of trained risk assessment investigators.

A critical problem facing investigators at a hazardous waste site is the nominal precision of the data available; the evaluations and recommendations of the assessments necessarily are based on data of indeterminate accuracy and precision. Furthermore, in order to complete the risk assessment, one must make assumptions regarding the identity, concentration, and distribution of chemical compounds within a site, and about the types and rates of exposures experienced by humans and animals at the site and in the environs. Extrapolation of the toxic effects observed in test animals to effects on humans and animals also is based upon assumptions—often being the extrapolation of effects of high dosages in test animals to the effects of low environmental dosages on humans. When performing a risk assessment, standard practice dictates that the investigator utilizes the most conservative assumption. However, as assumptions are pyramided, uncertainties are magnified. The sum of the assumptions is often described as the worst-case scenario; that is, the most conservative evaluation. Although this approach is valid for ruling out the worst case, there often is no method available with which to estimate the precision of the data or to comment on a situation representative of the most likely case. Therefore, when it is apparent that further study may not furnish more useful information or may not furnish information with sufficient precision, remediation of the worst-case condition may proceed immediately after the qualitative risk assessment.

A study of the risk assessment and risk management techniques employed at a site in New Hampshire in support of remedial activities underscored the complexity of risk analysis[3] and suggested techniques to improve the methodology. Sensitivity analysis, the study proposed, should be conducted to identify the major uncertainties within a data base. Where uncertainty is especially high, methods of reducing it through refinement of assumptions or generation of additional data should be investigated. If unacceptable ambiguity persists, investigators should seek alternative methods of site evaluation.

Because resources very often are scarce, even for addressing environmental hazards, it is necessary that the limited resources available for remediation be applied incisively and effectively. Attempts to reduce risks to zero or to implement remedies without knowledge of the risk level can multiply the cost of remediation without providing commensurate benefit. Accurate assessment of the current and future health risks of a given hazardous waste site is central to economical remedial action.

THE QUALITATIVE RISK ASSESSMENT

Through a qualitative risk assessment, data relating to a particular site are evaluated, and major mechanisms, pathways, and scenarios that may contribute to the level of risk at the site are identified. The qualitative risk assessment characterizes chemical hazards and formulates a preliminary estimate of chemical properties and site conditions. It provides an estimate of whether humans or animals in the environs can be affected adversely, and it predicts the nature and areal range of those effects. The qualitative risk assessment identifies specific factors that require further investigation, and documents the grounds for considering certain exposure routes and risk scenarios and eliminating others.

The qualitative risk assessment evaluates the hazard potential of a given waste material, scrutinizing each of the possible pathways—air, soil, surface water, and ground water—for its potential to facilitate exposure of receptors to the chemical components. A *complete exposure pathway* has these components:

- An available, functioning mobile waste source in a toxic and bio-available form. Important qualities of a waste substance are its physical state, toxicity, water solubility, vapor pressure, bioconcentration potential, and biodegradability.
- A transport mechanism capable of conveying the waste material to a potentially sensitive receptor. The common routes considered are air, surface water, ground water, and direct contact.
- A receptor within the limits of the effective range of the transport mechanism. Important factors are population density, land use, water use, and distance between the waste site and the receptor.

These three components are illustrated in Figure 4-1.

Frequently when working on a hazardous waste site, one finds conditions that demand immediate attention. In order to provide a suitable response,

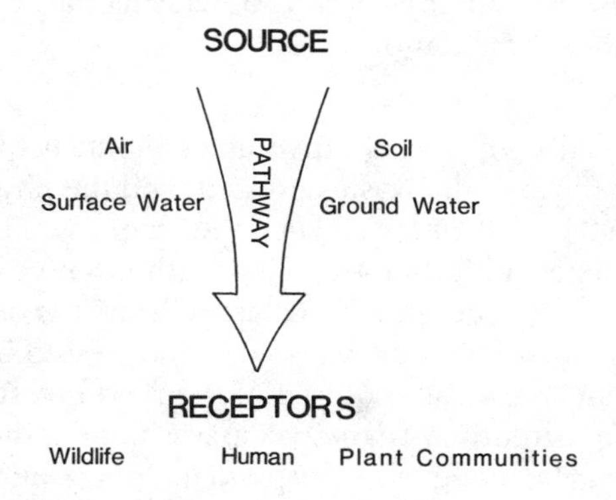

Figure 4-1. Essential components for complete exposure pathway.

the existing data base is evaluated by an abbreviated risk assessment. Subsequent remedial action is termed an interim remedial measure (IRM) or emergency remedial measure (ERM). The abbreviated risk assessment must furnish sufficient detail to prove that a legitimate technical foundation warranted the action.

In addition to identifying complete exposure scenarios, the qualitative risk assessment formulates the data requirements and the objectives of subsequent field investigations. With the qualitative risk assessment in hand, investigators can design a study that will define and quantify the major potential adverse effects of the hazardous waste site.

Formulation of Exposure Scenarios. The qualitative risk assessment evaluates independently the source, transport routes, and receptors. It then combines these three components into specific exposure scenarios. These components are reviewed and evaluated in combination; scenarios that do not meet the criterion of completeness, because they lack one or more of the three necessary components, are eliminated from further consideration, as they fall short of demonstrating a potential for damaging exposure. For example, improperly disposed toxic chemical compounds may become strongly bound to soil particles or compounds that degrade too quickly to be capable of mobilizing at toxic concentrations, and so cannot constitute a substantial risk to nearby populations. In these cases, the source is present, but the transport mechanism cannot supply the toxic waste to receptors. In another case, a toxic chemical may be released from a site but not represent a complete exposure scenario because of the absence of human or animal populations in zones where toxic levels of the material are present. Both of these cases represent uncontrolled hazardous chemical waste exposure scenarios with a low probability, or risk, of causing harm. They require no further evaluation for remediation.

Source. To proceed with a qualitative assessment, investigators require information about the site, the wastes present, and the physical, geological, and land use characteristics of the site and the environs. The minimum data requirements are listed in Table 4-1. The methods used to ascertain the conditions of the site were presented in Chapter 3, and the methods employed to define the characteristics of the waste were addressed in Chapter 2. It is important that the information required of the field investigation be evaluated prior to or concurrently with the risk assessment of the site. Thus, data collection efforts can be realigned to respond to data requirements specified by the risk assessment.

TABLE 4-1. Minimum data required for qualitative risk assessment.

Climate of area	Use of ground water in area
Geology of confining strata and formations	Location of proximate surface water bodies
Seasonal levels of ground water	Nature of process that generated wastes
Horizontal and vertical extent of waste material and residues	Volume and major hazardous constituents of wastes
Indication of type and depth of soil or other cover on site	Physical state of waste materials
Current and projected land use of site and environs	Physicochemical properties of waste materials
Human and wildlife population density in area	Toxicity of waste materials

To evaluate the source of risk, three attributes of the waste are scrutinized:

- The chemical, physical, and toxicological characteristics of the waste and waste components.
- The qualities that permit the waste to interact with transport media.
- The potential each waste component has for inducing toxic effects in humans and animals.

The objective of the source evaluation is to identify a set of compounds that are both mobile and toxic. These compounds are the focus of the risk assessment and serve as site-specific indicator parameters within the framework of a sampling and analysis program in which environmental media at on- and off-site locations are measured for chemical waste residues.

The essential first step in assessing the source is a review of the history of the site. The broader the information base is concerning the disposal history of the site, the more likely it is that an accurate assessment can be formulated of the effects of waste on the site. Production histories and process descriptions of the industries that occupied the site can be obtained by reviewing corporate documents, chemical engineering handbooks, and journals describing the processes prevalent during the production period in question. For example, if the site was a manufacturing area where wastes were deposited over a number of years, a thorough review of the processes employed at that location, the products manufactured, the raw materials, the wastes, the by-products, and the practices of waste management will yield vital information. Sometimes such information can be obtained by interviewing people who were employed at the facility if documents from the period of activity are unavailable.

Expending extra effort to secure reliable and complete data during the qualitative risk assessment can strengthen both the efficacy and the economy of the remediation. If a complete record of factual information cannot be developed, the investigators can infer connecting events. However, these inferences assume the worst case, and inject uncertainty into the evaluation. The effect upon the investigation is unfavorable because imprecision and bias are introduced; most often, the situation demands that the assumptions be biased conservatively on the side of safety. The result can be a risk assessment that prescribes remediation involving elaborate, costly measures even though investigators do not know whether site conditions warrant that level of remediation.

A blind survey of chemical residues can be conducted in lieu of a comprehensive site history. Such a survey is appropriate if little is known or can be reconstructed regarding the nature of the processes that produced the wastes. It may also be appropriate at locations such as municipal landfills where a wide range of wastes were deposited. This approach was devised by the U.S. EPA and is routinely utilized in remedial investigations. The waste material is divided into three organic fractions based on an extraction scheme, and is analyzed and quantified by gas chromatography/mass spectrometry to identify 129 specific organic and inorganic compounds, termed *priority pollutants.*

Caution must be exercised when utilizing the priority pollutant analysis, as the results often can be misleading. Occasionally, investigators assume that analysis of a particular waste material for the organic and inorganic priority pollutants by protocols approved by the U.S. EPA constitutes a comprehensive assay of all the possible chemical compounds present in the sample. However, the limitations of this approach must be borne in mind. By relying only on a blind analysis for priority pollutants, many toxic components of the waste material can be overlooked and eliminated from further consideration. Therefore, whenever possible, the analytical investigation should proceed based on the facts known about the chemical nature of the waste material developed from the site history. Investigators can further characterize the identity and quantity of these compounds by performing focused laboratory analyses of the waste materials.

After the compounds present at the source are identified, the properties controlling their movement and transport are evaluated. In this evaluation, the major factors considered are the physical state of the bulk waste and its matrix. The physical state of the waste material is important information to both the qualitative and the quantitative risk assessments. The waste may be in a solid, liquid, gaseous, or contained state, or it may be in a combination of these states. The amount of surface available to interact with the material's direct environment relative to the volume of the waste is a second important physical factor. The investigators determine the degree to which the waste material can complete a physical interaction with the environmental media it contacts, and then determine whether the current physical state of the waste enhances or diminishes its ability to be transported.

For example, a finely divided powdered waste will behave differently from a waste material with the same chemical composition but present as a nonporous solidified mass. The physical state of the powdered material will increase the amount of surface area available for interaction with infiltrating precipitation or ground water, as well as its ability to supply volatilized or particulate components to airborne releases. For liquid wastes, viscosity and density are important factors. A liquid waste with low viscosity and high specific gravity tends to flow along vertical barriers by the force of gravity as a discrete fluid, independent of ground water flow; such wastes also adhere to surfaces in thin layers. In contrast, the movement of a light liquid with high viscosity will depend upon the solubilizing properties of the water or air it contacts.

The physical state of wastes in containers is particularly difficult to interpret. It is necessary to determine whether the material in the container is leaking or has the potential to leak, or whether it is in a form that would prevent it from leaking even if the container were breached. Because most containers maintain their integrity for a defined length of time, both current and future physical states are estimated.

After the waste material, its components, and its matrix have been characterized, key physical and chemical properties are utilized to evaluate each chemical component. The properties considered are those that affect the transport and partitioning behavior of the chemicals in the environment. A number of source books contain compilations of physicochemical properties of chemicals that could be found in hazardous waste sites,[4] but data on many compounds are often nonexistent or difficult to retrieve.[5]

Two primary physicochemical properties of waste components evaluated are their propensity to leach from the deposited waste mass, or water solubility, and their potential to accumulate in living organisms, referred to as bio-accumulation. Mobility in surface and ground water is favored by compounds with high water solubility, but the tendency for bio-accumulation decreases with increasing water solubility. Because of the numerous factors that it controls, a compound's water solubility can be used to estimate other physicochemical properties. Additional properties evaluated are those that affect the persistence of a compound; a disposition to biodegrade, to photodegrade, or to hydrolyze can prevent a compound from reaching a receptor at a biologically significant concentration regardless of its mobility or apparent tendency to bio-accumulate. A summary of the pertinent properties of wastes is presented in the appendix to this chapter.

Those compounds whose properties of mobility and persistence suggest they may pose a risk are scrutinized, but the toxic effects of all compounds are evaluated in order to identify a set of compounds for consideration within the analytical, quantitative tier of the risk assessment.

The transport of wastes with ground water is an equilibrium partitioning process. The potential mobility of a waste material can be evaluated with bench-scale experiments using jar extraction, column extraction, reversed-phase thin-layer chromatography, and high pressure liquid chromatography.

Jar and column extraction techniques evaluate the effect of a soil matrix on the extractability of designated waste components. The simplest example of jar extraction is the Extraction Procedure test for hazardous waste characteristics, known as the EP Toxicity test in RCRA. The EP Toxicity test develops a ratio to compare leachate concentrations of selected parameters with their regulated concentrations. It permits a determination of whether a waste material is an environmental hazard. The objective of the jar extraction technique is to predict the concentration of a given waste component in leachate, and it is a worst-case estimate of leachability. The results can be used to determine leachate concentrations regardless of the waste concentrations in the matrix or, in cases when the matrix concentration is known, to determine an equilibrium or adsorption constant for the waste and the matrix being tested. Since the result calculated by jar extraction is an equilibrium in the liquid phase, the transport process is not actually evaluated. However, the measured ratio of the solute in the aqueous phase to its concentration in the waste matrix can be used to estimate a soil adsorption coefficient, and from that a retardation factor can be calculated. The retardation factor is used in certain models of ground water transport.

Column extraction is more difficult and utilized less often than jar extraction. Data from the column extraction technique can be used to predict the concentration of the material in leachate under field conditions and to calculate both an equilibrium constant and a retardation factor.

The chromatographic techniques permit a soil mobility index or retardation factor to be calculated. These measures indicate the rate at which the waste material is transported in ground water relative to marker compounds. The advantages of such experiments are that the exact composition of the waste need not be known, and the effect of the matrix on the mobility or extractability of particular waste components can be monitored.

The toxicity of each of the chemical compounds known or thought to be present at the hazardous waste site requires careful evaluation. It is usually necessary to develop a profile of each compound, listing acute and chronic toxic effects and data on carcinogenicity and mutagenicity. With these profiles in hand, the compounds from the hazardous waste site are compared one with another with respect to their potency and range of toxic effects, and they are listed in decreasing order of toxicity. Each compound is then categorized with respect to its acute and chronic LD-50 (the dose at which 50 percent of the test population is affected), and its potency as a carcinogen, if applicable. This list, which is used together with information about mobility and persistence characteristics, serves as a basis for choosing appropriate site specific compounds for inclusion in sampling and analytical programs.

The most comprehensive source of basic information on the toxic properties of chemicals is the Registry of Toxic Effects of Chemical Substances (RTECS).[6] The information in RTECS is a comprehensive collection of toxicological data, but it lists only the toxicological end-points of the compounds; it does not discuss the basis for the toxicity of the compounds. For that information, the investigator must refer to the original papers referenced by RTECS.

Additional information about the toxicological effects and environmental fate of chemical substances is available from the scientific literature, abstracts, and data bases. Papers are indexed by compound name or subject. A sampling of computerized data bases is presented in the appendix to this chapter.

Various divisions of the U.S. EPA have collected extensive information about chemical compounds commonly found at uncontrolled hazardous waste sites. This information, usually disseminated in published form for individual compounds, includes an assessment of an acceptable human dose or residual levels. In addition, as part of SARA, the Agency for Toxic Substances Disease Registry is required to prepare a toxicological profile of hazardous substances most commonly found at facilities on the National Priorities List. These profiles, as they become available, should be a valuable source of information for the development of risk assessments. Several additional sources of published toxicological information are listed in the appendix at the end of this chapter.

There are numerous means by which toxicological data, especially data on carcinogenic potency, are evaluated and extrapolated to humans. The various methods can produce results that are orders of magnitude different from each other when considering the same substance and dose. All of the techniques are based on theoretical considerations; no single method is universally applicable, and no single method has been proved accurate using animal models or epidemiological data. Therefore, no single toxicological profile should be considered to be uniquely authoritative. One should discuss the range of values generated by the various techniques when describing the degree of uncertainty inherent in a risk assessment.

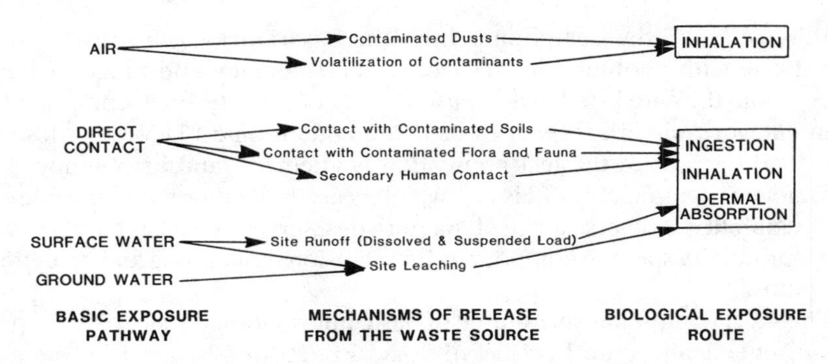

Figure 4-2. Potential transport mechanisms and exposure routes at hazardous waste site.

Transport Routes. The transport mechanisms at a hazardous waste site are scrutinized to identify those that facilitate the interaction of source waste materials with receptors in the environs. The mechanisms considered are:

- Leaching of soluble waste materials with ground water.
- Flow of surface water.
- Scouring action of air currents.
- Direct contact.

Each of these potential mechanisms is illustrated in Figure 4-2.

The transport of hazardous waste constituents with ground water is sometimes considered to be the pathway with the greatest long-term human health consequences, because aquifers supplying drinking water can be contaminated by chemical wastes for an indefinite time following a release. Therefore, when developing a qualitative risk assessment, it is important to determine whether ground water functions as an active waste transport mechanism. That is done by reviewing available data and by developing supplemental information through field investigations. The site features accounted are listed in Table 4-2.

Waste materials at an inactive landfill or lagoon can enter ground water in solution with direct infiltration of surface water or as a distinct, free-phase liquid. Wastes not present in the liquid phase can be dissolved and leached by the action of surface or ground water and transported off-site. Two dominant factors control the extent and rate at which infiltrating water or ground water can affect, leach, and transport mobile waste components: the depth of the bottom of the waste material relative to the height of the uppermost layer of ground water and the permeability or hydraulic conductivity of the waste materials and adjoining soils. For example, wastes in sandy soil in contact with ground water in an area that receives ample precipitation have a

TABLE 4-2. Site features reviewed in qualitative risk assessment.

Geology of area, especially confining materials	Ground water flow velocity
	Ground water use in the area
Depth to bedrock	
	Average annual precipitation
Physical dimensions	
	Type and depth of soil cover
Depth to ground water	
	Extent of vegetative cover
Seasonal ground water levels	

high potential for interaction and movement with the ground water. In contrast, waste materials in a clay soil removed from contact with the ground water have a far lower potential for interaction and movement.

The obvious and common approach to determining whether ground water is an active transport mechanism is to test for the presence of residues of the particular waste components in monitoring or residential wells at locations hydraulically downgradient of the waste site. The results of these tests are compared with those of similar tests at upgradient wells. During the qualitative study, only the presence of the components, not the concentration, is considered. If detectable concentrations of waste components not present at upgradient locations are found in ground water downgradient of the site, the ground water transport route is determined to be functioning. If a receptor on this route is identified, the ground water transport mechanism is further evaluated within the quantitative, tier two, risk assessment.

A second approach to evaluating the transport role of ground water is to establish a conceptual model of the system based on the hydrogeological properties of the area. For example, if available information indicates that the soils are highly permeable, and if the bottom of the landfill is situated within the ground water, investigators can deduce that the ground water route can function to transport waste components. In actual field conditions, such simplistic conditions without confounding factors rarely prevail. To assist the analysis of the ground water flow, the site's parameters are quantified into a mathematical model. (For additional information on modeling, see Chapter 7.)

Surface water can furnish a rapid and direct transport route for chemical substances, and wastes may be in dissolved or particulate form. In contrast to ground water, surface water can scour the earth and mobilize contaminants.

For the surface water transport route to be complete, waste materials must be within the influence of runoff, and the surface elevation and slope must be able to support the movement of the runoff at a velocity capable of carrying the wastes in suspension. Once moved by surface water, the contaminants can be deposited atop soils and provide a source for the direct contact or air routes, or they may remain with the water and be deposited into a surface water body such as a river or a lake.

Surface water transport may also act in conjunction with air or ground water transport. Ground water may discharge into surface water in the form of seeps, or as infiltration into rivers or lakes, and provide the initial source of waste. Alternatively, waste residues transported by air may be deposited on surface water and then be further transported. In both cases, the surface water is the secondary transport mechanism as well as a vehicle for exposure.

Waste components can be transported through the air if the waste materials are exposed to and affected by air currents. The wastes may be transported either in the vapor phase, if their vapor pressure favors their volatilization, or adhered to soil particles, if the soil is in a friable form. Wastes are also transported in the air when fugitive dusts are created. Heavy machinery, automobiles, or even foot traffic in an area of exposed waste material can generate dusts upon which waste can adsorb. If captured by the wind, these dusts can be transported into the environs of the site. Another method of airborne transport of wastes, uncommon but notable, is the movement of gases generated by the digestion of landfilled materials. Methane is commonly produced in landfills, and as it seeps to the surface, it can carry volatile waste components through the interstices of the soil and into the atmosphere.

Receptors. A *receptor* is any person, animal population, or natural or economic resource within the transport range of waste materials of a site. The objective in an evaluation of receptors is to examine critically the use of the land and resources proximate to the waste site, and to identify any land uses or resources that could be affected by exposure to the waste materials. The evaluation considers both current and future receptors.

The basis for evaluating receptors is a review of land use. The current and proposed uses of the hazardous waste site itself are examined. In order to identify potential receptors in the environs, a series of questions is juxtaposed against the known facts of current and future land uses. Typical queries are presented in Table 4-3. Residential, agricultural, and industrial areas are considered separately because people utilize each with different levels of intensity. Local and regional planning and zoning boards and agencies can furnish information on land use and population trends.

TABLE 4-3. Issues addressed in qualitative risk assessment to identify potential receptors.

ISSUES GERMANE TO THE SITE	
Current and projected land use.	Ability of owner to maintain property.
Nature of current and projected uses of adjoining land.	Economic value of property.
Owner.	Restrictions on access.
ISSUES GERMANE TO ADJOINING PROPERTIES	
Functioning transport routes and effective ranges.	Presence of rare or endangered species.
Source of potable water.	Specially designated wetlands.
Ground water use downgradient.	Important wildlife habitat.
Nearest surface water bodies.	Commercial fishing areas.
Prevailing winds.	Area's historical significance.
Current and projected population.	Unique natural or economic resources.
Agricultural uses.	
Nearby schools or residential institutions.	Land uses allowed by current zoning; future zoning.
Recreational uses.	Feasibility of zoning or other land use restrictions.

Once the issues in Table 4-3 have been addressed, investigators can propose a set of specific scenarios to identify potential receptors. Each of these scenarios is evaluated in relation to each of the transport mechanisms deemed functional.

The qualitative risk assessment provides documentation of an evaluation of all possible combinations of exposure and transport scenarios. Scenarios determined to function such that humans or animals could be exposed to hazardous chemicals are regarded as complete and evaluated further in the quantitative assessment. Scenarios deemed incapable of conveying hazardous wastes to the potential receptors are termed incomplete and not considered further. For the complete scenarios, a set of target compounds specific to the site is submitted for each transport and exposure route identified for further analysis of potential effects. Evaluators describe each of the scenarios in detail as a conclusion to the documentation of the qualitative assessment.

The record of the risk assessment effort presents the rationale for dismissing incomplete scenarios.

THE QUANTITATIVE RISK ASSESSMENT

In contrast to the descriptive qualitative risk assessment, the quantitative risk assessment estimates the degree of adverse effect due to release of specific wastes from a particular site. It is a tool used to calculate a numerical value for existing and future health risks associated with exposures by means of each identified complete pathway. To accomplish this end, the quantitative risk assessment:

- Specifies the identity and rate of release of contaminants from the source.
- Quantifies the rate of transport and dilution of releases.
- Quantifies the rates of exposure and uptake by receptors.
- Quantifies the degree of toxicological effect.

Data are developed on chemical residues in the various transport media to produce the quantitative risk assessment. Assumptions and estimates based on available information regarding the transport, fate, and toxic effects of the compounds at hand are also applied. A typical scenario developed to describe exposure from an aquifer utilized as drinking water would include leaching of wastes in contact with ground water, transport of those leached materials with the movement of ground water toward domestic wells, and human exposure by ingestion. A sample of ground water from a receptor's drinking water source is necessary to verify the concentration and identity of the contaminant in the water, and a profile of the receptor's water use is also needed. The exposure and uptake data are correlated with known toxicological effects of similar dosages. Finally, the quantitative assessment furnishes numerical values for the leaching and transport rates and the number of years over which the residues are expected to remain in the ground water, and an effect for the situation at hand is quantified.

In most cases, investigators lack the broad information required. The budget for a risk assessment often makes such a comprehensive survey of all media and all potential receptors impracticable. In most cases, too, residues from the waste site have not yet been transported to receptors. Therefore, many quantitative risk assessments must rely on available data, predictive models, and qualifying assumptions. The assumptions are designed to reflect worst-case conditions so that if they are in error, the error is on the side of safety. However, the assumptions still must be reasonable and probable under the circumstances. When risk assessments rely on numerous worst-case assumptions, the net effect can be the magnification of errors to a point where they describe a situation that is impossible under any circumstances.

To prevent the pyramiding of worst-case assumptions, each assumption is tempered with as much site-specific information as possible. For example, one standard assumption regarding direct contact exposure may be that a child between 5 and 18 years of age will play on a site 2 days each week for a total of 1 hour each day, that the activities and resultant exposures will continue through adulthood, and that, through hand-to-mouth activity, a child will ingest a total of 0.1 g/d and adults (ages 19 through 60) will ingest 0.05 g/d. These assumptions may be valid, but they should be evaluated relative to the conditions at hand: Are children in the area? Is the site likely to be attractive for play activities? Will the same child spend 2 days each week throughout childhood on the site? Will exposure continue through adulthood? If, based on what is known about the site, the standard exposure assumptions are inappropriate, adjustments must be made. The remainder of this chapter presents some of the methods and standard assumptions applied in the development of a quantitative risk assessment.

Waste Characterization. Evaluation of the source term of the quantitative risk assessment yields general information regarding the identity and physical and chemical characteristics of the waste, to produce an explicit characterization of the waste. Evaluators give primary consideration to accurate identification of the individual waste components and quantification of their respective concentrations. Because the quantitative waste characterization requires detailed analytical treatment, analyses are often limited to the site indicators specified in the qualitative assessment.

The variability of the waste and the effects of its matrix on the leachability of its components are also evaluated. The qualitative evaluation may have been based upon one or a few grab samples of the waste, but the quantitative evaluation must define the range of concentrations of each chemical component and the frequency of their detection. Current analytical techniques are designed to provide an estimate of the maximum amount of analyte in the waste. However, this may not represent the amount of analyte available for transport to off-site receptors, particularly as affected by surface or ground water; so the quantitative evaluation should include leachability tests to estimate the amount of waste components able to be dissolved and extracted from the waste by water.

The waste is also characterized relative to its total mass and volume. Most quantitative assessments evaluate the potential effects of releases of a waste over a lifetime, usually at least 70 years. Therefore, one must know the total amount of residue in the waste source to calculate whether enough material is available to provide a lifetime exposure at a given release rate. The information determined is the length of time the release would persist were the waste to remain in place.

Quantification of Release Rate. Once the waste source is quantified, transport of the waste components by each of the complete routes is measured or predicted. These release and transport rates then become the bases for quantitative exposure assessment.

The airborne transport route is often the most easily measured. Air sampling stations can be installed at the boundaries of the site or at the nearest established receptor location. Sampling can be conducted for both dust-borne and volatilized waste components, which can be analyzed at remarkably low detection limits. Air dispersion models evaluate these data and furnish estimates of the concentrations at off-site locations under various meteorological conditions. (Information about air sampling techniques is contained in Chapter 2.)

Release by the ground water route can be established and quantified in two ways: field measurement (addressed in Chapter 3) and transport modeling (presented in Chapter 7). The preferred method is to install a series of ground water monitoring wells at representative depths and locations. Samples of ground water are collected from the monitoring wells and analyzed for residues of each of the indicator compounds. Ideally, measurements are conducted at both on- and off-site locations as well as at the receptor. These data then form the basis for establishing the identities and concentrations of chemical components released from the source and affecting the receptors. An alternate method is to establish a conceptual model of the extent of leachate generated by the waste materials and subsequent inputs to ground water. The release is then evaluated relative to the volume and flow rate of the aquifer into which the releases take place, as well as the physicochemical characteristics of the waste components. Based on this information, estimates of the concentration of waste residues at various distances from the waste source, and thus exposures, can be predicted. A combination of the two approaches usually is employed.

Direct contact, although not a transport route, can be a significant means of exposure. Data describing direct exposure are generated from the collection and analysis of surface soil samples from the site and from receptor areas. The sampling should be sufficiently comprehensive to indicate the areal extent of the residues. For example, if a 40-acre (16-ha) site is considered the location for potential direct contact exposure, a representative characterization of the surface soils in that entire area, including soil concentration isopleths, should be reviewed.

If surface water is an element of a complete exposure scenario, the residues of waste components in both the surface water and the sediments are measured. A quantitative determination of the surface water route should also calculate the rate at which the waste materials are entering the surface water body and the rates at which those inputs are transported and diluted.

Estimation of Exposure and Uptake Rates. Evaluation of the receptor term, the third component of the quantitative risk assessment, includes two quantitative elements: the receptor count or census and the exposure and uptake estimate. The census requires identification and measurement of the population within the range of the waste-carrying transport routes to delineate the number of potential exposures via each of the routes considered. For example, characterization of the ground water route can establish that ground water users within a specific area hydraulically downgradient of the source are within the effective range of a plume of contaminated ground water. Therefore, areas upgradient of the site, along the perimeter of the plume or at distances downgradient where the plume becomes effectively dispersed or diluted are eliminated from further consideration, and the present and future population residing within the defined areas only are evaluated. Therefore, the aggregate risks associated with exposures from a particular site are presented as a function of the size of the population potentially effected.

The concentration of waste components in the various media and the existence of an exposed population having been estimated, the volumes and rates at which contaminated soil, water, or air are being taken up by the receptors are calculated. These values are estimated relative to the known toxicological effects of the doses ingested. The six major exposure and uptake routes considered in quantitative exposure assessments are: inhalation of ambient air, inhalation during showering or other domestic uses of ground water, ingestion through drinking water, ingestion through food, ingestion following incidental contact with contaminated materials, and absorption through the skin following direct contact.

Toxicological Evaluation. The final and sometimes most difficult step in completing the quantitative risk assessment is evaluation of potential toxicological effects related to exposures to the chemical compounds in question, otherwise known as the dose-response evaluation. This evaluation requires review of data on the toxicological effects of the compounds to derive estimates of the dosages and exposure periods at which deleterious effects reasonably may be expected to occur. Estimates of concentrations at which the majority of human or wildlife populations can be exposed without suffering adverse health effects are also formulated from these data. Those levels are compared with the exposures predicted by the quantitative exposure assessment.

A first approach in evaluating the quantified exposures is comparison of the exposures to existing standards and guidelines established by state and federal public health agencies. Useful standards include state and federal primary drinking water and ground water standards for organic and inorganic compounds in public drinking water supplies. When no standards are available, an Acceptable Daily Intake (ADI) is derived. The ADI is the amount of toxicant that is expected not to cause adverse effects to the general population following long-term repeated exposures.

These standards and criteria furnish values for concentrations representing safe exposures for the entire population. They assume lifetime exposures and consider infants and the elderly, the most sensitive receptor populations; and their derivation incorporates conservative safety or uncertainty factors. The standards are used to rule out potential effects. If a receptor-point concentration falls below a particular standard, then it can be concluded, from a human health standpoint, that exposure to the residue does not represent an unacceptable condition.

If receptor-point concentrations exceed the standards, or if standards are unavailable for the particular compounds, the exposures must be further evaluated. Two approaches generally are used. For exposures to noncarcinogens above the established standards, the data supporting establishment of the standard are reviewed, and the potential toxicological effects at the levels measured at the site are evaluated. A different approach is used if the compound being evaluated is a carcinogen. There are a number of techniques currently used to establish safe exposure levels to carcinogens, and scientists have differing views on their validity. The individual techniques are built upon different assumptions and mathematical extrapolation formulas, and rely on the results of animal cancer studies conducted at high dosage levels. Cancer incidence observed at those high levels is used to predict the likelihood of cancer in humans exposed to low levels of the test compounds. All of the techniques in use today are based on mathematical and theoretical grounds that cannot be validated, and the level of accuracy has not been demonstrated for any individual estimate. Depending on the assumptions or techniques used, the predicted cancer potency factors can vary by several orders of magnitude. Regardless of the limitations and uncertainties in the prediction of cancer risks based on animal-to-human extrapolation, cancer potency estimates, the unit cancer risks, have been calculated by the U.S. EPA Carcinogen Assessment Group for various known and potential carcinogens. The values are designed to furnish an upper-bound, worst-case cancer risk estimate for exposure to these compounds.

Use of the Quantitative Risk Assessment. Data from the quantitative risk assessment usually are presented in tables. Acute and chronic noncarcinogenic risks are given as percentages or multiples of the ADI. Quantitative estimates of carcinogenic risks are presented as the probability for an increase in cancer incidence should the exposure scenario occur on a continuous basis over a human receptor's 70-year lifetime. In order to address the effects of multiple exposures, the risk estimates are summed. The quantitative risk assessment also documents all data and assumptions used in generating the release, transport, exposure, and toxicological terms of the assessment. These data are presented for each scenario. For example, if the direct contact route is evaluated, the scenario is described, and each assumption used in the evaluation is presented with a discussion of the final conclusion of the effects of the exposures. The evaluation is further clarified by referencing and presenting the numerical data as raw calculations, as well as in tabular fashion.

A critical product of the quantitative risk assessment is an analysis of the reliability of the risk estimate. It is usually presented numerically as the likely range and limits of the risk value. It should also include a discussion of the major assumptions used and the appropriateness and uncertainty associated with each assumption.

The quantitative risk assessment can be a valuable aid in determining the effects of a particular chemical release and subsequent exposures. However, because of the number of variables used to estimate them, quantitative risk estimates are vulnerable to error. Evaluators must consider this potential error during the planning of the field investigation, as well as when generating assumptions describing the behavior of variables for which direct values are unavailable. A minimum requirement of the quantitative risk estimate is, then, a numerical as well as written discussion of the level of confidence associated with the estimate.

SUMMARY

A risk assessment measures the potential for chemical wastes and physical agents to escape from a source and communicate with receptors in the environment. It is a crucial component of many remedial action programs. The risk assessment interprets data about the source and articulates the degree of influence the waste site exerts on the environment. A risk assessment is the construct used to unify the three components (source, transport route, and receptor) associated with a release of chemicals into the environment, and it evaluates them as an integral process. It is employed to compare the effectiveness and applicability of various remedial designs.

An investigator initiates a risk assessment by marshaling and summarizing all available information for which the confidence level is acceptable. The range of possible exposure and risk scenarios is reviewed, and those inapplicable to the situation at hand are culled. As the risk assessment progresses, specific considerations are examined further. During the evaluation process, both the continuity of the waste source and feasible transport and exposure scenarios are reviewed. As the sufficiency of the data base is considered, specific additional data requirements are defined.

A risk assessment may be executed in two different levels of detail. The tier one study assesses the risk of exposure qualitatively. In it, the basic aspects of the waste site are accounted, and the analysis assesses whether waste materials can be transported to receptors and identifies the available pathways, determines whether any receptors can be affected by the wastes, and identifies transport and exposure scenarios. The qualitative risk assessment documents why the waste components, transport routes, or receptors excluded from the study were not considered. Its result may be the recommendation that a particular source or transport route be remedied without recourse to further analysis. The qualitative assessment can also conclude that the uncertainty of data precludes the development of useful information.

After completion of tier one, qualitative risk assessment, a second tier, quantitative risk assessment may be indicated. A quantitative risk assessment estimates the degree of adverse effect present upon release of specific waste components from a particular site. Whereas the qualitative risk assessment identified complete exposure pathways, the quantitative risk assessment formulates a numerical estimate of the existing and future health risks associated with each of the complete pathways. Actual measurements or projections derived from models are the tools of the quantitative risk assessment.

As many aspects of a hazardous waste site cannot be quantified, assumptions are postulated, and measures pertinent to the particular site are calculated. The calculated values are derived from data developed in the field investigations and gleaned from the literature on toxicology.

The quantitative risk assessment follows each component of the exposure continuum and generates for each an estimate of its magnitude. For example, to assess the exposure of a human population's potable water source via ground water transport, a typical scenario would include leaching of wastes in contact with ground water, transport of those leached materials with the ground water toward water supply wells, and human exposure to the contaminated ground water by ingestion. The data required for each element or link of the exposure scenario are quantified. It is essential to generate as much information from first-hand observation—primary data—as possible. In order to gauge the sufficiency of the quantitative risk assessment, an estimate of the level of precision required of the prediction is necessary.

A quantitative risk assessment can be a valuable tool in determining the effects of chemical release and exposures. Because of the number of variables in the estimate, the quantitative risk estimate is prone to error; and evaluators must consider that error when organizing the field investigation, and when making assumptions describing the behavior of the variables for which direct values are unavailable. The quantitative risk assessment must incorporate a discussion of the level of confidence associated with the estimate in addition to the numerical data.

NOTES

1. U.S. EPA Office of Emergency and Remedial Response. Draft Superfund health assessment manual. 1985.
2. Universities Associated for Research and Education in Pathology (9650 Rockville Pike, Bethesda, Md. 20814). Draft report: Executive scientific panel on the health aspects of the disposal of waste chemicals. Oct. 1984.
3. J. S. Evans, C. T. Petito, and D. M. Gravallese, Energy and Environmental Policy Center (John F. Kennedy School of Government, Harvard University, 79 John F. Kennedy St., Cambridge, Mass. 02138). Cleaning up the Gilson Road hazardous waste site: A case study. Feb. 1986.
4. C. H. Hansch and J. A. Leo. *Substituent constants for correlation analysis in chemistry and biology.* New York: John Wiley & Sons, 1979.
5. The reader is referred to W. J. Lyman, W. F. Reehl, and D. H. Rosenblatt. *Handbook of chemical property estimation methods—environmental behavior of organic compounds.* New York: McGraw-Hill Book Company, 1982. It is an excellent review of physicochemical properties and includes methods for their estimation. The approaches and formulas presented provide a basis for estimating necessary data where none are available, and the methods furnish an acceptable degree of accuracy.
6. National Institute of Occupational Safety and Health (U.S. Department of Health and Human Services, Public Health Service, Centers for Disease Control, National Institute for Occupational Safety and Health, Cincinnati, Ohio 45226). *Registry of toxic effects of chemical substances.* Nov. 1985. RTECS was compiled by a mandate of the Occupational Safety and Health Act (OSHA). The printed medium is updated annually. On microfiche, it is updated quarterly. RTECS is also available through computer information retrieval services.

APPENDIX

Pertinent Properties of Wastes

Water Solubility. Water solubility is the most important parameter affecting the environmental dynamics and fate of organic compounds. Solubility measures the maximum amount of a compound that can be dissolved in water at a specific temperature before two distinct phases appear. It has a major effect on mobility and bio-accumulation potential. An estimate of water solubility indicates the relative water contamination potential of a chemical; the concentration at which an organic chemical is transported with ground water is directly related to its water solubility. A compound's solubility is inversely related to its tendency to adsorb to soils and sediments, to its tendency to bio-accumulate, to its tendency to volatilize from water, and to its propensity for degradation by microorganisms.

Octanol-Water Partition Coefficient (K_{OW}). The quantity K_{OW}, a laboratory measurement with no units, is presented as the log of the ratio of the equilibrium concentration of a chemical in a two-phase liquid system composed of octanol and water. It is routinely used to estimate the tendency for an organic chemical to partition between two environmental phases of differing polarity such as an organic phase and an aqueous phase. K_{OW} is also used as a general estimate of the tendency of a compound to partition or bioconcentrate between water and the lipid fraction of fish, and as an estimate of its equilibrium partitioning to aqueous sediments. K_{OW} is inversely related to water solubility. The log K_{OW} values normally range from -3 to 7; highly water-soluble compounds, such as ethanol, have values of log K_{OW} < 1; hydrophobic compounds, such as certain PCBs and chlorinated dioxin congeners, have values of 6 to 7.

Soil Adsorption Coefficient (K_{OC}). The quantity K_{OC} measures the tendency of a chemical to partition between the solid and solution phases of a soil-water system. K_{OC} is similar to K_{OW}, and is expressed thus:

$$K_{OC} = \frac{\mu\text{g adsorbed/g organic carbon}}{\mu\text{g/ml in solution}}$$

K_{OC} accounts for the amount of organic carbon adsorbed to the soil. Therefore, partitioning is a function of the amount of hydrophobic environment present on the surface of the soil particle, and the soil simply serves as a matrix for the organic carbon. K_{OC} can be used to predict the extent that an organic solute will partition into soils when ground water moves through subsurface soils, to estimate the degree to which chemical components will become adsorbed to surface soils, and to estimate the partitioning to soils during runoff and the partitioning in aqueous sediments. Values for K_{OC} are not commonly reported in the literature, but they can be measured experimentally or estimated on the basis of water solubility, K_{OW}, or the Freundlich adsorption coefficient.

Vapor Pressure. Vapor pressure controls the rate and extent that a compound will evaporate when in contact with air. It is a pertinent measure at the interface of air with soil and air with water. Vapor pressure determines the extent that a compound can be transported in air following a release, and it controls the rate of volatilization from soils and from solution in water.

Acid Dissociation Constant (K_a). The constant K_a can represent the degree to which an organic acid or base adsorbs, partitions, dissolves, or is transported through environmental media and governs the amount of organic chemical in the ionized versus un-ionized form at a given pH. K_a is a direct function of the ratio of the ionized to the un-ionized form of the compound in place at a given time. Neutral (un-ionized) organic molecules have significantly lower solubility in water than their ionized forms; therefore, the ionized forms are more readily transported with the action of water, whereas the un-ionized forms are more strongly adsorbed or partitioned into nonpolar media.

Hydrolysis. Hydrolysis is associated with a compound's chance of becoming water-soluble and with its potential for detoxification; it is the process by which an organic molecule reacts with water to become cleaved. Hydrolysis involves a chemical process that divides the compound into two simpler and usually more water-soluble forms. During hydrolysis, water (a nucleophile) reacts with an electrophilic center on the molecule; only organic compounds with hydrolyzable eletrophilic centers can become hydrolyzed.

Biodegradability. Biodegradability is the ability of a compound to become metabolized by microorganisms, and it governs the persistence of that compound in the environment. Biodegradation can result from utilization of the material as a sole carbon source or from the co-metabolism of the material with other carbon sources. Materials resistant to biodegradation remain as residues and often bio-accumulate. The biodegradability of a compound affects the distance the compound can travel if released from a hazardous waste site. Biodegradable material can be released from a hazardous waste site with less adverse consequence than can persistent compounds.

Photodegradability. Many organic compounds absorb visible and ultraviolet wavelengths of light energy and thereby degrade into different chemical forms. Photodegradable compounds are less persistent in the environment than compounds unaffected by light if sunlight is available to them. The presence of chemical photoinitiators (compounds that accelerate the rate of photolysis) and quenching materials (compounds that reduce the rate of photolysis) in the environment is difficult to predict.

Reactivity. Reactivity describes the potential for a material to react spontaneously at a given temperature or pressure, or following the addition of water, air, heat, or incompatible materials. Many highly toxic and carcinogenic compounds are hazardous because of their reactive nature. The reactivity of materials at a hazardous waste site must be considered because it indicates whether they may explode or react violently; highly reactive materials have the potential of affecting populations in the site's environs. The concentration of a compound, the nature of its matrix, and the presence of other materials have an effect on reactivity. The persistence of reactive materials once they leave the site must be considered. Because of their chemical instability, many highly reactive compounds have short environmental half-lives and when released may react to become chemically decomposed to a nonhazardous material.

Sample Listing of Computerized Data Bases

ANALYTICAL ABSTRACTS: The Royal Society of Chemistry, The University, Nottingham, England NG7 2RD; 0602 507411

BIOSIS PREVIEWS: BioSciences Information Service, 2100 Arch Street, Philadelphia, PA 19103; (215) 587-4800, (800) 532-4806

CA SEARCH, CHEMNAME, CHEMSIS, CHEMZERO: Manager, User Education, Chemical Abstracts Service, P.O. Box 3012, Columbus, OH 43210

CHEMICAL EXPOSURE: Chemical Effects Information Center, Bldg. 2001, Oak Ridge National Laboratory, P.O. Box X, Oak Ridge, TN 37830; (615) 574-7772

CHEMICAL REGULATIONS AND GUIDELINES SYSTEM (CRGS): CRC Systems, Inc., 4020 Williamsburg Ct., Fairfax, VA 22032; (703) 385-0440 x120

DISSERTATION ABSTRACTS ONLINE: Dissertation Publishing, UMI, 300 North Zeeb Road, Ann Arbor, MI 48106; (800) 521-0600, (313) 761-4700

ENVIROLINE: Environmental Information Center, Inc., 292 Madison Avenue, New York, NY 10017; (212) 949-9494

ENVIRONMENTAL BIBLIOGRAPHY: International Academy at Santa Barbara, Environmental Studies Institute, 2060 Alameda Padre Sierra, Santa Barbara, CA 93103; (805) 965-5010

ENVIRONMENTAL HEALTH NEWS: Occupational Health Services, Inc., 400 Plaza Drive, P.O. Box 1505, Secaucus, NJ 07094; (201) 865-7500

FEDERAL REGISTER ABSTRACTS: National Standards Association, 5161 River Road, Bethesda, MD 20816; (301) 951-1310

HEILBRON: Chapman and Hall, Ltd., 11 New Fetter Lane, London EC4P 4EE, United Kingdom; 01-583-9855

KIRK-OTHMER ONLINE & WILEY'S ENCYCLOPEDIA OF POLYMER SCIENCE & ENGINEERING: Engineering/Online John Wiley & Sons, Inc., 605 Third Avenue, New York, NY 10158; (212) 850-6360

MEDIS: Mead Data Central, P.O. Box 933, 9393 Springboro Pike, Dayton, OH 45401; (800) 227-4908

MEDLINE: U.S. National Library of Medicine, 8600 Rockville Pike, Bethesda, MD 20209; (301) 496-3147
NTIS (NATIONAL TECHNICAL INFORMATION SERVICE): U.S. Department of Commerce, 5285 Port Royal Road, Springfield, VA 22161; (703) 487-4600
POLLUTION ABSTRACTS: Director, Database Services, Cambridge Scientific Abstracts, 5161 River Road, Bethesda, MD 20816; (800) 638-8076, (301) 951-1400
SCISEARCH: Institute for Scientific Information (ISI), 3501 Market Street, University City Science Center, Philadelphia, PA 19104; (215) 386-0100, (800) 523-1850
TSCA INITIAL INVENTORY: TSCA Assistance Office, U.S. Environmental Protection Agency, Office of Pesticides and Toxic Substances, 401 M Street, S.W., Washington, DC 20460, (800) 424-9065, (202) 554-1404
WATER RESOURCES ABSTRACTS: Office of Water Research and Technology, U.S. Department of the Interior, Washington, DC 20240; (202) 343-8435
WORLD ENVIRONMENT REPORT: Business Publishers Inc., 951 Pershing Drive, Silver Springs, MD 20910; (301) 587-6300

Sources of Published Toxicological Information Related to Hazardous Waste Sites

Health effects assessments
U.S. EPA, Office of Health and Environmental Assessment
Recommended maximum contaminant level goals
U.S. EPA, Office of Drinking Water
Federal water quality criteria
U.S. EPA, Office of Water Regulations and Standards
Documentation for development of toxicity and volume scores for purpose of scheduling hazardous wastes
U.S. EPA, Office of Solid Waste
Compilation of chemical, physical, and biological properties of compounds present at hazardous waste sites
U.S. EPA, Office of Waste Program Enforcement
Water-related environmental fate of 129 priority pollutants
U.S. EPA, Office of Water
Water quality standards handbook
U.S. EPA, Office of Water
Draft superfund health effects assessment guidance document
U.S. EPA, Office of Solid Waste and Emergency Response
Support documents for state and federal water qualtiy standards
American Conference of Governmental Industrial Hygienists
Documentation for occupational safety and health standards
Occupational Safety and Health Administration

Chapter 5
The Role of the Laboratory in Remediation Work

The modern environmental laboratory is a complex collection of remarkably sensitive instruments, powerful data gathering and processing computers, and trained scientists. Analysis results are routinely determined in the range of parts per million, per billion, and even per trillion. In the realm of these minuscule quantities, there are few absolutes. For example, no water taken from the environment is pure. Contaminants are always present, and their quantity is referred to as *background.*

An investigation of a hazardous waste site attempts to identify and quantify all significant contaminants at the site. The analytical tools available to investigators in the completion of this task are diverse and far-ranging. To begin, the potential contaminants are first delineated; next, a sampling and analytical program is established. Thus, the field program meshes with the laboratory analytical program.

There are a number of potential pollutants present at a hazardous waste site. Both the pollutants to be monitored at the site and the analytical protocols to be used to identify and quantify those substances are considered simultaneously, and this is done early on, as the investigation is outlined. The site investigator selects pollutants that have recognized protocols and are economical to analyze; the laboratory chemist can assist the site investigator in selecting protocols that will achieve the goals of the investigation. The following factors are important during the selection process:

- Reasonable and specific levels of detection are chosen for the pollutants in question. Low detection limits—parts per billion rather than parts per million, say—are not required for samples in a highly contaminated matrix. If significant contamination is present, there is no point in reaching low sensitivity.
- Indicator parameters—for example, total organic carbon, chlorides, specific conductance—are selected to monitor the variability of contamina-

This chapter was developed by David R. Hill of OBG Laboratories, Inc.

tion within a specific location. When noticeable changes are found, additional, more specific analyses are scheduled to achieve further definition.
- Analytes that are regulated and thus have published levels of acceptance or are naturally occurring, or both, are selected. If standards for the selected primary pollutants have not been published, the laboratory scientist is faced with the difficulty of determining whether or not the analyte is a potential problem at the site.

The primary task of the laboratory is to generate data that are technically acceptable and are of good enough quality to enable site investigators to evaluate the potential problems at hand. The investigators' evaluation and ultimate conclusion can only be as good as the data supplied them by, among others, the environmental laboratory. Therefore, the laboratory supervisor should be included in establishing the protocols and criteria of the investigation in sample collection, sample preservation, analytical protocols, and quality control and quality assurance (QA/QC).

Each of these topics—sampling, analysis, and QA/QC—is presented in this chapter. The vocabulary used in the laboratory is sometimes arcane; therefore, a glossary of laboratory terms is supplied in the appendix to this chapter. It includes formulas used to compute the descriptive statistics of the QA/QC program.

SAMPLING

Sampling at a hazardous waste site begins with the initial hydrogeological work, but the first involvement of the laboratory occurs when a sampling program is developed. To evaluate a site accurately for potential contaminants, samples are collected from a variety of matrices: surface waters, ground water, soil, sludge, sediment, leachate, air, and biota. (Techniques of field investigation were addressed in Chapters 2 and 3.) An experienced chemist can advise the field crew leader on the best sampling schemes to develop the necessary data base. If samples are not representative of their original environment, or if the integrity of the constituents changes between the time of sampling and the time of analysis, the samples are worthless. From the perspective of the laboratory supervisor, the sampling program should consider:

- *Background level of the substances deemed contaminants:* Background levels must be known in order to select the matrices required to meet the objectives of the investigation.
- *Training and qualifications of sampling teams:* In most cases, the sampling at hazardous waste sites is performed in remote locations. Experienced crews can make decisions in the field when confronted with situations not

previously broached in the training sessions specific to the investigation at hand. Their experience can avert delays and insure that proper procedures are employed for the investigation.

- *Calibration of field equipment:* Because an investigation relies on data gathered in the field in order to support the validity of the samples collected, the field equipment must be accurately calibrated. Otherwise, the instruments may vary from day to day, and the data will be inconsistent and defective.
- *Selection of samples to support and define potential problems:* Improper sample selection may not define the extent and magnitude of the contamination.
- *Methods of sample collection:* The samples must be representative, and cross-contamination must be precluded.
- *Decontamination procedures:* No external contamination may be brought to the site, and no contamination may be carried from it.
- *Maintenance of chain of custody:* Chain of custody is mandatory if litigation is possible. Custody from collection of sample to delivery to laboratory must be maintained in order to enter the samples into a court proceeding as evidence.

Once the crew is adequately prepared, the success of sampling in the field depends upon careful attention to many factors: good housekeeping, the collection of representative samples, proper handling and preservation of samples, and maintenance of a chain of custody, as discussed in the following paragraphs.

Good housekeeping refers to an organized sampling routine, including the full recording of all pertinent events in field notebooks and the proper labeling of all forms and containers. These procedures are written out and reviewed with the samplers in order that they may be followed. Most sampling guidelines are specified by the enforcement, compliance, monitoring, or program offices of the U.S. EPA.[1] The following are examples of good sampling practices:

- Before the sampling equipment is used, it is checked to insure that it is clean and in proper operating condition.
- Before a sample is collected, all the sample bottles are inspected to insure that there is no contamination. Foul bottles are cleaned, and if the bottles cannot be cleaned, they are replaced.
- Whenever the sampling event does not proceed as planned, any abnormality encountered and corrective action taken are recorded. This information is useful to evaluate the reliability of the equipment and to reexamine

potentially faulty data derived from the samples collected. In addition, the supervisor of the sampling program reviews the field log and highlights any abnormalities in weather conditions or the status of the site sampled.

Sampling and analytical programs are expensive, but the success of any investigation depends upon the samples collected. Care should be taken to insure that the samples are representative of the area under investigation. Also, whenever samples are collected, a sufficient volume is taken to permit quality assurance testing through split samples or spiked samples, and all utensils are cleaned between samples to avoid cross-contamination. Some of the techniques employed to collect representative samples are:

- When ground water is analyzed, the well is evacuated to remove standing, stagnant water (the volume of water removed being three to five times the volume of the well), and it is sampled when it returns to its original elevation. Care is exercised to exclude sediment or particles of organic matter, as organic and inorganic compounds may adsorb to certain materials of the sampling containers or equipment.
- In the collection of surface water, the mouth of the sampling container is placed below the surface of the water and into the current to avoid trapping floating material and to avoid contamination. In addition, the investigator's hands are kept as far from the mouth of the jar as possible. When it is practicable to do so, the sampling vessel is forced through the entire cross section of the stream.
- A depth-integrated sampler is used to collect discrete subsurface lake samples at depths of 4 ft (1.2 m) or deeper. Its operation is monitored to witness that it opens at the proper time. If there is doubt about the validity of a sample, it is discarded, and another is taken.
- Soil or sediment may be collected in a number of acceptable ways: grab, split spoon, boring, excavation. Grab, for surface collection, and split spoon, for subsurface collection, are the most common.

(Additional information on specific sampling techniques used in the field is presented in Chapters 2 and 3.)

A variety of factors affect the choice of containers and caps used for taking samples: resistance to breakage, size, weight, interference with constituents, cost, and availability. Two types of materials are widely used for collecting samples: glass and plastic. Samples are often collected in amber-colored bottles to minimize their degradation by light. Sometimes, special considerations are made for certain classes of compounds. In the collection of samples for extractable organics, for example, the cap should be lined with Teflon™ or aluminum foil. This precaution prevents contamination

TABLE 5-1. Recommended sampling volumes and containers.

CONTAINER	PARAMETERS	VOLUME
Glass bottle with Teflon™-lined cap	Oil and grease Phenols	2 L 1 L
Glass bottle with Teflon™-lined cap	PCB/pesticides Extractables (base/neutrals/acid) organics	1 L
Plastic bottle and cap	Inorganics Solids Metals Cyanide Nutrients	1 L
Duplicate glass bottle and Teflon™ septum cap	Volatile halogenated organics Trihalomethanes	40 ml

from the cap, liner, or sealer. Investigators take care to collect one parameter—or one analytical scan—per container because the sample jar must be rinsed with the extracting solvent to remove organics adsorbed to its walls. Samples are not filtered because organics can adsorb to containers, filter media, or both. In another example, samples of purgeable organics are collected with no head space. Otherwise, the parameter's insoluble components will accumulate in the air space above the sample, and when the container is opened for analysis, the volatilized components will escape. For general information on containers and volumes, see Table 5-1.

Because the consequences of an uncontrolled hazardous waste site investigation are difficult to predict, litigation is a possibility for several years after the remedial investigation and feasibility study have been completed. Therefore, an accurate record must be maintained of sample collection, transport, analysis, and disposal. Chain of custody procedures, which track the samples from the field to the laboratory and final disposal, are instituted and followed throughout the sampling and analysis.

The project coordinator must be prepared to produce documentation that traces the samples through each step. (See Figure 2-1.) On the chain of custody form, samplers note meteorological data, equipment employed, evacuation techniques, any calculations, physical characteristics of samples, date, time of day, location, and any extraordinary events that occurred during sampling. Samples are physical evidence and are handled according to procedural safeguards. The National Enforcement Investigation Center of the U.S. EPA defines custody of evidence as: in actual physical possession, in view after being in physical possession, in a locked repository, or in a secure and restricted area.

The individual who takes the samples completes the custody form, packages the samples, and seals the package with evidence tape—tape that makes tampering conspicuous. The sealed package is transported to the laboratory by the sampler or by a courier service. If a courier is used, the package must have an air bill or bill of lading that has been completed by the sampler and signed by both the sampler and the courier. Upon its arrival at the laboratory, the sample custodian—an individual assigned to accept and log samples—signs the courier's transport bill. The sample custodian has several duties and responsibilities upon receiving the sample, as the person who: records receipt of samples; inspects sample shipping containers for the presence or absence of custody seals; assesses container integrity; records the condition of shipping and sample containers in laboratory logs; signs the chain of custody form and bill of lading from the courier; verifies and records agreement or disagreement of information on the sample documents; labels the sample with a unique laboratory sample number; places the samples in storage, secure storage if desirable; and, finally, schedules the samples for analysis.

ANALYSIS

In order to develop the analytical protocols—the procedures applied to analyze the samples—laboratory personnel synthesize information from a number of sources, examining the compounds of concern, the necessary or feasible detection limits, and the matrices in which the contaminants are found. In addition to these physical facts, the laboratory supervisor reviews the objectives of the particular remedial program and recommends an analytical scheme. In many cases, the laboratory's efforts are oriented toward problem definition and potentially long-term monitoring.

Analytical protocols are among the most constantly changing laboratory practices. These protocols are repeatedly improved and redefined to incorporate changes in knowledge about chemical substances. However, the laboratory chemist has little choice in the selection of protocols for the analysis of potentially hazardous substances, as government agencies dictate the protocols permitted. Before any analysis is performed, it is necessary to determine which government institution will receive and review the results. Prior to analysis, agreement between the laboratory and the controlling government agency is reached to insure that the chosen protocols will satisfy the objectives of the remediation program.

The U.S. EPA has published analytical methods that have been thoroughly tested, protocols that have been subjected to peer review and evaluation studies. These protocols describe in detail the equipment, reagents, quality control, and sequence of steps necessary to produce valid results. Methods of analysis are selected from several established sources (see bibliography to this chapter for pertinent references). State agencies with the responsibility to administer remedial investigations have the option to incorporate the U.S. EPA protocols, as most do, or they may develop and require their own, as the State of California did. The procedures should be performed by experienced chemists who are able to interpret the results.

The exact composition of the hazardous waste samples is unknown when they reach the laboratory and must be readied for analysis. The objective of the protocol is to identify a constituent selectively, to remove any possible interference, and to calculate the quantity of material in the sample. In most cases, alternatives protocols can be selected to accommodate the potential problems posed by different matrices or other interfering factors, the required detection limits, and the cost. Table 5-2 presents some of the alternatives available for analyzing groups of pollutants.

The analytical chemist selects the best alternative among the published protocols to achieve the objectives of the investigation. (The decision-making process is diagrammed in Figure 5-1.) The chemist must consider selectivity of the instrument and the chemical characteristics of the compounds in question. Familiarity with the theory of the protocols and instru-

TABLE 5-2. Examples of alternative analytical techniques.

ANALYTES	ALTERNATIVE TECHNIQUES
Metals	Direct aspiration—flame Graphite furnace Inductively coupled plasma spectrometry
Purgeable organics (aqueous)	Purge and trap Solvent extraction Direct injection
Purgeable organics (soil)	Sparge method Solvent extraction Dynamic or static head space
Extractable organics (aqueous)	Liquid-liquid extraction Adsorption followed by extraction
Extractable organics (soil)	Solvent extraction Thermal desorption Wrist action shaker Continuous extractor Liquid-liquid extraction
Cation/anion	Colorimetric Titrimetric Potentiometric

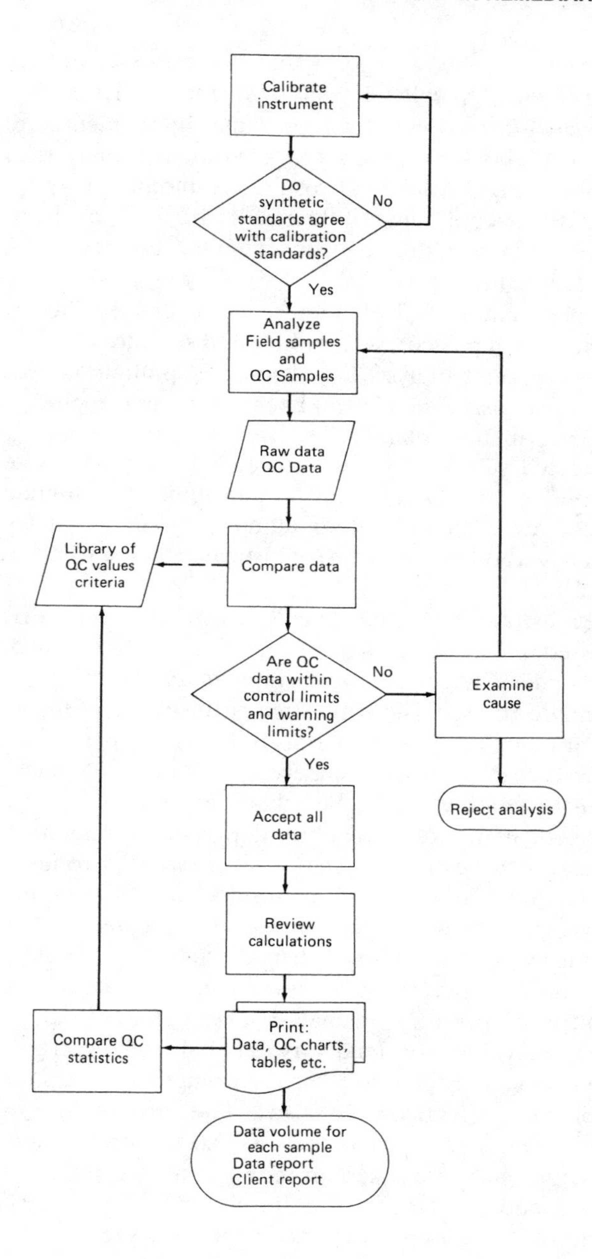

Figure 5-1. Flow chart of analytical process.

ments is essential, as is knowledge of their limitations. Table 5-3 lists the analytical instruments typically chosen for various analyses.

To achieve acceptable qualitative and quantitative results, the chemist incorporates good laboratory practices, including the analysis of sufficient quality control samples. Analytical results are monitored frequently, and decisions about the acceptability of the results are made as the analyses are performed. Several factors that are consequential to accurate and precise analysis are listed in Table 5-4.

Not all samples require full characterization. During the investigation phase of a remediation project, several hundred samples are collected. It is impracticable to conduct analyses for all potential pollutants—it is too time-consuming and expensive, as it would require several techniques; so the analyses are performed in phases. The first step in a phased program is analysis of selected samples. At this stage, indicator parameters or site-specific compounds are sought to reduce the volume of redundant data and to accord a broad review of the contamination of the site. At each subsequent phase of the analysis, additional, more sophisticated techniques are conducted on fewer samples.

An example illustrates this point. At a PCB disposal site, the original scope of work suggested that all samples from 30 sites be analyzed for 200 different hazardous substances by gas chromatograph/mass spectrometer. U.S. EPA protocols mandate the specific sophisticated instrument and require substantial time for completion; the analysis is, in sum, expensive. To expedite the work and reduce the cost, yet still develop sufficient information, investigators decided to employ techniques that were compound-specific. The samples were screened with a gas chromatograph equipped with a flame ionization detector, an instrument that has the capacity to identify 90 percent of the compounds in question. The results from the gas chromatograph scans were reviewed, and where concentrations were greater than 1 part per million, analysis by gas chromatograph/mass spectrometer was scheduled. In addition, one sample from a clean zone was scheduled for gas chromatograph/mass spectrometer analysis to verify the absence of pollutants. Effectively the analytical workload was reduced from 2000 gas chromatograph/mass spectrometer scans to 500 gas chromatograph scans and 156 gas chromatograph/mass spectrometer scans. The cost of the analyses was reduced from $1 million to approximately $400,000. The decision to proceed with a phased approach was based on the regulatory standards established for remedial alternatives. For example, the action level for PCBs in soil is 50 ppm; therefore, soil at 10 ppb was not further analyzed, as the increased sensitivity of detection did not justify the additional cost.

Paramount to the reviewer and ultimate user of the data is the detection

TABLE 5-3. Major instruments required in an environmental laboratory.

ANALYTICAL INSTRUMENT	COMPOUNDS ANALYZED
Gas chromatograph/mass spectrometer/data system (GC/MS/DS)	Organic priority pollutants and hazardous substances
GC-electron capture detector	Organic and organochlorine compounds, pesticides, PCBs
GC-flame ionization detector	Organic compounds
GC-thermionic detector	Organic compounds
GC-nitrogen/phosphorus detector	Organosulfur and organophosphorus compounds
GC-conductivity detector	Organic compounds, PCBs, volatile compounds
GC-photoionization detector	Aromatic and unsaturated organic compounds
Total organic carbon analyzer	Organic and inorganic carbon
Total organic halogen analyzer	Organochlorine compounds
High performance liquid chromatography	Cleanup of extracts prior to injection
Gel permeation unit	Size exclusion cleanup of extracts
Atomic absorption spectrometer	Metals analysis
Infrared spectrophotometer	Functional groups on organic compounds
Inductively coupled plasma spectrometry	Metals

TABLE 5-4. Important procedural considerations during analysis.

Calibrate instruments initially with standard solution at five concentrations.

Verify calibration with U.S. EPA reference samples.

Analyze samples.

Analyze duplicate samples, 1 for every 20 samples to monitor precision.

Analyze spiked samples, 1 for every 20 to monitor accuracy.

Continue calibration with reference sample every 10 to 15 analyses to verify stability of instrument.

If results are beyond control limits during analysis, recalculate and reanalyze as necessary.

Document results, quality control data, and any corrective measures.

Sign and date work sheet, and place in secure evidence file.

limit achieved by the protocol. In most cases, the protocols have proven sensitivities for particular compounds. However, confounding influences caused by a matrix in which a suspected substance is bound are considered in the selection of detection limits:

- In an aqueous solution, a high concentration of extraneous organics might mask the substance sought.
- If there is a high concentration of unknown substances in the water, the sample may have to be diluted; the detection limit is, therefore, increased.
- An aqueous matrix may form an emulsion, and the proportion of the sample recovered will be reduced.
- A high level of organics obviates the possibility of concentrating the sample; detection limits, then, are raised.
- The detection limits of a sample from soil are always higher because there is less of the sample present than in other matrices.

The techniques available to effect sample cleanup—separation of the sample from its matrix—are primitive when compared with the sophistication of the instruments of analysis. The cleanup of a sample not only removes the sources of interference, it also removes part of the compound of interest.

Having characterized the contamination, the laboratory manager submits that definition to the supervisor of the remedial work. The parties involved in the remediation effort then make decisions about the scope of the remediation, and that information becomes the basis for development of alternatives for remediation.

Chain of Custody. The hand-to-hand custody of samples practiced in the field continues in the laboratory, as it is maintained throughout the preparation and analysis of the samples. Custody sheets permit multiple entries because several people handle the samples throughout the analytical scheme. (See Figure 5-2.) In the sequence of events in the laboratory, the analyst retrieves the sample from secure storage and then documents the reason for possession of the sample, notes the date, and signs the form. After analysis, the sample is returned to storage, and the custody form is again dated and signed. Each individual who possesses the sample follows these procedures. During analysis, the analyst initials the form and indicates the date of all intermediate and final steps.

Upon completion of analysis, the project officer or a surrogate assimilates all the field and laboratory notes and develops an evidence file for the project. This file is arranged in chronological order for ease of review, and it

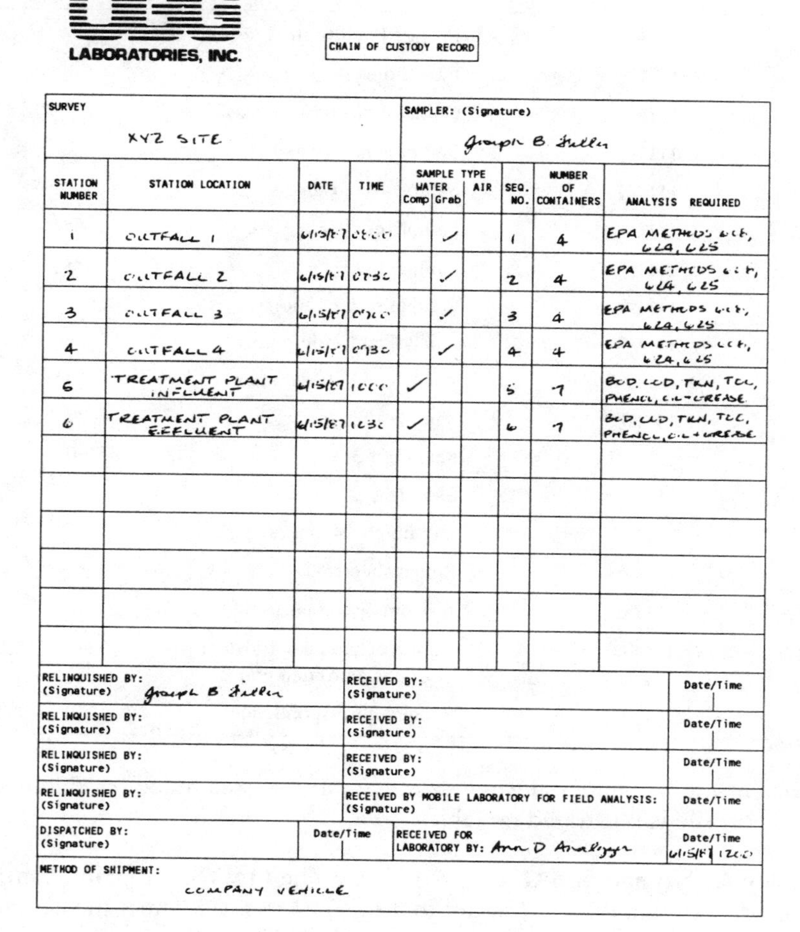

OBG LABORATORIES, INC.

CHAIN OF CUSTODY RECORD

SURVEY	SAMPLER: (Signature)
XYZ SITE	Joseph B. Fuller

STATION NUMBER	STATION LOCATION	DATE	TIME	SAMPLE TYPE WATER Comp	Grab	AIR	SEQ. NO.	NUMBER OF CONTAINERS	ANALYSIS REQUIRED
1	OUTFALL 1	6/15/87	0800		✓		1	4	EPA METHODS 608, 624, 625
2	OUTFALL 2	6/15/87	0830		✓		2	4	EPA METHODS 608, 624, 625
3	OUTFALL 3	6/15/87	0900		✓		3	4	EPA METHODS 608, 624, 625
4	OUTFALL 4	6/15/87	0930		✓		4	4	EPA METHODS 608, 624, 625
6	TREATMENT PLANT INFLUENT	6/15/87	1000	✓			5	7	BOD, COD, TKN, TOC, PHENOL, OIL+GREASE
6	TREATMENT PLANT EFFLUENT	6/15/87	1030	✓			6	7	BOD, COD, TKN, TOC, PHENOL, OIL+GREASE

RELINQUISHED BY: (Signature)	RECEIVED BY: (Signature)	Date/Time
Joseph B. Fuller		
RELINQUISHED BY: (Signature)	RECEIVED BY: (Signature)	Date/Time
RELINQUISHED BY: (Signature)	RECEIVED BY: (Signature)	Date/Time
RELINQUISHED BY: (Signature)	RECEIVED BY MOBILE LABORATORY FOR FIELD ANALYSIS: (Signature)	Date/Time

DISPATCHED BY: (Signature)	Date/Time	RECEIVED FOR LABORATORY BY: Ann D Analyzer	Date/Time 6/15/87 1200

METHOD OF SHIPMENT: COMPANY VEHICLE

OBG Laboratories, Inc.
Box 4942 / 1304 Buckley Road / Syracuse, New York 13221 / (315) 457-1494
Oakdale Medical Building / 700 Harry L. Drive / Johnson City, New York 13790

Figure 5-2. Chain of custody form for analysis.

TABLE 5-5. Illustration of document inventory (evidence) file.

DOCUMENT CONTROL NUMBER	TYPE	NO. OF PAGES
1111-1	Project file inventory sheet	1
1111-2	Field notes	30
1111-3	Chain of custody records	7
1111-4	Shipping manifests	27
1111-5	Sample log-in sheets	40
1111-6	Sample control records	40
1111-7	Sample tickets	500
1111-8	Sample traffic reports	127
1111-9	Analytical traffic reports	127
1111-10	Analytical data summary	10
1111-11	Sample # 2	20
1111-12	Sample # 3	20
1111-62	Sample # 50	20
1111-63	Lab notebook pages	37
1111-64	Bench sheets	50
1111-65	Instrument log pages	13
1111-66	Copies of mass spectral data, graphs, chromatograms	43
1111-67	Related correspondence	4

is inventoried, numbered, and stored for future reference. A document inventory file is illustrated in Table 5-5.

Quality Assurance and Quality Control. The QA/QC program is implemented and maintained to assure that the data reported meet the demands of their intended use, and it is governed by U.S. EPA, state, and local government mandates. The program sets guidelines and direction for all the physical, chemical, and microbiological measurements performed by the laboratory. A laboratory's QA/QC guidelines support its surveillance program and analytical efforts. They reflect the most economical approach and both guarantee and certify that all data collected, stored, reported, and used by the laboratory are scientifically valid, defensible, and of known precision and accuracy. Thus, the caliber of the data generated by the laboratory is documented.

The measurement of physical, chemical, and microbiological properties of pollutants in various environmental matrices involves uncertainties that cannot be eradicated completely. In addition, random (indeterminate) and systematic (determinate) errors are inherent in all analytical methods because of uncertainties in measurements. Additional errors, often unrecognized, are introduced by juxtaposing chemical reactions and other undesirable physical and chemical effects. In many instances, absolute values cannot be attained directly. Error can, however, be reduced to tolerable limits by examination and control of the significant variables, and uncertainty is minimized through use of statistical methods.[2] The QA/QC estimates the accuracy (the probable "true value") and precision (the range of the error in measurement) for the various analytical methodologies by analyzing blanks, duplicates, spikes, and synthetic standards. The major effort of the QA/QC program is development of a workable, day-to-day QA/QC model. It furnishes control charts and control limits that measure the laboratory's daily performance.

After sufficient quality control data are collected from analyses, statistical methods are used to evaluate the quality of the data by calculating control and warning limits. The data produced from these tests are maintained by a computer-assisted data management system. That data system has the secondary function of relating QA/QC data to analytical performance in daily as well as varying time frames.

The quality control system calculates, stores, segregates, interprets, monitors, and retrieves each item of QA/QC information daily. Each sample from the field carries a unique identifying code. Each QA/QC sample is also assigned a code identifying it as a blank, duplicate, spike, or synthetic known, and is placed in a data base; thus a permanent record of every quality control sample is established.

This data base is used as the starting point for statistical assessment of the quality control data, which helps investigators to understand the information from the analytical protocols. Specific statistical programs generate precision (X-bar and R) and accuracy (P-bar) quality control charts for each pollutant. A minimum of 15 duplicates and spiked samples or synthetic-known analyses is required to generate a control chart. These charts provide a graphic representation of the QA/QC information and are used by laboratory personnel for daily monitoring of the accuracy and precision of the various analytical methods. They are the tools needed to detect qualitative variations in the analytical methods used to quantify environmental pollution. They provide a continuous indication of the state of an analytical procedure with respect to quality, and assist investigators in deciding when and how to take corrective action.

For statistical reports and evaluations, the terms *accuracy* and *precision* are not synonymous. *Precision* expresses the agreement or reproducibility of a set of replicate results among themselves without assumptions about prior information concerning the true result, whereas *accuracy* implies correctness of results. The precision is usually expressed in terms of the standard deviation *(s)*, variance *(s^2)*, or range. Good precision often is an indication of good accuracy; however, high precision can be achieved with poor accuracy if systematic errors are present in the method or instrument used. The mean $\pm$ 2s includes 95.4 percent of the values, and the mean $\pm$ 3*s* includes 99.7 percent of the data in a normal distribution curve. (See Figure 5-3.) Common practice sets warning limits at $\pm$ 2*s* and control limits at $\pm$ 3*s* on either side of the mean. Therefore, there is a 95.4 percent confidence level for the warning limits, and a 99.7 percent confidence level for the control limits. The qualitative relationships between the upper and lower control limits and between the upper and lower warning limits and the mean are shown in Figure 5-3. The larger the number of replicate analyses, the greater is the statistical confidence that the true mean lies within control or warning limits on either side of the experimental mean.

The control limits on quality control charts are the paramount criteria for assessing the significance of variations in the analytical results. For instance, when the plotted quality control indicators of percent recoveries, relative percent error, and so forth fall within the control limits, the conclusion is that the analytical methods used are within control bounds. If, however, a quality control indicator value falls outside the control limits, an assignable cause is present. Thus, control limits are action limits. They enable the laboratory

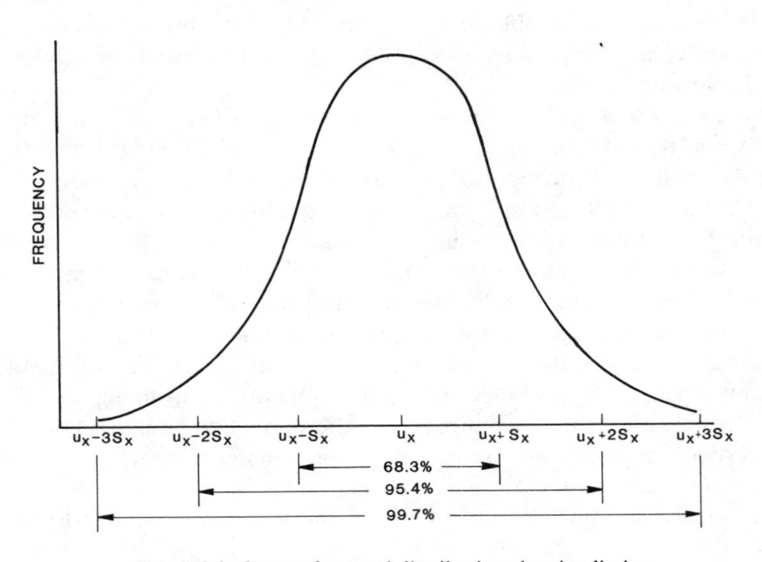

Figure 5-3. Curve of normal distribution showing limits.

supervisor to perceive deviations in analytical procedures and to take corrective action before producing erroneous results or results that exceed the absolute maximum tolerable limits.

The statistical techniques that generate data for the quality control charts of Figure 5-4 involve complex mathematics. The control limits and the warning limits are calculated from the quality control data of duplicate

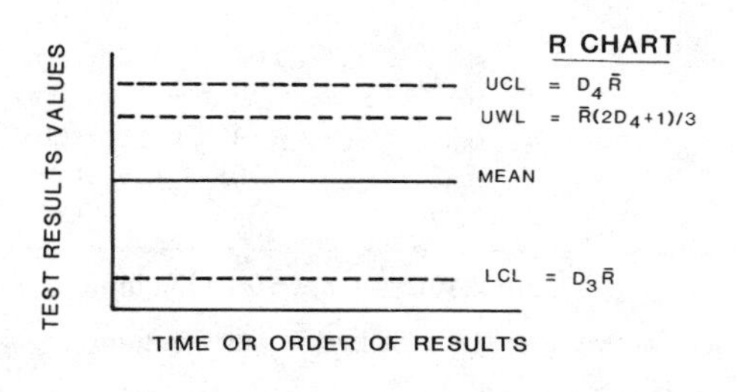

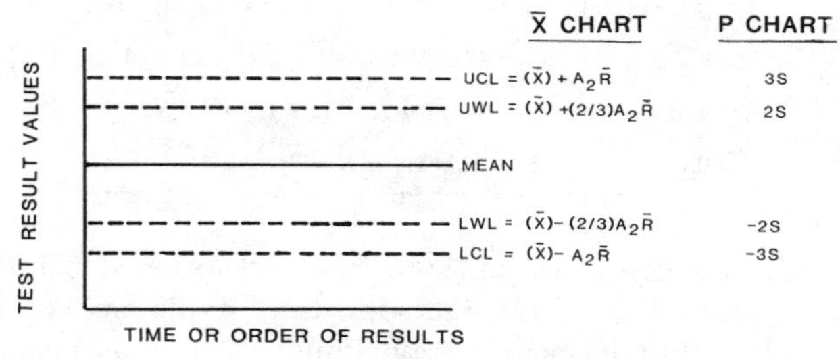

Figure 5-4. Essentials of control charts.

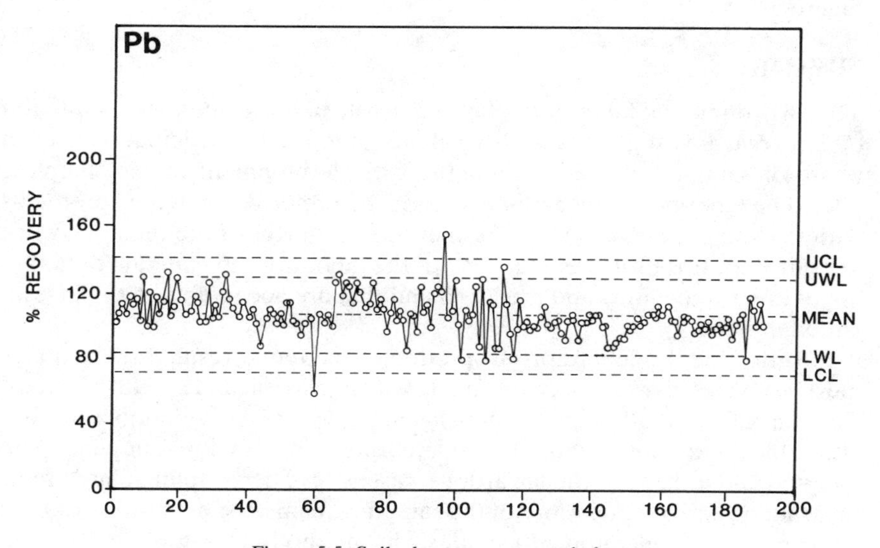

Figure 5-5. Spiked recovery control chart.

TABLE 5-6. Statistical factors and equations for calculating quality control chart lines.

OBSERVATIONS IN SUBGROUP (n)	FACTOR A_2	D_2	D_3	D_4
2	1.88	1.13	0.00	3.27
3	1.02	1.69	0.00	2.58
4	0.73	2.06	0.00	2.28
5	0.58	2.33	0.00	2.12
6	0.48	2.53	0.00	2.00
7	0.42	2.70	0.08	1.92
8	0.37	2.85	0.14	1.86

UCL = Upper control limit; UWL = Upper warning limit

LCL = Lower control limit; LWL = Lower warning limit

UCL for $\overline{X} = |\overline{X}| + A_2\overline{R}$; UWL for $\overline{X} = |\overline{X}| + (2/3)A_2\overline{R}$

LCL for $\overline{X} = |\overline{X}| - A_2\overline{R}$; LWL for $\overline{X} = |\overline{X}| - (2/3)A_2\overline{R}$

UCL for $R = D_4\overline{R}$; UWL for $R = \overline{R} + (2/3)(D_4\overline{R} - \overline{R})$

LCL for $R = D_3\overline{R}$; LWL for $R = \overline{R}(2D_4 + 1)/3$

analyses by using the equations and statistical factors listed in Table 5-6. The statistical factors A_2, D_2, D_3, D_4, and so forth of Table 5-6 were calculated such that the control limits carried a maximum risk of making an error only 0.1 to 0.3 percent of the time. A spiked recovery control chart is shown in Figure 5-5.[3]

SUMMARY

The environmental laboratory plays a key role in the successful completion of remedial investigations and feasibility studies. It participates in each phase of sample collection beginning with development of the sampling plan. The laboratory supervisor discusses the preferred analytical protocols with the site investigator to insure that the objectives of the remedial work are satisfied. It is ultimately the task of the laboratory to generate data that are of known accuracy and precision and that are acceptable to the investigation at hand.

In order for the laboratory to produce acceptable results, the samples must be collected from the environment with consistent and careful methods. The U.S. EPA and other government agencies dictate the techniques that are allowable, and those methods are incorporated into the sampling plan adopted and utilized at the hazardous waste site. The sampling plan directs the training and qualifications of the sampling teams, the calibration of field equipment, the selection of samples, the methods of sample collection, decontamination procedures, and maintenance of chain of custody from collection in the field to delivery to the laboratory. Throughout the sampling period, the field crew gives attention to good housekeeping practices, collecting representative samples, correct handling and preservation of samples, and, at all times, chain of custody. Sampling, if performed correctly, is the basis for the information developed by the laboratory. If it is performed incorrectly, the resources expended collecting the samples in the field will have been wasted.

In deciding which analyses will be performed on the samples collected from a particular site, the laboratory chooses among protocols that have been tested and approved by government regulatory agencies, notably the U.S. EPA. The various available protocols are related to the objectives of the investigation, and those that best address the study's objectives are selected. The detection limits of the analysis also are matched to the level of information that must be developed to achieve the study's objectives. It is sometimes both economical and acceptable practice to conduct the analysis in phases. With such an approach, a proportion of the total number of samples is analyzed just to characterize the contaminants at the site, and once the general qualities of the site are known, additional samples are subjected to more detailed, quantitative analysis.

In order to produce consistently accurate and precise data, the laboratory must attend continually to the protocols and instruments used. Because the quantities under study are minute, statistical methods are applied to interpret the data from all analyses. A computer-assisted management information system monitors the results of each instrument and indicates whether it is within acceptable bounds of precision and accuracy. The system produces a quality control/quality assurance model. Thus, the laboratory is able to supply the site investigator with data of defined reliability. All the records of the analysis, notably the chain of custody documents and the QA/QC data, are organized and stored to confirm the accuracy and precision of the results.

The laboratory is an important tool in investigation of a hazardous waste site. It should be involved from the initial stages of planning to assist the project investigators in developing the strategy of study. Without quality analytical data, including the correct sampling methods, the investigation will fall short of its goals.

BIBLIOGRAPHY

American Public Health Association, American Water Works Association, Water Pollution Control Federation. *Standard methods for the examination of water and wastewater.* 16th Ed. 1985.

American Society for Testing Materials. *Annual book of standards.* Section 11, Vols. 11.01 and 11.02. Water and environmental technology. 1984.

Guidelines establishing test procedures for the analysis of pollutants under the Clean Water Act. 40 CFR 136. 26 Oct. 1984.

Hazardous Waste and Consolidated Permit Regulations. 40 CFR 261.

U.S. Department of Health, Education and Welfare (now Department of Health and Human Services). *NIOSH manual of analytical methods.* Vols. I-V. Aug. 1978.

U.S. EPA. Organics analysis of multi-media/multi-concentration contract laboratory program. Jan. 1986.

———. Industrial Technology Division. Method 1624 and 1625 isotope dilution GC/MS. Washington, DC: Jan. 1985.

———. Methods for chemical analysis of water and wastes. Mar. 1983. EPA-600/4-79-020.

———. *Handbook for sampling and sample preservation of water and wastewater.* Sept. 1982. EPA-600/4-82-028.

———. *Test methods for the evaluation of solid wastes.* SW-846. Vols. I-III. July 1982.

———. *Handbook for analytical quality control in water and wastewater laboratories.* Cincinnati: 1979. EPA-600/4-79-019.

———. *Procedures manual for ground water monitoring at solid waste disposal facilities.* Aug. 1977. EPA/530/SW-WI.

U.S. EPA, MERL. *Methods for benzidine, chlorinated organic compounds, pentachlorophenol, and pesticides in water and wastewater.* Cincinnati: Sept. 1978.

———. *Sampling and analysis procedures for screening of fish for priority pollutants.* Cincinnati: Aug. 1977.

U.S. EPA Region VIII. *Determination of 2,3,7,8-TCDD in soil and sediment.* Feb. 1983.

NOTES

1. Sample preservation methods and holding times are for the parameters referenced in the National Pollutant Discharge Elimination System (NPDES) and the Primary Drinking Water Standards.

2. U.S. EPA. *Handbook for analytical quality control in water and wastewater laboratories.* Mar. 1979. EPA-600/4-79-019. U.S. EPA. *Manual of analytical quality control for pesticides and related compounds in human and environmental samples.* Jan. 1979. EPA-600/1-79-008.

3. Taken from: U.S. EPA. *Handbook for analytical quality control in water and wastewater laboratories.* Mar. 1979. EPA-600/4-79/019. Hart P. Samson and C. Rubin. *Fundamentals of statistical quality control.* Reading, Mass: Addison-Wesley. p. 40.

APPENDIX

Glossary of Laboratory Terms

Absolute error: Difference between experimental result and true value.

Accuracy: Nearness of analytical result or set of results to true value. Usually expressed as *error, bias,* or *percent recovery.*

Average or arithmetic mean: Summary statistic calculated thus:

$$X = \left[\sum_{i=1}^{n} X_i \right] / n$$

X_i = each observation; n = number of observations.

Blank: Sample of clean material that represents the matrix of samples; introduced into analytical procedure to identify background contamination from equipment used in the procedure.

Calibration curve: The line developed during analysis of a series of standard solutions that span the linear range of the instrument.

Cleanup: Separation of a contaminant from its matrix in preparation for analysis.

Control limit: Level at which data generated are within the 99.7 percent confidence level of the natural limit distribution curve. This level is commonly set at three times the standard deviation ($3s$).

Duplicate samples: Samples derived when a field sample is divided into equal amounts. Duplicate samples are samples treated as unique samples.

Field blank: Sample of material taken from clean area of field site; furnishes data describing background level of contamination.

Percent recovery (PR): (1) Recovery of synthetic standard of known value:

$$PR = \frac{\text{(Observed value)}}{\text{True value}} \times 100$$

(2) recovery of known amount of analyte (spike) added to sample of known value:

$$PR = \frac{\text{(concentration of spike + sample)} - \text{sample}}{\text{concentration of spike}} \times 100$$

Percent relative error (PRE): The absolute error divided by the true value and multiplied by 100. If the true value is unknown, *PRE* measures the range of a replicate analysis divided by the mean of the replicate value. For duplicates, *PRE* is calculated thus:

$$\text{PRE} = \frac{100|X_2 - X_1|}{(X_2 + X_1)/2} = (100|X_2 - X_1|)/\overline{X}_j$$

Precision: See *percent recovery.*

Random error: (also, *indeterminate error*): Errors inherent in analytical method because of uncertainties in measurements.

Range (R_i^2): Difference between the highest and lowest values of a set of data. For n sets of duplicate values, (X_2, X_1),

$$R_i i = |X_2 - X_1|.$$

$\overline{R}$*:* Average range of n sets of duplicate values; calculated thus:

$$\overline{R} = \left[\sum_{i=1}^{n} R_i \right].$$

Reference sample: A sample of known and verifiable concentration used to check the standard generated by the laboratory.

Relative error: Absolute error divided by true value. See also *percent relative error.*

Spike: Sample containing a known amount of contaminant of concern. The field sample and the spiked sample are analyzed and compared to determine the amount recovered. See *percent recovery.*

Split sample: Samples derived when a single sample is divided into two or more containers. These multiple sample volumes of the identical sample may be analyzed at different times for QA/QC.

Standard deviation(s): Descriptive statistic used to describe dispersion of a data set. The s for n values is:

$$s = \left[\frac{\sum_{i=1}^{n} (X_i - \overline{X})^2}{n - 1} \right]^{1/2}$$

Synthetic standard: Solution prepared independently of the laboratory analyzing field samples; contains exact amount of contaminant of concern. *Known* samples may contain several contaminants. *Synthetic knowns* are used to verify the laboratory's prepared standards.

Systematic error: (also, *determinate error*) Errors that can be reproduced independent of individual samples. These errors are usually associated with equipment or method. They are positive or negative.

Variance (s^2 or CV): Descriptive statistic. The *relative standard deviation (*s^2*)* or *coefficient of variation (CV)* is:

$$s^2 = CV = 100s/\overline{X}$$

Warning limit: Level at which data generated are within the 95 percent confidence level of the normal distribution. This level is commonly set at two times the standard deviation ($2s$) from the mean.

$\overline{X}$: Arithmetic mean; average.

Chapter 6
Health and Safety at Hazardous Waste Sites

Remedial work at hazardous waste sites often poses safety and health concerns distinct from those of traditional industrial workplaces because hazardous materials are present and because the site may not be fully controlled. Uncertainty prevails with regard to substances that may be present:

- The variety, quantity, and identity of compounds present are not known precisely.
- The physical states and matrices are not defined.
- The antagonistic, additive, or synergistic interactions of the compounds are unknown.
- The ages and sources of the wastes may not be known.

Finally, there is the ever present possibility that contamination on the site could affect workers or receptors in the environs.

To compensate for the uncertainty, extraordinary safety measures are designed and implemented. A comprehensive safety plan is instituted at every hazardous waste site, tailored to address all the known specific hazards of the particular site. Moreover, the program contains measures designed to furnish protection against any potential, unknown hazards.

In order for a potential threat to present an actual risk to either human health or the environment, three constituents are necessary:

- *Source:* A threat, such as an accumulation of chemical contamination, must be present.
- *Transmitter:* There must be a channel available for the source to reach a receptor. For example, ground water is a vehicle that distributes chemical substances in the soils.
- *Receptor:* Life forms—humans, wildlife, flora—must be in a position to be affected by the danger.

This chapter was developed by Ruth Wegmann of O'Brien & Gere Engineers, Inc.

TABLE 6-1. Classification of potential hazards during remediation of uncontrolled waste sites.

Chemical exposure (via any route)	Biological hazards (for example, infectious wastes or poisonous plants or animals)
General safety hazards (for example, treacherous terrain, heavy equipment)	Electrical hazards
Fire and explosion	Temperature (for example, heat stress, exposure to cold)
Oxygen deficiency	
Radiation	Noise

(This concept was introduced in Chapter 4.)

Sources of hazards are listed in Table 6-1. During development of the safety plan, each of these hazards is addressed.[1]

SOURCES OF HAZARD

Any outdoor work site can present hazards. Rough terrain, slippery surfaces, sharp objects, heavy equipment, and unstable surfaces pose dangers. In addition, the agility and mobility of workers on a hazardous waste site are often encumbered by protective clothing and equipment. Yet, the single most dangerous and variable source of risk at a hazardous waste site may be exposure to chemicals.

Chemical substances can be in gaseous, liquid, and solid phases; they can be absorbed by the skin, enter through skin punctures, or be inhaled or ingested. The point of injury from the toxic chemical may be at the locus of contact, the substance may have a systemic effect, or it may travel in the body to a target organ or area. It is, therefore, mandatory that persons developing the safety protocols grasp well all information about chemicals present on the site, their concentrations, and potential noxious effects in order to provide complementary countermeasures.

Fires and explosions are possible at hazardous waste sites, and their sources are varied. Combustion can be caused by chemical reactions; the ignition of flammable or explosive chemicals or of other materials, such as structures, following an explosion; the agitation of shock- or friction-sensitive materials; or the sudden release of materials under pressure.[2] Explosion and fire can occur spontaneously, but they often are a corollary to human intervention on the site.

In addition to the dangers from the shock of explosions and the heat of fires, there can be release of toxic chemicals into the atmosphere during combustion. Both site workers and residents of the environs are then at risk. The potential for explosions and fire are reduced by these measures:[3]

- Field monitoring for explosive atmospheres or flammable vapors using LEL (lower explosion limit) meters.
- Segregating potential sources of ignition from flammable materials.
- Utilizing nonsparking, explosion-proof equipment.
- Careful handling of all materials to prevent agitation or release of chemicals.

Fires and explosion can create chaos at a hazardous waste site. Specific procedures are instituted for use in such circumstances, to manage the emergency, primarily, but also to maintain order.

A chemical reaction can consume ambient oxygen, or the air can be displaced by another gas. Any atmosphere that has less than 19.5 percent oxygen by volume is considered oxygen-deficient. (Normal air contains approximately 21 percent oxygen.) The potential for a deficiency of oxygen exists in, for example, confined spaces of structures or tanks, or in low-lying areas of the site where heavier gases could settle. If such conditions might be created, the atmosphere is monitored with an oxygen meter. During the rare times when workers must operate in an oxygen-deficient atmosphere, fully sealed, self-contained breathing apparatus or supplied-air respiratory equipment is used to furnish air.

Radioactive materials that may be present at a hazardous waste site emit three basic types of harmful ionizing radiation: α (alpha), β (beta), and γ (gamma). The α particles transfer their energy for a very short distance, and they are usually blocked by clothing or the outerlayers of skin; thus, their ability to penetrate is negligible. However, α particles can pose an internal hazard; toxic effects can occur following their inhalation or ingestion. The β radiation may exist at varying energy levels. It is capable of causing "beta burns" to the skin and damage to blood vessels, and matter emitting β radiation may also be inhaled or ingested. Thus, β radiation poses both an internal and an external hazard. The γ radiation is quite able to penetrate; it passes easily through clothing and human tissue, and it can also cause serious and permanent injury to the body.[4]

Maximum permissible levels of external and internal radiation have been established by OSHA.[5] Monitoring is conducted with various instruments, including thermal luminescent detector badges and personal radiation detectors; Geiger counters are also used. The best protection against radiation hazards is offered by respiratory equipment, which prevents the inhalation of radiation-bearing matter, and thorough decontamination and washing, which prevent prolonged contact with the skin or accidental ingestion. Under controlled conditions, protection measures are based on considerations of time, distance, and shielding. If radiation levels on the site are above 2 rem/h over the background level, all site activities are ceased, and the site and the danger are assessed by a health physicist before any work is resumed.

If infectious wastes or biological wastes from hospitals, medical laboratories, or research facilities are contained in the site, special precautions are necessary. Meticulous decontamination and cleaning of any exposed body parts and of equipment are crucial for protection from biological hazards. In addition, if the site supports poisonous plants or animals including poison ivy, snakes, or insects, workers may be subjected to danger. Workers are alerted to such perils to enable them to identify the dangers in the field; in addition, protection from contact is furnished to them, and first aid treatment is available to counter exposure.

Dangers from electricity are created in a number of ways: when overhead power lines or buried cables are present at the site, when equipment is electrically driven and cables are lying on the ground, and during lightning storms. Some of these hazards can be managed. Conscientious use of low-voltage equipment that is bonded (the static charge is equilibrated between the items) and grounded with watertight and corrosion-resistant connecting cables mitigates the dangers. It is also prudent to observe weather conditions.

The physical fitness, age, weight, and adaptation of the workers to the local climate are considered as the work schedule is developed. Most remedial work is performed during the construction season—the warmer months of the year—and workers often are required to wear airtight protective clothing that prevents the natural release of body heat and moisture generated during physical activity. Therefore, heat stress is an ever present danger at hazardous waste sites. The symptoms of heat stress are rashes, cramps, discomfort, and drowsiness; unless it is relieved, the individual may experience fainting, heat stroke, or even death. To protect workers from heat stress, only a minimal amount of protective equipment must be worn, and work and rest periods are judiciously scheduled. Workers are also trained and encouraged to replace body fluids.

Workers exposed to cold can suffer frostbite or hypothermia. The field team leader is responsible for monitoring ambient conditions, including both temperature and wind chill, and decisions made in scheduling the work consider these factors. When workers must be exposed to cold temperatures, appropriate clothing is required, and warm shelter is made readily available.

The exposure of workers to noise is restrained by OSHA regulations.[6] Any workers exposed to noise at or above 85 dBA (8-hour time-weighted average measured in decibels on the A scale, the scale emphasizing frequencies within the range of human hearing and used for all OSHA noise exposure regulations) are notified of their jeopardy by the field team leader. When such levels are experienced, it is the responsibility of the field team leader to administer a continuing and effective hearing conservation program. Should the exposure to noise exceed 90 dBA, protective equipment is provided. In addition to the annoyance, discomfort, and possible hearing damage that may result from excess noise, communication is disrupted. Without clear communication available at all times, workers may be unable to warn each other of danger. Disposable or reusable earplugs or ear muffs protect hearing. In selecting the hearing protection equipment to be utilized at a hazardous waste site, worker comfort and safety as well as noise reduction are consequential considerations.

PROVISIONS OF A HEALTH AND SAFETY PLAN

The federal government commands stringent safety measures at hazardous waste sites, and state governments also are active in regulating safety. A number of government guidelines are listed in Table 6-2. A site-specific health and safety plan, as part of a comprehensive health and safety program, puts these guidelines into operation. The plan is followed by all consultants, such as those who execute the remedial investigation, those who design the feasibility study, the contractors who implement the remedial measures, and all persons who visit the site. In addition to government regulation, the rules of private corporations have a bearing at hazardous waste sites. Corporate safety measures generally include safety checklists, directions for movement of vehicles on plant premises, medical and first aid provisions, general safety policies, and handbooks. All persons who work at the site or who visit are expected to follow these measures also.

In response to the requirements of SARA, OSHA promulgated interim final standards regarding the health and safety of workers engaged in hazardous waste operations.[7] These regulations were published and took effect on December 19, 1986. The final standards were subject to revision and public comment. No schedule has been set for promulgation of the final standards. The topics addressed by the regulations are listed in Table 6-3.

TABLE 6-2. Sources of federal safety rules.

OSHA Requirements, 29 CFR 1910.120: Protection of Workers in Hazardous Waste Operations
EPA Order 1440.2: Health and Safety Requirements for Employees Engaged in Field Activities
EPA Order 1440.3: Respiratory Protection
EPA *Occupational Health and Safety Manual*
OSHA Requirements, 29 CFR 1950-1956: Provisions for State Safety Plans
Interim Standards Operating Safety Guide (revised September 1982, EPA Office of Emergency and Remedial Response)

TABLE 6-3. Safety topics addressed by OSHA regulations.

Safety and health program	Informational programs
Site characterization and analysis	Material handling
Site control	Decontamination
Training of personnel	Emergency response
Medical surveillance	Illumination
Engineering controls, work practices, personal protective equipment	Sanitation
	Excavation
Monitoring	Contractors, subcontractors

The new OSHA regulations apply to workers in the following situations:

- Response to the discovery of hazardous substances under CERCLA, including initial investigation at CERCLA sites before the presence or absence of hazardous substances has been ascertained.
- Major cleanup actions under RCRA.
- Operations involving hazardous waste storage, disposal, and treatment facilities.
- Hazardous waste cleanup operations designated by state or local governments.
- Emergency responses to the release or threatened release of hazardous substances; emergency operations following such releases.

Anyone who employs people engaged in hazardous waste operations is required to develop and implement a safety and health program for those employees. The program is to identify, evaluate, and control safety and health hazards, and it is to provide response measures for emergencies. At a minimum, the program is to address the topics listed in Table 6-3. The program is also furnished to any subcontractor involved with the hazardous waste operation.

The *health and safety plan* is a written document that applies the health and safety program to a particular project. The plan is prepared prior to the remedial investigation and responds to the specific dangers present at the hazardous waste site. The plan's pertinent measures are practiced when advance visits are made to the site, such as when initial sampling is conducted; it is assiduously followed during the full-scale remediation. New information about the site's hazards, such as that developed after analysis of the sampling performed during the field investigation, is incorporated into the plan in order to define better the hazards and requisite protection measures.

The health and safety plan addresses the components of the health and safety program listed in Table 6-3, as well as the following topics:

- Organization of personnel and programs.
- Site background and history.
- Evaluation of hazards.
- Site delineation.
- Protection of workers' safety; monitoring.
- Decontamination.
- Emergency and contingency provisions.

Each of the topics of a safety plan is addressed below. Several government publications contain directions for the establishment of safety programs. (See bibliography to this chapter).

It is the responsibility of the safety and health officer specifically assigned to the hazardous waste site remediation project to train workers and visitors to observe the correct safety procedures; the task is sometimes delegated to the field team leader or to the contractor in charge of the remediation. The health and safety plan contains provisions that match the sequence of remediation efforts, and it progresses in step with these efforts. Thus, personal protective equipment, safety training, decontamination, and emergency procedures are kept current with the work at hand.

To insure that the health and safety plan is followed, workers are told of the potential hazards on the site and the importance of safety; they are trained to utilize the safety protocols. Information about safety is essential for both workers and visitors to the site. Visitors also are trained to recognize hazards and to observe the safety protocols of the safety plan. The topics included in the training are listed in Table 6-3. Additional important issues are:[8]

- The chemistry of hazardous materials.
- Toxicology.
- Industrial hygiene.
- Hazard evaluation.

It is also necessary for all workers to become familiar with the layout of the site in order to be aware of hazardous areas, evacuation routes, emergency assembly areas, the location of emergency equipment, and the routes to be used by emergency response teams. In addition, emergency procedures are exercised regularly, usually monthly, through drills. The drills serve two purposes: first, during each drill, the emergency equipment is checked to insure that it is available and functional; second, the fact that these drills are held is documented as part of the record of preparation for the remediation effort.

Project Organization. Organization is the keystone of the safety program. The safety and health protocols must operate both during routine operations and during emergencies. While effective organization permits efficient routine operations, unequivocal organization is mandatory in times of emergency. At those times, the health and safety plan imposes order on an otherwise chaotic situation. The plan identifies important people and the lines of coordination and command.

Project organization begins away from the project site. Senior-level management of the corporation or government agency responsible for the remedial effort defines the project's objectives, delineates the chain of command, delegates responsibility, furnishes resources, and monitors and evaluates the progress of the effort. These managers usually benefit from expert technical

TABLE 6-4. Central on-site personnel involved in hazardous waste site remediation work.

TITLE	GENERAL DESCRIPTION	SPECIFIC RESPONSIBILITIES
PROJECT TEAM LEADER	Reports to upper-level management. Has authority to direct response operations. Assumes total control of site activities.	• Prepares and organizes the background review of the situation, the work plan, the site safety plan, and the field team. • Obtains permission for site access and coordinates activities with appropriate officials. • Insures that the work plan is completed and on schedule. • Briefs the field teams on their specific assignments. • Uses the site safety and health officer to insure that safety and health requirements are met. • Prepares the final report and support files on the response activities. • Serves as the liaison with public officials.
SITE SAFETY AND HEALTH OFFICER	Advises project team leader on all aspects of health and safety on site. Recommends stopping work if any operation threatens worker or public health or safety.	• Selects protective clothing and equipment. • Periodically inspects protective clothing and equipment. • Insures that protective clothing and equipment are properly stored and maintained. • Controls entry and exit at the access control points. • Coordinates safety and health program activities with the scientific advisor. • Confirms each team member's suitability for work based on a physician's recommendation.

and legal counsel and support services from consultants and other advisers. Among the specialists are physicians and hospital and emergency care personnel; they, together with authorities in the fields of engineering, public relations, industrial hygiene, law, and toxicology, provide recommendations on the design of both the general work plan and the safety plan. Auxiliary specialized personnel may be required to participate in the work also, including scientific advisors, logistics officers, photographers, financial agents, and security officers. Table 6-4 lists the functions of people essential to the remedial work, their general duties, and specific responsibilities.[9]

The authority and the responsibility of the principal personnel of the remediation effort are clearly delineated in order to expedite the smooth, orderly, and safe execution of the work. This chain of command is indispensa-

TABLE 6-4. ***(continued)***

TITLE	GENERAL DESCRIPTION	SPECIFIC RESPONSIBILITIES
		• Monitors the work parties for signs of stress, such as cold exposure, heat stress, and fatigue. • Monitors on-site hazards and conditions. • Participates in the preparation of and implementation of the site safety plan. • Conducts periodic inspections to determine if the site safety plan is being followed. • Enforces the "buddy" system. • Knows emergency procedures, evacuation routes, and the telephone numbers of the ambulance, local hospital, poison control center, fire department, and police department. • When necessary, notifies local public emergency officials. • Coordinates emergency medical care.
FIELD TEAM LEADER	May be the same person as project team leader and may be a member of the work party. Responsible for field team operations and safety.	• Manages field operation. • Executes the work plan and schedule. • Enforces safety procedures. • Coordinates with the site safety officer in determining protection level. • Enforces site control. • Documents field activities and sample collection. • Serves as a liaison with public officials.

(continued)

TABLE 6-4. ***Continued***

TITLE	GENERAL DESCRIPTION	SPECIFIC RESPONSIBILITIES
COMMAND POST SUPERVISOR	May be the same person as the field team leader. Responsible for communications and emergency assistance.	• Notifies emergency response personnel by telephone or radio in the event of an emergency. • Assists the site safety officer in a rescue, if necessary. • Maintains a log of communication and site activities. • Assists other field team members in the clean areas, as needed. • Maintains line-of-sight and communication contact with the work parties via walkie-talkies, signal horns, or other means.
DECONTAMINATION STATION OFFICER(S)	Responsible for decontamination procedures, equipment, and supplies.	• Sets up decontamination lines and the decontamination solutions appropriate to the type of chemical contamination on site. • Controls the decontamination of all equipment, personnel, and samples from the contaminated areas. • Assists in the disposal of contaminated clothing and materials. • Insures that all required equipment is available. • Advises medical personnel of potential exposures and consequences.
RESCUE TEAM	Used primarily on large sites with multiple work parties in the contaminated area.	• Stands by, partially dressed in protective gear, near hazardous work areas. • Rescues any worker whose health or safety is endangered.
WORK PARTY	Depending on size of field team, any or all of field team may be in work party, but work party should consist of at least two people.	• Safely completes the on-site tasks required to fulfill the work plan. • Complies with site safety plan. • Notifies site safety officer or supervisor of unsafe conditions.

ble in setting precedents, upholding decisions, assuming liability, and delegating duties. Safety is a factor within this chain of command because channels of communication that permit the clear, accurate, and timely exchange of information are prerequisite to the promotion of a safe working climate.

Two systems of communication are required at a hazardous waste site: one for internal communications among personnel on the site and the other for communication between the site and the project personnel, authorities, and services off-site. During the remediation operations, communication on the site can be difficult because of the level of activity and noise, the dimensions of the work area, and the use of protective clothing and equipment. Oral communication can be impeded by background noise and equipment designed for respiratory and hearing protection. Therefore, prearranged auditory and visual cues are employed, and communication devices such as two-way radios and noisemakers are utilized. Communication between the site and project managers and the civil authorities, surrounding population, government agencies, and others (including personnel who work at the site and are also residents of the locale) are important, too. Through this channel are exchanged both routine information, such as changes in work schedules and the status of the remediation, and critical information regarding emergencies.

Background and History of the Site. Information describing the hazards present at the site is the key to surmounting them, and historical records are an important source of such information. By learning a site's history of use and management and the sequence of events that have led to a crisis, investigators can begin to define the hazards and can surmise transitory states of the wastes. In addition, potential obstacles to certain remedial alternatives can be projected. Background information is found in corporate documents and municipal and other public records. Previous reports and assessments describing the site or the uses and processes it supported are valuable. Monitoring data characterizing the water, soil, sediments, air, or biota gathered during the field investigation also can furnish insight into the identity of contaminants and the extent of contamination.

Evaluation of Hazards. Once the extent, severity, and degree of risk existent on a site are defined, health and safety protocols can be established. The breadth of the protocols complements the intensity of the estimated hazards. The safety officer evaluates the potential sources of danger and prepares suitable safety protocols as a response. Certain dangers, such as oxygen deficiency, are unique to the site and affect only the workers performing the investigation and remediation. Others, such as chemical exposure, may affect both the workers and the population in the site's environs. The geographical extent of the danger is addressed by the protocols. For example, oxygen tanks are provided for workers who might suffer oxygen deficiency while in a situation of oxygen-deficient air during the remedial efforts, and evacuation procedures coordinated with municipal emergency officials are planned for residents whose safety might be jeopardized by, for example, fugitive toxic emissions. Factors considered in developing the protocols are the topography and hydrogeology of the site, weather conditions, artifacts on site, availability of storage facilities, accessibility of power supplies, and accessibility of emergency medical facilities. For each bona fide health and safety threat, specific routes of exposure are defined. Then, protection from each source of every threat is devised.

Site Delineation. A hazardous waste site is veritably different from a typical industrial workplace and requires special attention to make it suitably safe. Specific areas are designated for various functions, and the workers are trained to observe assiduously the delineation. Designated areas are:

- Work area (contamination zone).
- Entry and egress point.
- Decontamination area.
- Emergency and first aid facilities.
- Field office trailer.
- Shower/locker room facilities.

Figure 6-1 illustrates how these areas might be configured.

A fence placed on the perimeter of the work area marks the contamination zone. The fence line designates the area in which the hazardous materials are present; that area is a health and safety risk to workers, and within it all requisite protective equipment is used at all times.

Entry and egress are permitted through one point on the perimeter of the

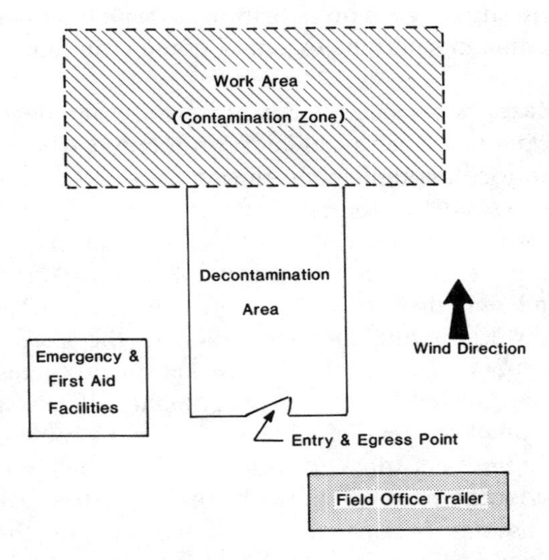

Figure 6-1. Typical layout of hazardous waste site undergoing remediation.

contamination zone. At that point, the decontamination area begins. All personnel and equipment pass into or from the work area (contamination zone) only through the designated entry and egress point and directly into the decontamination area. The health and safety officer or a surrogate monitors this gate and registers all personnel who enter or leave the contamination zone. That agent permits entry only to those personnel fully dressed in all required protective clothing and equipment. The procedures used at this gate are outlined in Table 6-5.

To avoid distribution of the contaminants from the work area into the clean areas, all workers implement decontamination procedures in a decontamination area located contiguous to the contamination zone. The decontamination procedures insure that contamination is not transferred from the protective clothing or equipment utilized in the work area to the person removing them or to outside areas or personnel. Decontamination procedures are strictly ordered and are designed to be specific to the level of protection being utilized. At the conclusion of the decontamination procedures within the decontamination area, the workers are considered clean and are permitted into the "clean areas"; that is, beyond the work area and the decontamination area. Shower and locker room facilities are made available to the workers if the level of contamination is such that a shower and complete change of clothes is a required part of the decontamination procedure.

Emergency and first aid facilities are located adjacent to the decontamination area because they necessarily must be near the decontamination and work areas to provide efficient and effective health and safety services to the on-site personnel. These facilities are usually housed in a trailer that also stores additional protective clothing and equipment and provides emergency showers and eye wash stations, first aid equipment, fire extinguishers,

TABLE 6-5. Entry and exit procedures between contamination zone and decontamination area.

ENTRY
1. Dress in required safety clothing and equipment.
2. Notify site health and safety officer (or surrogate) of intended operation.
3. Undergo review by safety and health officer or agent for prerequisite personal protective equipment and clothing.
4. Record entry time and name.
5. Enter contamination zone only through entry and egress point.

EGRESS
1. Exit only through designated entry and egress point.
2. Implement warranted decontamination procedure.
3. Sign out and record time.

and air horns or other warning signal devices. The health and safety plan for the project and emergency information, including the telephone numbers of off-site emergency personnel, are also located in the trailer, together with communications equipment for both on-site and off-site contact.

The administrative center of the remediation work is a field office trailer. The project team leader, the health and safety officer, and the other principals of the endeavor are headquartered there.

Protection of Workers' Safety. A medical program is established for personnel at waste sites to address the special health hazards and high levels of stress they may experience. This program is essential because it assesses and monitors the workers' health and fitness both before and during remedial work activities. The medical program also functions to provide emergency treatment and to maintain accurate medical records. See Table 6-6 for a summary of the program's responsibilities.[10]

The personal protective equipment worn by workers at a hazardous waste site remediation effort is designed to provide the minimum level of defense appropriate to the contaminants at hand. Care for this equipment includes several elements: selection and use of equipment appropriate to the site's hazards, including limitations on its use, such as hot weather conditions; inspection, maintenance, storage, and decontamination of the equipment; proper fit of the equipment and training in the correct procedure for donning and doffing it; consideration of the duration of work tasks; monitoring of the equipment during use; and evaluation of the effectiveness of the program that supervises use of the personal protective equipment.

Protection at hazardous waste sites is categorized into four levels:

- *Level A:* Highest level of respiratory, skin, eye, and mucous membrane protection. This level is used if: (a) the chemical substance has been identified at concentrations that warrant using fully encapsulating equipment,

TABLE 6-6. Components of site medical program.

- SURVEILLANCE
  Pre-employment screening
  Periodic medical examinations and follow-up examinations when appropriate
  Termination examinations
- TREATMENT
  Emergency
  Nonemergency (case by case)
- RECORDKEEPING
- PROGRAM REVIEW

or (b) the chemical substance presents a high degree of contact hazard that warrants using fully encapsulating equipment.

Work performed in a confined or poorly ventilated area also requires Level A protection until conditions change and a lower level of protection is indicated.

- *Level B:* Highest level of respiratory protection, but lesser level of skin and eye protection. This is the minimum level recommended during initial visits to the site and then until the nature of the hazards has been determined to demand less protection.
- *Level C:* Criteria for using air-purifying respirators are met, but skin and eye exposure is unlikely; hazardous airborne substances are known and their concentrations have been measured.
- *Level D:* No special safety equipment required except that typical of any construction site.

The health and safety officer establishes the level of protection required and determines whether the level should be advanced or reduced. The items of safety equipment required for the various levels are listed in Table 6-7.

TABLE 6-7. Personal safety equipment required for various hazard levels.

	LEVEL			
	A	B	C	D
Positive-pressure (pressure demand), self-contained breathing apparatus (SCBA)* or pressure-demand supplied-air respirator with escape SCBA*	■	■		
Full-face, air-purifying respirator*			■	
Fully encapsulating, chemical-resistant suit	■			
Chemical-resistant clothing		■	■	
Chemical-resistant inner gloves	■	■	■	
Chemical-resistant outer gloves	□	■	■	□
Chemical-resistant overboots	■	■	■	
Boots or shoes with steel toe and shank	■	■	■	■
Thermal luminescent detector badge (radiation)	□	□	□	
Personal radiation detector	□	□	□	
Hard hat	□	■	■	■
Coveralls	□	□	□	■
Two-way radio communications (intrinsically safe)	■	■	■	
Escape mask			□	
Safety eyewear				■

* = OSHA/NIOSH approved.
■ = Recommended
□ = Optional

The chemical-resistant clothing for Levels B and C includes overalls and long-sleeve jackets, or hooded one- or two-piece chemical-resistant splash suits, or one-piece disposable chemical-resistant suits. All joints where the items of clothing overlap with boots and gloves are securely taped with reinforced duct tape to insure a sealed barrier to the contaminated environment.

The contaminants at a hazardous waste site may interact with the materials of the protective clothing or equipment. For example, corrosive, acidic, or basic wastes require disposable coveralls that are specially treated to provide protection from such chemicals. Rubber gloves and boots may be damaged when in contact with certain solvent wastes. Equipment manufacturers are the source of accurate and reliable information about items of protective equipment and the chemical materials with which the equipment is compatible. In other words, the manufacturer is the source of information about the range of appropriate protective uses of the equipment.

Filtering cartridges for air-purifying respirators are selected to compensate for specific air quality conditions. Cartridges have been designed for protection against organic vapors, pesticides, asbestos, dusts, radionuclides, mists, fumes, and other specific gases or compounds. Special cartridges can be used that combine two or more protective features. If such combination cartridges are unavailable, it is often possible to stack single-purpose cartridges on the respirator; however, when cartridges are stacked, it becomes more difficult for a worker to draw air through them.

Safety clothing and equipment provide protection and are worn comfortably only if they fit well. Ill-fitting gloves, boots, or suits are a safety hazard. Without a firm fit, respirators provide no protection; therefore, each worker who will use a respirator is tested to insure that the fit is fast, and that a complete seal is formed around the face, thus barring any inadvertent inhalation of the contaminated atmosphere.

To benefit fully from the safety equipment, workers are instructed about all aspects of the health and safety plan that affect them, including the use of personal protective equipment. They are trained to dress for maximum protection. When Level A or Level B is in force, workers are taught proper breathing techniques with the positive-pressure self-contained breathing apparatus. They are also told how to care for and maintain their protective clothing and equipment, as those items require decontamination and upkeep in order to provide dependable service. All equipment and articles of protective clothing are inspected regularly to confirm their suitability for continued use.

The air of a hazardous waste site is monitored prior to establishment of the health and safety plan to determine the level of respiratory protection required for workers. Potential airborne contaminants are identified and quantified. Gross assessments of the measurements are made in the field;

TABLE 6-8. Considerations for instituting air quality monitoring during remediation of a hazardous waste site.

Extent of contaminants of site	Work operation in progress • Schedule • Location
Physical state of contaminants	Materials being moved
Accessibility	Weather conditions
Availability of trained personnel	Short-term vapor emissions
Availability of monitoring equipment	Level of dust generated
Availability of analytical laboratory	

definitive and usually long-term analyses on air samples are performed by a qualified environmental laboratory. The monitoring locations are chosen to suit the contamination conditions of the site. During remediation, samples may be taken from within the work zone, from the decontamination area, or from the workers; alternatively, the samples may be taken at the perimeter of the site, especially to audit possible fugitive emissions that may reach the environs. Factors that affect the frequency, location, and method of monitoring are listed in Table 6-8. These factors determine the monitoring equipment best suited for a particular site.[11]

Decontamination. Decontamination procedures are instituted to prevent contamination of any area beyond the work zone and decontamination area. Distinct decontamination procedures, dependent on the level of protection employed, are developed for workers, for their personal protective equipment, and for tools and machinery. All wastes or waste streams generated by decontamination activities are collected and disposed of appropriately. The specific decontamination procedures vary with the level of protection employed.[12] (See Table 6-5 for the general procedure used by workers when entering or leaving the work zone.) Workers are attended by decontamination line personnel—employees who help the workers wash, scrub, rinse, and remove protective equipment.

Decontamination of workers and personal protective equipment generally progresses through the following steps:

1. Tools, sampling devices, containers, monitoring instruments, radios, clipboards, and all other items of equipment used in the work zone are deposited on a drop cloth of plastic.
2. Outer boots and outer gloves are scrubbed with detergent solution and rinsed with abundant clean water.

3. The tape that sealed the boots and gloves to the outer protective suit is removed. These boots and gloves are removed, and they along with the sealing tape, are placed in separate containers lined with plastic.

   If a worker is leaving the work zone to change a cartridge or the entire respirator, this is the point at which the worker does so. Following the exchange, new outer gloves and boot covers are donned, the joints are taped, and the worker returns to duty.
4. The protective suit is removed and deposited in a plastic-lined disposal container.
5. Respirators are removed and cleaned with detergent solution by personnel assisting with the decontamination process.
6. Inner gloves are removed and deposited in a plastic-lined disposal container.
7. The worker showers and changes clothes to complete the decontamination process.

Tools and equipment, including heavy machinery used on the site, are decontaminated by decontamination line personnel wearing the proper personal protective equipment and clothing. Gross accumulations of contaminated soil are swept or scraped off. All surfaces that have contacted the contaminated wastes are steam-cleaned or washed with detergents and rinsed.

Emergency and Contingency Provisions. Emergencies at hazardous waste sites are more likely to occur and more difficult to manage than those at conventional construction sites. Well-defined procedures and well-trained personnel insure that such emergencies are supervised correctly. OSHA requires that certain elements compose an emergency response plan, and those minimum requirements are listed in Table 6-9. Moreover, emergency

TABLE 6-9. Minimum requirements in emergency response plan specified by OSHA.

Pre-emergency planning	Decontamination procedure
Personnel roles: • Lines of authority • Training • Communication	Emergency medical treatment and first aid
Emergency recognition; emergency prevention	Emergency alerting and response procedures
Safe distances; places of refuge	Critique of response and follow-up action
Site security and control	Personal protective equipment; emergency equipment
Evacuation routes; evacuation procedure	

response personnel—both on-site and off-site—require training in the correct procedures to employ when handling emergencies.

During emergencies, the project team leader is responsible for all response actions and directs those measures. Whenever an emergency occurs, all personnel are evacuated from the immediate area because they are safer at a remove from the problem area and cannot then interfere with efforts to manage the crisis. When evacuation of the site is required, decontamination procedures are carried out to the fullest extent practicable, but they are not permitted to interfere with the safe egress of all personnel from the immediate area of danger. The sequence of response actions in an emergency is illustrated in Figure 6-2.

When an accident or medical emergency occurs, the persons at risk may be transported from the site for care. The victim is decontaminated immediately whenever possible and to the degree practicable. The dimensions of the illness or injury dictate the extent of decontamination that is feasible. When it would interfere with necessary medical treatment, decontamination is postponed. The victim is wrapped in blankets, plastic, or another barrier material to reduce the possibility of spreading the victim's contamination. All persons who will provide emergency medical treatment are alerted to the fact that decontamination was not performed, and they are instructed in the correct decontamination procedures. If decontamination may be performed, it is done promptly, and the individual's protective clothing and equipment are washed and rinsed, or removed.

The following situations may be the source of an emergency:

- *Major release of hazardous materials:* The immediate response to a release of hazardous materials follows the general order: (1) containment; (2) cleanup; (3) disposal. Releases into unprotected areas are contained with absorbent booms, blankets, pads, or granular materials. Following containment, the spilled material is recovered if possible or cleaned using additional absorbent materials. The contaminated absorbent materials are collected and contained along with any contaminated soils, and all polluted materials are disposed of in an approved hazardous waste facility.[13]
- *Major chemical exposure:* Whenever anyone on the waste site is exposed to a hazardous chemical, the health and safety officer is notified immediately, and first aid is administered straightaway. The health and safety officer then prepares information to describe how the individual should be treated, including a copy of the material safety data sheet, a form that details the chemical, physical, toxicological, and fire hazard potential for a compound and lists the general safety precautions applicable when handling the material. The patient and the safety information are dispatched to a medical facility for care.

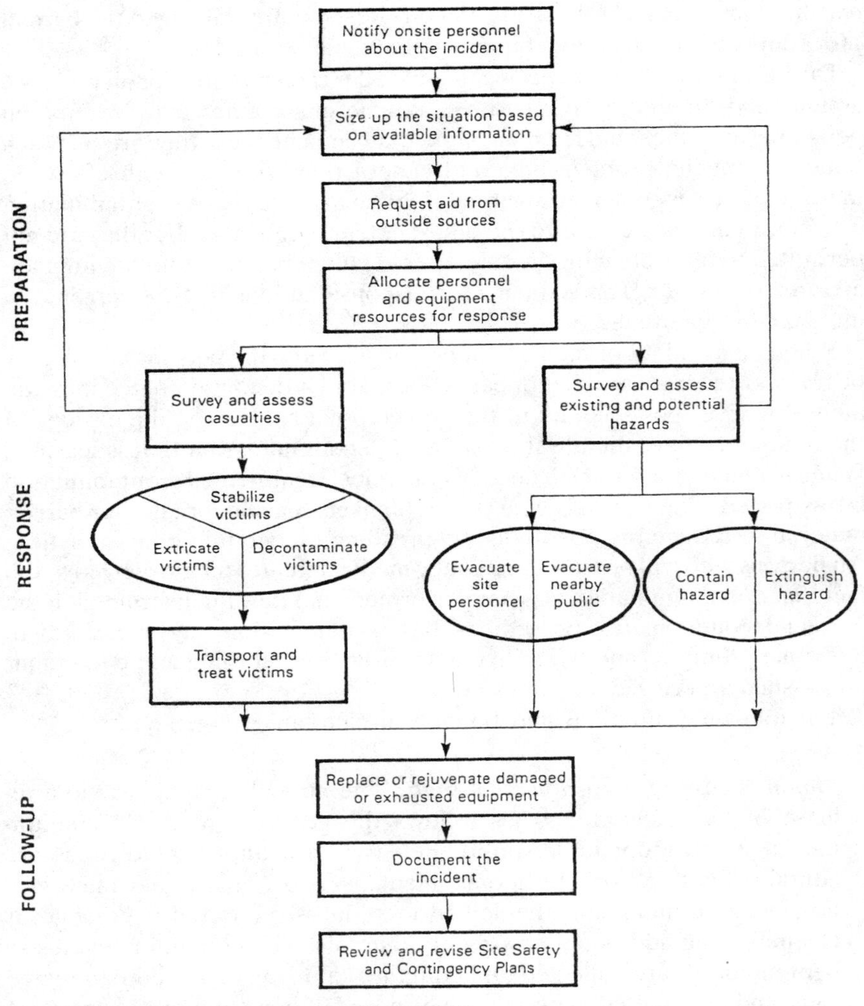

Figure 6-2. Flowchart of emergency response. Source: *Occupational Safety and Health Guidance Manual for Hazardous Waste Site Activities*, NIOSH/OSHA/USCG/EPA, 1985.

- *Medical crisis:* The demand for medical treatment at a hazardous waste site can range from bandaging minor cuts to life-saving first aid and immediate transport to expert medical care. Individuals with specialized training in first aid are identified to site personnel, and all workers are alerted to the location of first aid kits and other emergency supplies, as well as emergency phone numbers used to summon outside assistance.
- *Fire or explosion:* All personnel working on the hazardous waste site are informed of the locations of all fire-fighting equipment, and everyone is trained in the proper use of the equipment. No one is permitted to attempt to extinguish a fire individually, unless it is apparent that it is sufficiently minor and can be quickly doused. For all other fires, the area is evacuated immediately, and the designated emergency response team—composed either of the site personnel or personnel from an off-premises fire company—is summoned.
- *Accident involving equipment:* Workers operating around heavy equipment may be injured. Whenever such an injury occurs, the worker is given first aid treatment immediately and is transported to a nearby hospital. A single individual is assigned the responsibility to call for any ambulance or other emergency vehicle required; that same person is accountable for directing the vehicle to the victim.
- *Flood:* Flooding at a hazardous waste site poses a particularly dangerous situation because the waste from the site can be spread into the environs. If the National Weather Service issues a flood warning affecting the site, all personnel are evacuated. If time permits, all equipment is removed, and power lines are disconnected. If practicable, the heavy equipment at the site can be used to construct berms or levees to divert the flood waters.

If an emergency threatens the health or safety of a community in the environs of the site, the authorities in the area and the public are so informed by the project team leader, with the assent of senior-level management. If necessary, residents are evacuated. For any evacuation, the health and safety officer, the project team leader, and senior-level management coordinate the action with the proper public agencies.

At certain times of crisis at a hazardous waste site, the health and safety officer is required to notify relevant federal and state agencies. On the federal level, the Occupational Safety and Health Administration, the U.S. EPA, the Coast Guard, and the Department of Transportation often require that they be notified of the emergency and of the response to the emergency.[14]

Arrangements After an Emergency. Following every emergency, all procedures are reviewed, and emergency equipment that was used is restored to readiness. It is of such importance to have all emergency preparations in place that no one is permitted on the site until all emergency equipment is again serviceable. Each emergency is investigated, its circumstances are recorded, and the completion of restoration is then reported to project administrators and pertinent federal and state agencies. It is imperative that these reports be accurate, objective, and complete; like chain of custody forms, they are signed by each person who contributes to them or reviews them. These reports subsequently can serve as tools in the safety training given site workers and may provide the information necessary to determine the efficacy of project policies and procedures. The emergency reports provide a permanent record that can be utilized by insurance companies to assess liability and by government agencies in their review of the conduct of the project.

CASE EXAMPLE: DEVELOPMENT OF A SITE SAFETY PLAN

A major chemical company located on the East Coast owned two large waste storage lagoons containing 15 million gal (57,000 m^3) of oily sludge materials. These lagoons had been used since 1947 for the storage of still bottoms and acid washes. The company engaged a consultant to determine how the materials could be pumped from the impoundments, transported to a refinery, and reclaimed for use as a fuel. Whenever hazardous materials are moved, a safety plan is necessary to address all the phases of the task. Therefore, a safety plan unique to the site and project was formulated. The consultant acted as the health and safety officer for the site. Hazards at the site were posed by the use of heavy equipment, the remoteness of the lagoons from other site facilities such as first aid and emergency response, and hot, humid weather anticipated for the length of the work schedule.

In order to develop a sound and applicable safety program, the work plan of the project was reviewed first. All aspects of the work—time schedules, labor requirements, equipment to be used, location of the work, lines of communications, and materials to be handled—were studied. All existing information related to the project was also reviewed. This information included data in previous safety plans, project reports, and letters and memoranda regarding the site.

In addition, the process that generated the waste was reviewed in order to identify the likely components of the waste. The still bottoms and acid washes were by-products of an oil refinery that recovered benzene, toluene, xylene, and naphtha solvents. For safety reasons, it was especially important to characterize thoroughly the nature of the materials to be handled. The available material safety data sheets describing the constituent materials were scanned; the responsible corporation was telephoned to determine if it could provide further assistance or direction to avenues of additional research.

TABLE 6-10. Data describing wastes present in lagoons (case example).

Flashpoint: 64-78°F (18-26°C)

Amount volatile: 25% (by total weight loss)

pH: <2

Major components:
- Benzene
- Chlorobenzene
- Methylene chloride
- Toluene
- Dichlorobenzene
- Naphthalene
- Sulfides

With these data, a firm understanding of the wastes' chemical, physical, and toxicological properties was developed.

Data on hand indicated that the waste materials in the impoundment had stratified into three distinct layers: the bottom was hard and crumbly; the middle was viscous and rubbery; and the top was a liquid, oily sludge. Both lagoons were covered with a layer of water ranging from 1 to 5 ft (0.3 to 1.5 m) thick. The review of the chemical and physical properties of the materials yielded the information presented in Table 6-10.

Having evaluated the available data, the safety consultant traveled to the site in order to view the conditions firsthand and to examine records available only at the site. It was imperative that the consultant review the site in person to insure that the safety plan be developed in a practical manner. During the tour of the site, the waste lagoon and its immediate surroundings were photographed, and details of the layout of the site were noted and sketched for reference. (Note: During any initial visit to the site containing hazardous wastes, it is necessary to wear Level B protection. Once the site has been classified in terms of its health hazard, the level of protection may be modified.) Of particular interest were both the active railroad tracks, which separated the lagoons from the remainder of the site, and the strong odor associated with any exposed waste materials on the banks of the lagoons, an odor caused by the sulfur in the compounds. Specific items registered during the site visit are listed in Table 6-11.

Knowledge gained in the site visit was combined with information garnered from the review of the background data. The safety plan was then outlined, utilizing the framework presented in this chapter. Applicable federal, state, and local regulations were followed. The responsible corporation, as is often the case, had its own guidelines for truck routes and for fire and emergency response. Those guidelines were incorporated into the safety plan.

TABLE 6-11. Site details considered in development of site safety plan (case example).

Topography	Site for decontamination area
Waterways and water bodies	Space for emergency first aid facility
Odors	Utility access points
Roads on-site	Environs
Paths and trails	• Industries • Residences
Access and evacuation routes	• Group quarters
Site for field office trailer	

In order for any safety program to be effective, the employees who will work on the site must be familiar with the provisions of the program, and they must be educated in the proper use and care of the safety equipment. The obvious solution is training. However, there were practical problems in training the large number of workers who would be involved and with the distance the health and safety officer had to travel to furnish the training. In addition, worker turnover was expected, and all new workers had to be trained adequately before starting work on the site. Otherwise, the safety plan could not be fully implemented, and it would be ineffective.

It appeared, then, that the training program would have to be given not only at the initiation of the project but repeatedly during the course of the work. In practical terms, such a schedule would have demanded that the health and safety officer travel to the site for each session, transporting the safety demonstration equipment each time. Such travel would have been time-consuming and expensive.

An alternative was developed that insured adequate training and manageable costs. A videotape of the training demonstration was made. In order to create the tape, a script was written, and an audiovisual firm was engaged to shoot the tape. The health and safety officer made the presentations in the tape, and another individual was recruited to demonstrate use of the safety clothes and equipment. The actual taping took less than one day. When the tape was edited, supplementary information was added: tables, outlines, and site photographs to emphasize key points. This videotape incorporated all the information and demonstrations normally presented in live training programs.

Every employee and each person involved in the reclamation project viewed the videotape before being permitted to enter the site, even to perform the initial field work. The name of each viewer, the date, and the time of each viewing session were recorded, and these records became part of the documentation of the training program. For the initial viewing session, the safety consultant was on hand to answer questions. By noting where the workers had the most questions, the consultant could prepare in advance supplementary explanatory material for future sessions.

Once the training tape had been prepared, the complete safety plan was drafted. The outline developed earlier was the basis for the draft plan. This draft was reviewed by both the consultant's management and the management of the corporation. The safety plan was modified in line with pertinent comments, and it was issued as a procedures document of the reclamation project.

The plan was then ready for implementation, but a major difficulty had to be overcome: the subcontractors' workers resisted wearing the required safety gear. A consultant can do little to enforce safety measures; his or her role is to provide technical assistance and expert advice, not to furnish any enforcement. The consultant advised the workers of the importance to their personal health and safety of wearing the appropriate clothing and of using the safety equipment stipulated in the plan. They also were told of the liability to their employers if they did not conform to the contractor-approved safety plan. However, it was, ultimately, the responsibility of the subcontractors to enforce the safety procedures for their employees. Because the use of appropriate safety gear was mandatory, the workers were finally persuaded to comply.

Each phase of safety plan development was documented. All communication with contractors, subcontractors, the responsible corporation, and personnel who assisted in writing the safety plan was filed to insure that a complete record would exist. Moreover, all phases of the safety plan's development were communicated to all contractors, subcontractors, corporate personnel, and the consultant's management.

Thus the safety plan was developed for one hazardous waste site reclamation project. Its implementation was not without difficulty, due to both economic constraints and personnel snags, but because safety was essential, every facet was executed completely. No safety practice and no documentation were omitted.

SUMMARY

No plan of remedial action at a hazardous waste site can be considered complete or sufficient without careful regard and particular attention to the special conditions surrounding the health and safety of workers and other personnel. Hazardous waste sites can present health and safety hazards more serious than those found at conventional construction areas. Therefore, health and safety are crucial concerns throughout the remedial work effort.

The liability for the safety of all workers on the site is a prominent concern of project managers; safety issues are amply addressed in the comprehensive health and safety program for the project, and the protocols are stringently implemented.

The basic requirements are:

- An accurate evaluation of the type and extent of contamination.
- Meticulous design and implementation of an effective, site-specific health and safety plan.
- Constant awareness of and preparation for all possible exigencies.

Evaluation of the hazards present on the site and the extent of contamination requires a tour of the site; sampling and analyses to describe accurately its chemical contamination; and an understanding of the history of use of the land. These data are evaluated in the context of present weather conditions, accessibility, and other pertinent parameters.

With that information in hand, a specific health and safety plan is formulated to address existing and potential dangers. The health and safety plan is critical for several reasons. Many of the regulations listed in Table 6-2, as well as right-to-know provisions,[15] require the verifiable provision of safety information and training for all workers on a hazardous waste site. Furthermore, the health and safety plan serves as a record of compliance with OSHA requirements, including:

- Project organization.
- Site background and history.
- Hazard evaluation.
- Site delineation and control.
- Monitoring programs.
- Safety equipment, protocols, and training programs.
- Decontamination procedures.
- Emergency and contingency provisions.

A number of specific safety points also deserve attention in the plan. Foremost among them is authority for the supervision and implementation of the safety protocols; the plan establishes lines of coordination and communication. Another important item is selection of the protective clothing and equipment; both overprotection and lack of protection jeopardize the health and safety of workers. Finally, a project contingency plan describes the actions to be taken in the event of fire, explosion, chemical release, medical problem, or other emergency. At such extraordinary times, especially, the chain of command and communication are observed closely to provide the most efficient and comprehensive resolution of the difficulty.

BIBLIOGRAPHY

American Conference of Governmental Industrial Hygienists, Inc. (ACGIH). *Guidelines for the selection of chemical protective clothing,* 2d Ed. Mar. 1985.

American Industrial Hygiene Association (AIHA). L. R. Birkner. *Respiratory protection—a manual and guideline.* 1980.

Ferguson, J. S. and W. F. Martin. Overview of occupational safety and health guidelines for Superfund sites. *American Industrial Hygiene Association Journal.* 46(1985):175-180.

National Institute of Occupational Safety and Health (NIOSH), OSHA, U.S. Coast Guard (USCG), and U.S. EPA. *Occupational safety and health guidance manual for hazardous waste site activities.* Oct. 1985.

NIOSH. *Personal protective equipment for hazardous materials incidents: A selection guide.* 1984. Publication No. 84-114.

National Safety Council. *Fundamentals of industrial hygiene.* 2d Ed. 1979.

Recovery for exposure to hazardous substances: the Superfund [subsection 301(e)] study and beyond. *Environmental Law Reporter.* 14(1984):10098-10141.

U.S. EPA and Federal Emergency Management Agency. *Planning guide and checklist for hazardous materials contingency plans.* 1981.

Walsh, J., J. Lippitt, and M. Scott. Costs of remedial actions at uncontrolled hazardous waste sites—impacts of worker health and safety considerations. Conference paper. June 82-June 83. 1983. EPA-600/D-84-019.

NOTES

1. National Institute for Occupational Safety and Health (NIOSH), Occupational Safety and Health Administration (OSHA), U.S. Coast Guard (USCG), and U.S. EPA. *Occupational safety and health guidance manual for hazardous waste site activities.* Oct. 1985. Table 1.
2. *Occupational safety and health guidance manual for hazardous waste site activities.*
3. *Occupational safety and health guidance manual for hazardous waste site activities.*
4. National Safety Council. *Fundamentals of industrial hygiene.* 2d Ed. 1979.
5. 29 CFR 1910.96, Ionizing Radiation.
6. 29 CFR 1910.95, Occupational Noise Exposure.
7. 29 CFR 1910.120.
8. *Occupational safety and health guidance manual for hazardous waste site activities.*
9. *Occupational safety and health guidance manual for hazardous waste site activities.* Table 3-2.

10. *Occupational safety and health guidance manual for hazardous waste site activities.* Section 5.

11. A useful reference on the subject is the *Occupational safety and health guidance manual for hazardous waste site activities* of NIOSH, OSHA, USCG, and U.S. EPA.

12. *Occupational safety and health guidance manual for hazardous waste site activities.* Appendix D.

13. Federal regulations require notification of the National Response Center when reportable quantities of hazardous substances are released. See CERCLA, 40 CFR 302. Certain states also regulate the release of hazardous substances.

14. 40 CFR 265 Subpart D for owners and operators of hazardous waste facilities; 40 CFR 302, Releases of Hazardous Substances; 29 CFR 1904, Occupational Injuries.

15. 29 CFR 1910.1200, the Hazard Communication Standard. Some states also have "right-to-know" laws.

Chapter 7
Ground Water Models: Tracking Contaminant Migration

The physical and chemical properties of ground water can be simulated with models. Hydrogeologists and engineers use ground water models to facilitate their understanding, evaluation, and management of ground water contamination due to hazardous wastes. Ground water models are practical because they have the capability to simulate hydrogeological systems and to test remedial alternatives rapidly, systems and alternatives whose complexity renders them expensive or impossible to investigate by direct field methods. Remedial measures can be expeditiously and economically tested by using ground water models to forecast the response the remedial measure will induce in the hydrogeological system. The ability to predict the response of a ground water system significantly enhances the remediation and management of hazardous waste sites.

Ground water models may be physical or mathematical in design. They evaluate processes in the zone of saturation (vadose zone) or the unsaturated zone, or the interaction of the two zones. The most widely known and used ground water models are those applied to the zone of saturation. This chapter surveys ground water modeling and its application to hazardous waste investigations and remediation. The focus is on the logic and method of applying models. Detailed discussions of the theory and mechanics of modeling are not presented because the literature abounds with discourse about specific models and modeling.[1-12] For the purposes of this book, the theory of ground water modeling is introduced, and the goals of model use, the varieties of levels and types of models available, and the advantages and limitations of models are reviewed as applied to hazardous waste sites. Three case studies are presented as vehicles to carry the logic of ground water model applications. These studies discuss both the process of applying models to specific problems and the advantages and limitations of the particular applications.

This chapter was developed by Guy A. Swenson III of O'Brien & Gere Engineers, Inc.

THEORY AND LOGIC OF MODELING

The scientific method is a recognized, systematic approach to the development of conceptual models—theories—that describe physical, chemical, and biological processes. The basic steps of the scientific method are:

1. *Review data:* Data are collected and reviewed.
2. *Develop conceptual model:* A conceptual model, a theory, is developed to explain the observed data.
3. *Test conceptual model:* A hypothesis—a prediction of present or future conditions, or both—is proposed to test the conceptual model. Additional data are gathered, and they are evaluated to determine whether the hypothesis is supported or contradicted.
4. *Modify conceptual model:* To account for any contradictory data found during Step 3, the conceptual model is modified.

Steps 3 and 4 are repeated, and the conceptual model is refined to the satisfaction of the scientist.

The process that hydrogeologists and engineers use to develop an understanding of hazardous waste sites and ground water contamination parallels the approach of the scientific method:

1. *Review data:* Data describing the contaminants of concern and the hydrogeological conditions influencing ground water flow are collected and reviewed.
2. *Develop conceptual model:* A conceptual model, a theory, of the hazardous waste site is developed. This conceptual model enunciates the current understanding of the physical, chemical, and biological processes governing movement of the contaminants from the hazardous waste site in the ground water system.
3. *Test conceptual model:* A hypothesis of existing or future conditions is fabricated, and either additional site data are collected or a ground water model is applied.
4. *Modify conceptual model:* The conceptual model is altered to account for contradictory information identified in Step 3.

Steps 3 and 4 are repeated until the conceptual model is refined to the extent necessary for the hazardous waste study.

USES AND METHODS OF MODELING

Ground water models have been developed to address the extensive variety of hazardous waste problems and physical, chemical, and biological processes that govern ground water flow and the transport of contaminants in ground water. Ground water models have been developed to simulate one or more of the following phenomena:

- *Ground water flow:* to address the physical processes that govern ground water movement.
- *Solute transport:* to combine the hydraulics of ground water flow models with the physical, chemical, and biological processes that control the migration and degradation of chemical constituents dissolved in the ground water.
- *Phased transport:* to incorporate the physical, chemical, and biological processes that control the migration and disposition of chemical constituents that exist in a separate liquid phase within the ground water regime.

Ground water models that embrace two or more of these categories are able to abstract a variety of complex hazardous waste problems.[13]

Three methods of conducting ground water modeling have been or are in use:

- *Physical models* use reduced-scale versions of the real world. They are often called *sand boxes,* as they generally consist of a receptacle containing sand or another porous material. Although such models can be useful in evaluating certain processes because they mimic the real world, the constraints of scale, the limits on the complexity of such models, and the time required for such modeling are significant disadvantages to their use. They have largely been replaced by the other modeling methods.[14]
- *Analog models* basically mimic hydraulic processes with electrical or thermal systems. Resistor-capacitor networks and other analog models were common in the 1950s and 1960s but have been generally replaced by mathematical models because of the cost and time disadvantages of analog models.[15]
- *Mathematical models* rely on the solution of an equation or sets of equations to describe the processes of ground water hydraulics and contaminant migration.

With recent advances in computer capabilities, many mathematical problems are now readily solved; therefore, mathematical models have largely replaced physical and analog models. Mathematical models are more versatile, available, and efficient than earlier methods. The majority of ground water models currently in use are mathematical. These models can be grouped into three categories: analytical, numerical, and statistical.[16–26]

Analytical models use analytical equations (equations that can be solved algebraically) to describe hydrogeological processes. Such equations are readily evaluated without the use of a computer; however, they represent idealized conditions, as many simplifying assumptions are necessary. *Numerical models,* both finite-difference and finite-element, employ numerical techniques to solve complex mathematical equations describing the processes controlling ground water flow and contaminant migration. Although such equations can evaluate more complex hydrogeological problems than analytical equations, numerical modeling is typically more time-consuming and costly and requires the use of a computer. *Statistical models* rely on statistical associations and probabilities to evaluate hydrogeological problems. These models have been developed to address the complexity of hydrogeological systems and the typically limited data base available to support a ground water model. Statistical models are hybrids composed of an analytical or numerical model and a statistical framework. Analytical and numerical models allow only single values for each input parameter; consequently, a complex hydrogeological system requires an extensive data base and a detailed model. Statistical models, on the other hand, rely on probability and statistics to account for a limited data base or a complex system.

APPLICATION OF GROUND WATER MODELS

The use of a model is dictated by the goals of a hazardous waste project. The goals to which modeling contributes can be grouped generally into three categories:

- Reconnaissance.
- Interpretation.
- Prediction.

In practice, modeling efforts often involve two or more of these objectives.

Reconnaissance studies test the preliminary conceptual model and identify areas of significant data deficiencies. They are performed when the available hydrogeological information is limited or when only a highly simplified conceptual model is required. Reconnaissance model studies can facilitate rapid, economical reviews of hazardous waste sites and the potential effects of contaminant migration from the site. Reconnaissance modeling also effectively identifies deficiencies in an existing data base and provides guidance for future field studies. For example, in testing the preliminary conceptual model, it may become apparent that data on the hydraulic conductivity of an aquifer are insufficient or that the hydraulic potential of an aquifer near a discharge boundary is not well understood.

Interpretive studies typically follow extensive data collection efforts, and they test and refine the conceptual model. The resulting refined conceptual model furnishes a basis for the description of the existing hydrogeological system and for predictions of future contaminant migration. The use of ground water models for interpretive studies is clearly an application of the scientific method.

Predictive studies forecast the future response of a hydrological system to existing stresses and to future conditions. Predictive studies can forecast future contaminant migration to provide a basis for determining potential risks or the necessity for site remediation. These studies can also evaluate the future response of a hydrogeological system to proposed remedial measures and thereby evaluate the effectiveness and design of remedial alternatives.

Ground water modeling efforts generally proceed according to the following steps:

1. Problem definition and review of objectives.
2. Collection and evaluation of site data.
3. Development of the conceptual model of the site:
   (a) Definition of the nature and extent of the hydrogeological system.
   (b) Identification of the physical and chemical processes governing the hydrogeological conditions and the migration and transformation of any contaminant.
4. Identification of the appropriate mathematical equations and simplifying assumptions—boundary and initial conditions, for example—to be used to represent the physical and chemical processes important to the site problem and to the remedial project's objectives.
5. Selection or development of the mathematical or computer code or codes for the appropriate model.
6. Identification of the model input parameters as they relate to the site.
7. Sensitivity analysis of the model and calibration of the model to determine its accuracy and validity.
8. Utilization of the model as a tool to achieve the objectives of the study.

CASE STUDIES

It is most enlightening to illustrate the logic and method of applying ground water modeling through examples. Ground water models are a tool of hydrogeologists and engineers, but it is essential that others involved with management of hazardous waste site remediation understand the rationale and techniques of their applications. Three case studies are presented here; several others are put forth in the literature.[27–36]

Reconnaissance Study. The owner of an industrial landfill that had been closed for 20 years performed an initial hydrogeological investigation of the

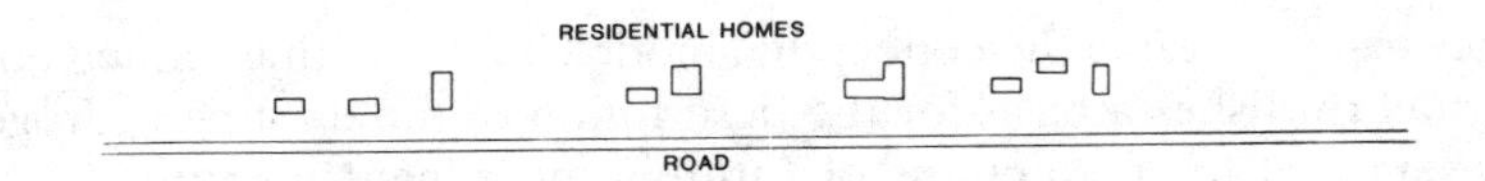

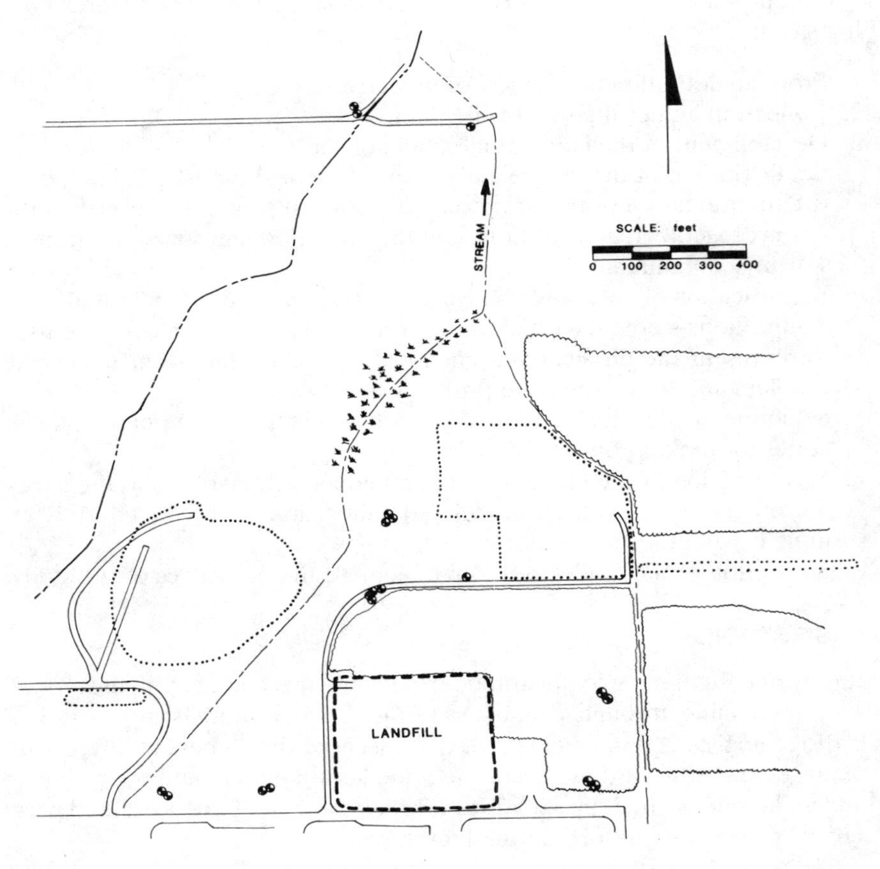

Figure 7-1. Landfill and area of potential migration.

immediate landfill site. That study detected elevated concentrations of both organic and inorganic contaminants in a fractured bedrock aquifer, an aquifer used as the source of water supply for nearby residences; see Figure 7-1. The concentrations of contaminants in the ground water on the site exceeded levels considered acceptable for drinking water supplies. The landfill owner commissioned a study to delineate the extent of ground water contamination and to evaluate remedial alternatives that could mitigate any hazard. The landfill owner, however, was immediately concerned with the potential for the contaminants to migrate from the landfill and to affect the potable water wells. In order to provide a rapid assessment and not to arouse public concern needlessly, investigators performed a reconnaissance modeling study without acquiring additional data outside the landfill's premises. The specific objectives of the reconnaissance study were, first, to determine if the contaminants would migrate beyond the landfill site and adversely affect the residential wells and, second, to estimate the time for the contaminants to reach the wells. The data that were available consisted of regional ground water reports and the initial hydrogeological investigation of the landfill and proximate area. Although that first investigation furnished some detail, the information from the site's monitoring wells was generally limited to the perimeter of the landfill.

As the first step in this assessment, a preliminary conceptual model of the area was developed using the available data. The known facts are listed in Table 7-1.

TABLE 7-1. Data base used to evaluate ground water contamination, reconnaissance study.

Aquifer known to be source of potable water.	Hydraulic gradient proximate to site 0.002 to 0.028 ft/ft (0.06 to 0.85 cm/cm) toward the north.
Regional geology suggesting aquifer extends beyond landfill.	Bulk hydraulic conductivity of fractured bedrock aquifer estimated to be 3.7×10^{-4} to 6.3×10^{-7} cm/s.
Aquifer composed of fractured siltstone and sandstone.	Residential wells downgradient of site situated along a road 2800 and 850 ft (850 and 260 m) from landfill.
Ground water flow expected only in bedrock fractures.	Concentration of contaminants adjacent to landfill in ground water known.
Thickness of aquifer unknown, but existing monitoring wells penetrated 30 ft (9 m) of aquifer.	

The restricted data base did not permit the hydrogeological properties of the bedrock aquifer and the process of contaminant migration to be well defined in the conceptual model. A quick review, taking into consideration the range of flow velocities of the ground water, indicated that at a low rate (3.5×10^{-5} ft/day [1.2×10^{-8} cm/s]), no contaminants would affect the residential wells 2800 ft (850 m) away from the landfill within 100 years; at a high rate (3.0 ft/day [1.1×10^{-3} cm/s]), the wells would be affected within 20 years. However, this simplified assessment gave insufficient data upon which to make a judgment about the best course of action regarding the potential effects on residential wells. Therefore, the investigation team turned to ground water modeling to provide adequate information upon which to base sound decisions.

An appropriate ground water model had to be selected. Although certain hazardous waste problems require that site-specific models be developed, the time and cost of developing one was unacceptable in the case at hand. Therefore, previously developed models were evaluated for application to the site. The restricted data base was insufficient to support a complex numerical model. A simplified analytical model could have been supported by the data base, yet it would have lacked the means to evaluate the variability of the hydrogeological conditions except in providing a best-case/worst-case/most-likely-case analysis. Therefore, a statistical model was chosen. The model selected was one developed by Thomas A. Prickett & Associates for use in porous media. This model combined an analytical solute transport equation[37] with a statistical package that accounted for the variability of the hydrogeological conditions. The statistical model generated results that identified the probability of a specific concentration of a contaminant occurring at specific distances from the source of contamination.

In using the statistical model, a range of values was initially assigned to each variable in the model. The range of values for this site was formulated from the available hydrogeological data and from the literature on ground water; see Table 7-2.

The model accounted for the variability of the hydrogeologic parameters by randomly selecting an input value from within the range assigned to each variable. The analytical equation was solved with the randomly selected values, and the results were tabulated. By performing numerous calculations with randomly selected input values, a statistical table of results was developed. This table indicated both the frequency and the probability of occurrence of contaminant concentrations at different distances from the contaminant source.

Before the statistical model was applied, assumptions inherent in the conceptual model or the model code were identified. The nature of the analytical solute transport equation, the nonhomogeneous complexion of

TABLE 7-2. Reconnaissance model: range of input values.

	HIGH	LIKELY	LOW
Darcy velocity (ft/d)	3.0×10^{-2}	2.8×10^{-3}	3.5×10^{-6}
Porosity (Dim)	0.10	—	0.10
Longitudinal dispersivity (ft)	100.	—	20.
Transverse dispersivity (ft)	60.	—	1.
Aquifer thickness (ft)	30.	—	30.
Contaminant loading (lb/d)			
Chloride	12.8	5.3	1.6
VHO	0.0198	0.0085	0.0024
Retardation			
Chloride	1.	—	1.
VHO	5.	—	5.
Distance x (ft)	2800.		2800.
Distance y (ft)	0.	0.	
Time (years)	20 (current), 50	—	20 (current), 50

the hydrogeology of the landfill area, and the limited data base prompted these assumptions. Without an awareness of these conjectures by all parties, the model could have been misapplied or the results misinterpreted. However, if actual conditions did, in fact, depart materially from these assumptions, the modeling results would have been voided. The assumptions were:

- The data upon which the conceptual model was based were limited in the information they furnished about the bedrock aquifer adjacent to the site. Only general hydrogeological information was available about the bedrock aquifer beyond the property boundary. It was essential to assume that the available hydrogeological information was representative of actual on-site and off-site conditions; otherwise, the contaminant migration could not have been evaluated, and the project would not have been completed.
- The analytical solute transport equation dictated that the modeler assume a closed ground water system. For the site at hand, precipitation did recharge the aquifer, and the only discharge, a minimal one, was through the residential wells. The necessary assumption of a closed system injected a conservative bias into the modeling effort.
- A variety of contaminants had been identified in the ground water adjacent to the site. Chlorides and VHO (volatile halogenated organics) were selected as representative contaminants for the modeling effort because they represented the most conservative contaminants that could be transported.
- The analytical solute transport equation required that the source of contamination be a point with respect to the distance of contaminant

migration. Because the locale of concern with respect to contaminant migration was 2800 ft (850 m), the 200-ft (61-m) radius of the landfill was sufficiently small to view it as a point source.

- The landfill had not been altered at the time of this investigation; therefore, it was assumed that the source of contamination delivered a constant load.
- The flow of ground water from the landfill was generally toward the north and northwest. No information was available to determine the direction of flow beyond the landfill property line, but the regional topography sloped toward the north, and it appeared reasonable to assume that the flow of ground water off-site was to the north.
- Because the exact direction of flow was unknown and contaminant concentrations are highest directly downgradient of a source, it was assumed that the observation points, the residential wells, were located directly downgradient of the source. This assumption was conservative, as an observation point located away from the centerline of the plume would display lower contaminant concentrations.

The theory regarding the dispersion and retardation of contaminant transport is currently under debate, yet the mathematical equation used in Prickett & Associate's model has been demonstrated in the literature to yield rather accurate representations of actual contaminant migration. Therefore, it was assumed that the terms of the equation describing dispersion and retardation represented actual dispersion and retardation.

The reconnaissance model was calibrated in a typical manner: the model was run, and the results were compared with data gathered from the site. The conceptual model and ground water model were modified until there was an acceptable correlation between the model's results and the site data. From Figure 7-1, it is apparent that the monitoring wells form, generally, concentric rings around the locus of the landfill with radii about 400 and 1700 ft (120 and 520 m) from the landfill's centroid. Because an analytical data base existed for the monitoring wells, two calibration points, at $x = 400$ and $x = 1700$ ft (120 and 520 m) from the source of contamination, were selected. Both calibration points were assumed to be located directly downgradient of the source of contamination. For model calibration, a time period of 20 years was used, as that was the length of time since the landfill had begun to be used. To calibrate this statistical model, the maximum contaminant concentration in each monitoring-well ring was compared with the result of the ground water model. The model's result indicated, with a 99 percent probability, that the concentration would be equal to or below a certain value. The model was modified by varying the loading rates of the contaminants, as information on this variable was based on quite limited data, and variation in the loading rate directly affected contaminant concentrations. Table 7-3

TABLE 7-3. Model calibration: maximum observed value and model values at 99 percent confidence level.

	CHLORIDE (ppm)		VHO (ppb)	
DISTANCE	WELL	MODEL	WELL	MODEL
400 ft (120 m)	80	7400	4400	4400
1700 ft (520 m)	1680	1680	7.5	364

displays the highest levels of concentration observed at the monitoring wells. It also presents the results of the model from the final calibration.

Table 7-3 demonstrates that the model results were not successfully matched to the data base. The lack of successful calibration may have reflected the fact that the data base for the conceptual and statistical model was limited, and, thus, the models did not match actual site conditions. For example, the modeling of a fractured bedrock aquifer by a porous media model may lack sufficient accuracy to account for the observed data. The observed chloride concentrations may suggest that either the source was no longer active, or the chlorides were released as contaminant slugs rather than as a constant source. In either case, the assumption of a constant source would not have been compatible with the actual site conditions.

The collection of additional data or further attempts to calibrate the model were not possible because of scheduling constraints. Were the ground water model a conventional analytical or numerical model, such poor calibration would have jeopardized the modeling effort. However, because the model was statistically based and evaluated probability of occurrence, it was still possible to utilize it. Therefore, the model was calibrated so that the results yielded at the 99 percent confidence level were greater than or equal to the observed maximum values. (Refer to Table 7-3.) This calibration, though not so accurate as anticipated, was deemed sufficient for this reconnaissance study.

The statistical model was run to simulate time 20 years and 50 years since the initiation of the landfilling operations. The 20-year time was simulated to evaluate the existing conditions, the 50-year time to evaluate future conditions. The difference, 30 years, is considerable but was chosen because of the limited data base of the model and the need to be conservative in assessing the potential migration of the contaminants off-site. The regulatory guideline for VHO was 100 ppb; the drinking water standard for chloride was 250 ppm. The product of the modeling is presented in Table 7-4.

TABLE 7-4. Model results.

DISTANCE: $x = 2800$ ft (850 m)				
	CHLORIDE		VHO	
TIME	A	B	A	B
t = 20 years (present)	875 ppm	90%	50 ppb	99.5%
t = 50 years	1650 ppm	67%	550 ppb	92%
Regulatory guideline	250 ppm		100 ppb	

NOTES: A = 99% probability that the concentration is below this value
B = Probability the concentration is below regulatory standards or guidelines.

The model results identified a range of potential levels of contamination that could be expected at the observation points beyond the landfill site, and they provided the probability of occurrence of specific concentrations. Although the potential concentration of VHO at $x = 2800$ ft (850 m) included values very much in excess of the 100 ppb regulatory guideline, there was only an approximately 5 to 10 percent probability that the guideline would be exceeded. For chloride, there was a 10 to 30 percent probability that the concentration would exceed the drinking water standard of 250 ppm, but a worst-case evaluation would result in a concentration greatly in excess of the standard.

The information from the model furnished a quantitative assessment of the variable conditions of the site. Therefore, the management decision-making process was approached with substantial confidence. The modeling effort indicated that residences 2800 ft (850 m) from the landfill had less than a 10 percent probability of significant contamination from VHO in the near future. The probability of chloride affecting the residential wells was greater, but still below 30 percent. The conclusion was that no imminent risk existed, but the landfill owner was advised to undertake a feasibility study in order to evaluate alternative remedial measures to prevent potential future deleterious effects.

Interpretive Study. An industrial facility was located above a high-yielding sand and gravel aquifer. Prior to environmental restrictions, the facility had discharged industrial waste water containing 1,1,1-trichloroethane into a leach field. The aquifer became contaminated, and a neighboring municipal well used for potable water was subsequently closed. The owner of the facility ceased discharges and performed a hydrogeological investigation to

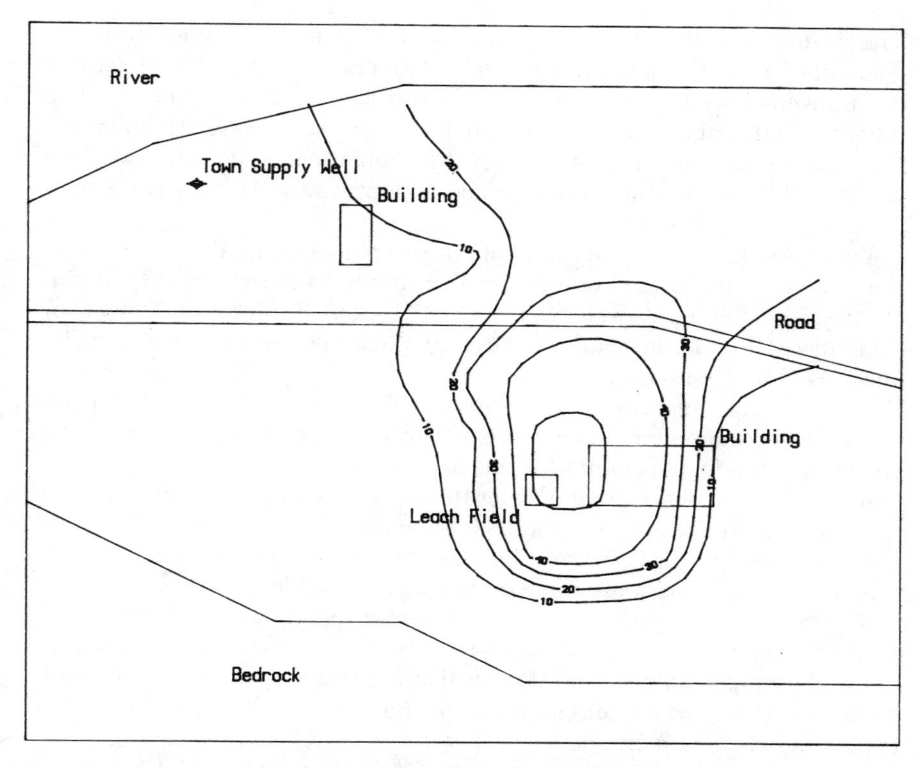

Figure 7-2. 1,1,1-Trichloroethane plume in sand and gravel aquifer.

delineate the horizontal and vertical extent of the contamination. A plume of contamination originating from the industrial facility was identified and is illustrated in Figure 7-2. The discharge had ended five years prior to the investigation, and the aquifer had a high hydraulic conductivity; therefore, the contaminant plume was expected to have migrated away from the site. Nevertheless, data indicated that the plume was still located at the source. The apparent discrepancy between the conceptual model and the site data had to be resolved prior to the evaluation of remedial options. A ground water model was developed to evaluate the conceptual model in light of the data base developed during the investigation.

The detailed hydrogeological investigation had defined the extent of ground water contamination and the direction and rate of ground water flow both on-site and off-site. The investigation also provided the basis for the conceptual model. According to the conceptual model, the unconfined alluvial sand and gravel aquifer was found to be about 70 ft (20 m) thick and located in a bedrock valley. The saturated thickness was about 50 ft (15 m).

The ground water flow was north toward a river and had an average hydraulic gradient of 2.4×10^{-3} ft/ft (0.77 m/m). An aquifer performance test defined the transmissivity of the aquifer as 124,800 gal/day/ft (1550 m^2/d). The presence of the municipal well downgradient of the source of the contamination had accelerated the migration of the contaminants, but after the well was closed, the contaminants continued to flow toward the river at a rate of about 2.7 ft/day (0.8 m/d).

The shape and movement of a contaminant plume primarily depend upon the velocity of the ground water flow. The source had been active for about 500 days, and the supply well had been sealed about 30 days later. At a rate of 2.7 ft/day (0.8 m/d), the center of the contaminant plume was expected to be about 1400 ft (425 m) from the site. Figure 7-2 indicates a plume centered at the source. The objective of the modeling effort was to simulate the migration of the contaminant plume in order to test and, as needed, to modify the conceptual model to account for this discrepancy.

In many situations, a ground water flow model is initially developed, and then a contaminant transport model is appended to the flow model. At this site, the ground water flow was relatively uniform toward the river, and with this uniformity, it was determined that a transport model only was sufficient. There are few well-documented, tested, and readily available transport models. Because an extensive data base was available and the objective was to test the conceptual model, a statistical model was not sufficiently detailed. Numerical transport models are available, but because the accessible information indicated that ground water flow was relatively uniform, it was more economical to use an analytical version of the Random Walk model developed by Prickett, Naymik, and Lonnquist.[38] The Random Walk model simulates contaminant transport in a two-dimensional hydrogeological system. Its input variables are listed in Table 7-5.

Utilizing the data of Table 7-5, the migration of the contaminant plume to date was simulated. If the conceptual model were correct, the simulation should have reflected the initial rapid migration of the contaminant to the municipal well; after the well was closed and the discharge terminated, the velocity of the plume's migration should have shown a decline, and the plume's center should have moved toward the river. See Figure 7-3. Parts (a) and (b) of the figure illustrate the results of the ground water model and show the contaminant rapidly reaching the municipal well. Once the municipal well was closed, the modeling showed that the contaminant plume migrated away from the industrial facility as a single mass, as illustrated in part (c) of Figure 7-3. However, as noted earlier, the data collected from the site indicated that the contamination continued to issue from the industrial facility. Therefore, neither the ground water model nor the conceptual

TABLE 7-5. Input parameters for interpretive ground water model.

PARAMETER	VALUE	
Transmissivity	124,800 gal/day/ft	1550 m^2/d
Specific yield	0.2	
Hydraulic conductivity	2,600 gal/day/ft^2	0.12 cm/s
Porosity	0.3	
Longitudinal dispersivity	60 ft	18 m
Transverse dispersivity	20 ft	6 m
Retardation coefficient	2 ft	61 cm
Regional ground water flow *(x)*	0.94 ft/day west	29 cm/d west
Regional ground water flow *(y)*	2.57 ft/day north	78 cm/d north
Particle mass	1.8 lb	0.8 kg
Source discharge	22,400 gal/day	85 m^3/d
Supply well discharge	576,000 gal/day	2,180 m^3/d

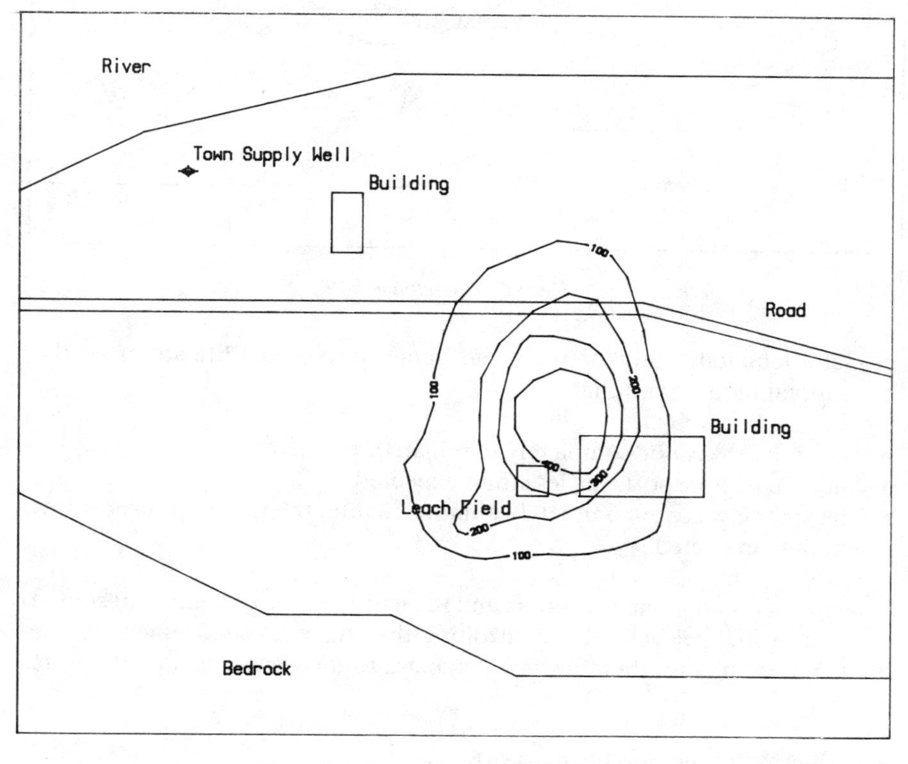

Figure 7-3. Plume of 1,1,1-trichloroethane.

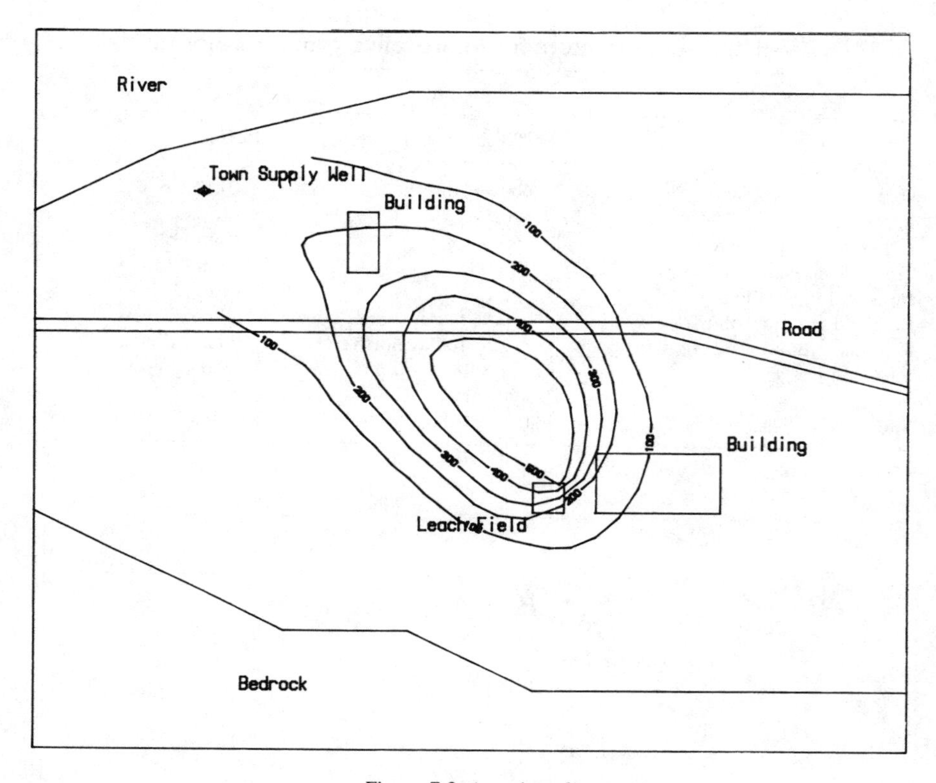

Figure 7-3. *(continued).*

model adequately simulated the site data. Possible modifications of the conceptual model were that:

- The source was not completely eliminated.
- The rate of transport was less than expected.
- The source was not completely eliminated, and the rate of transport was less than expected.

The rate of contaminant transport and the nature of the contaminant source were the principal variables controlling the shape and movement of the contaminant plume; therefore, two changes to the conceptual model were evaluated:

- Reduction in the rate of transport.
- Consideration of a potential continuous low-level source of contamination.

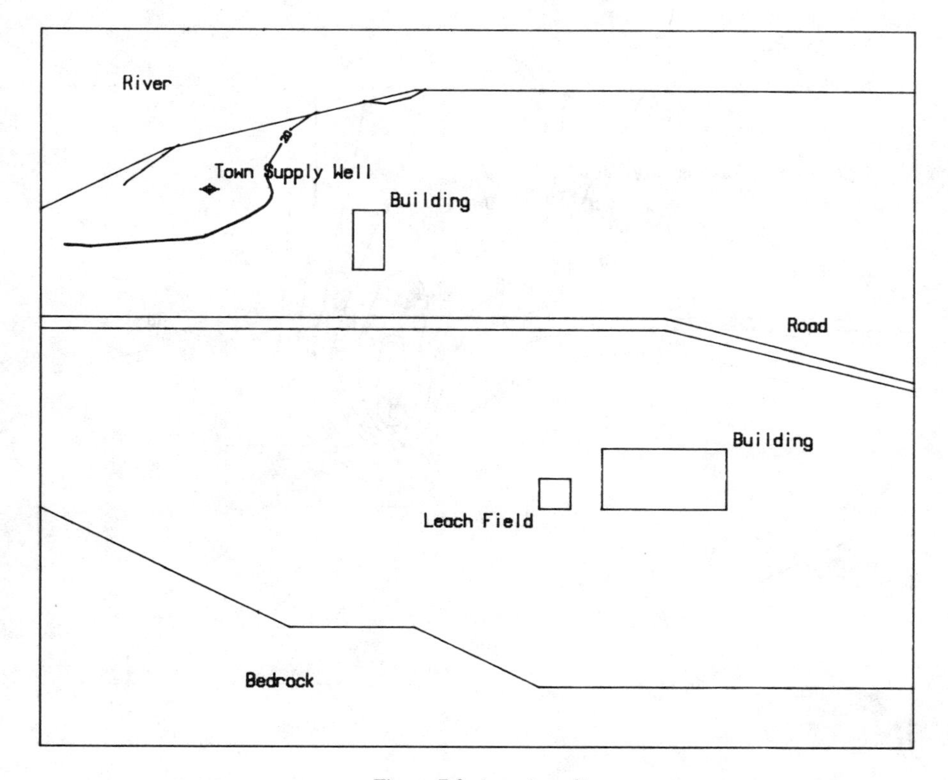

Figure 7-3. *(continued).*

The contaminant transport rate could have been reduced by using a lower hydraulic conductivity for the aquifer or a greater contaminant retardation. Because the hydraulic conductivity was well established through an aquifer performance test, the retardation rate was chosen for modification, and it was increased from 2 to 10 ft (0.6 to 3.0 m).

The results of the model utilizing a retardation coefficient of 10 ft (3.0 m) still did not reflect actual conditions, as they indicated that the contamination would not have reached the supply well, as it had, in fact, done. Therefore the presence of a low-level constant source subsequent to the original discharge was evaluated. Figure 7-4 depicts a plume at the time of the modeling effort moving under the new assumed circumstances. These results compared favorably with the site data and suggested that a remnant or secondary source of contamination persisted at the site. The modified ground water model now simulated known site conditions.

The apparent discrepancy between the site data and the conceptual

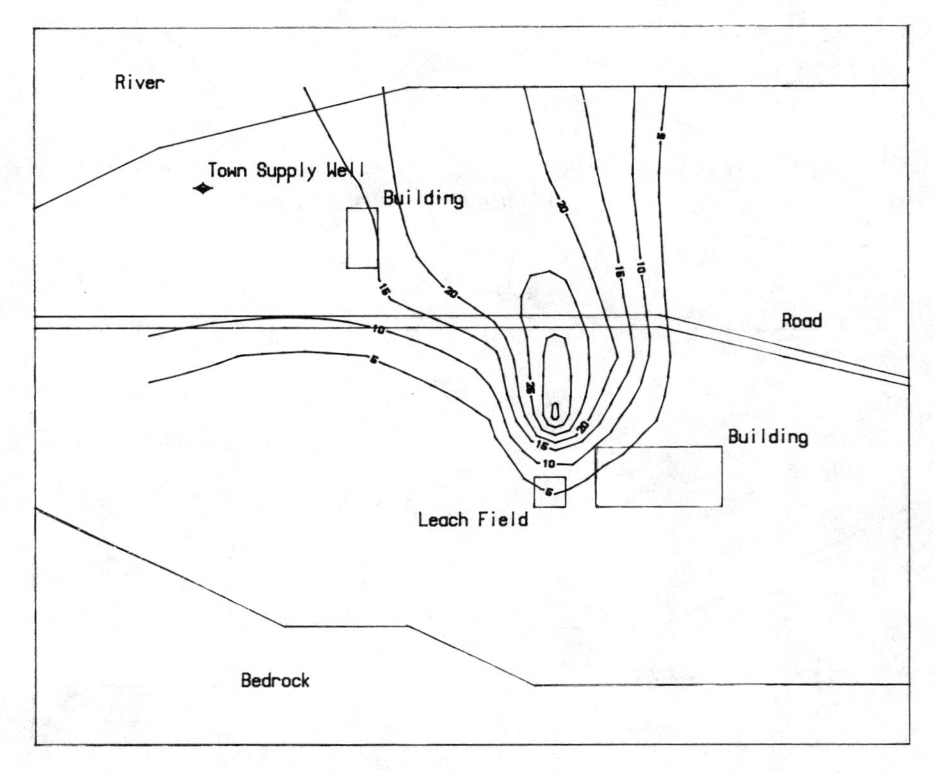

Figure 7-4. Simulation of constant low level source of contamination.

model having been resolved, the model was then employed in a predictive study to evaluate the remedial option of installing a ground water recovery well. The predictive simulations indicated that an on-site recovery well would control and collect the contaminated ground water in a more effective manner than an off-site well. Based on this predictive study, a recovery well was installed on the industrial property; see Figure 7-5.

Predictive Study. An industrial and municipal landfill was the subject of an extensive investigation that disclosed that hazardous waste had been deposited in the landfill, and organic chemicals had contaminated the ground water in the vicinity. A 48-inch (122-cm) low permeable cap and a circumferential slurry wall linked below grade to a clay formation with low conductivity were proposed to inhibit the migration of the organic chemicals. However, the regulatory agency overseeing work on the site had to be convinced that such containment would mitigate the ground water contamination. Therefore,

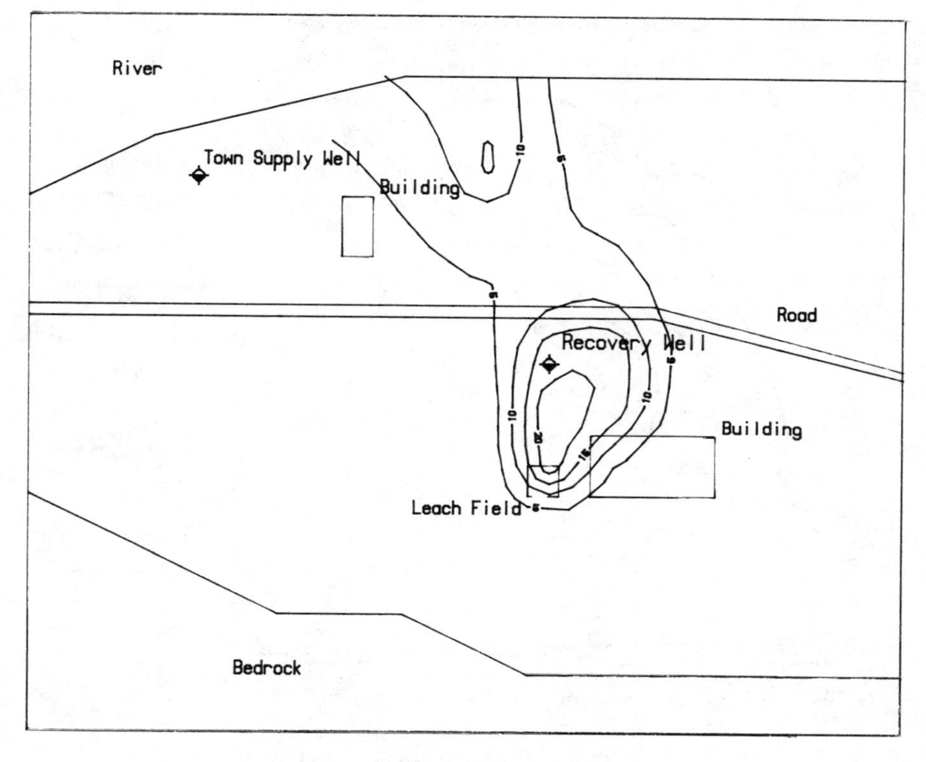

Figure 7-5. Effects of recovery well on contaminant plume.

the future flow of the ground water and transport of contaminants had to be predicted. Ground water modeling was used to provide the technical justification necessary to support the proposed remedial alternative.

The hydrogeological investigation completed at the site provided a satisfactory data base for a conceptual model. The landfill material had been deposited on the surface of a sand quarry. Immediately beneath the landfill material was a medium-grained sand unit ranging from 25 to 30 ft (7.5 to 9 m) thick. Beneath the sand unit was a lacustrine clay unit with low hydraulic conductivity, less than 10^{-7} cm/sec, which was 15 to 70 ft (5 to 21 m) thick. (See Figure 7-6.) Ground water flowed principally in the sand unit in southerly and southwesterly directions. The hydraulic conductivity of the sand unit, about 100 gal/day/ft^2 (4.7×10^{-3} cm/s), was determined through an aquifer performance test on the sand unit. Laboratory and in situ permeability tests on the clay unit provided the basis for estimating the hydraulic conductivity of the clay.

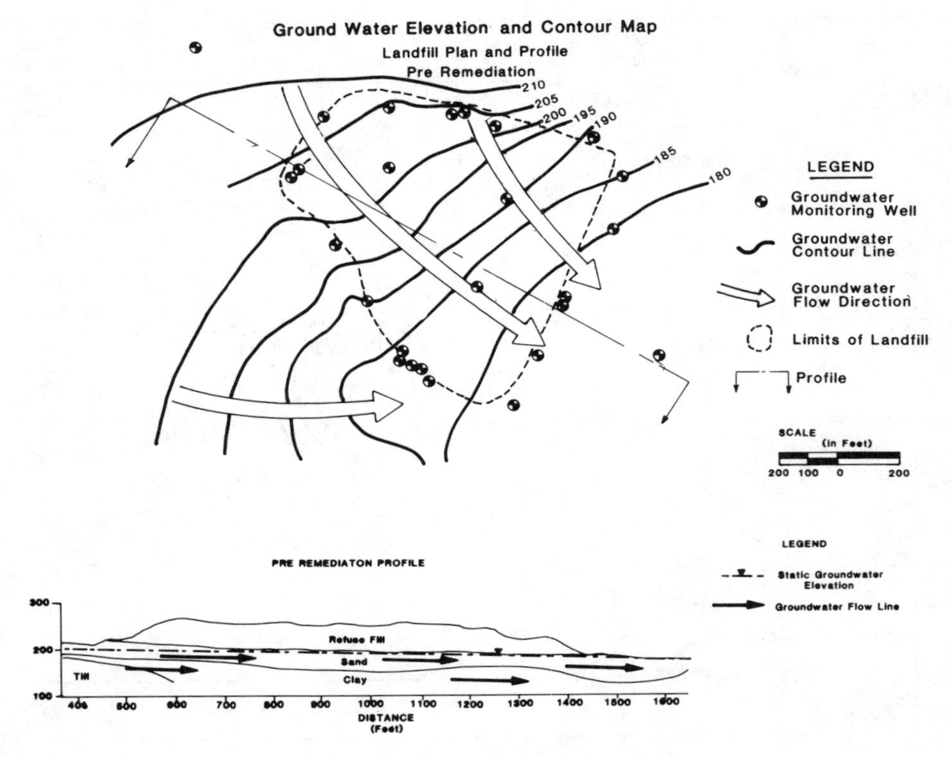

Figure 7-6. Hydrogeological conditions of landfill.

Encapsulation (addressed in detail in Chapter 9) is a practical remedy because it reduces the flow of ground water into, through, and from a landfill. Because encapsulation relies on the control of ground water flow rather than the treatment of the contaminants, flow models were considered more appropriate than solute transport models to predict the future ground water flow and contaminant transport following remediation. Ground water would flow around and into the encapsulated landfill in all three spatial directions, so the model had to be able to assess multidimensional flow. Three-dimensional ground water flow models are available, but their use is time-consuming and expensive. In addition, the position of the regulatory agency did not suggest that an elaborate modeling effort was necessary to support the remedial proposal. Therefore, a relatively simple approach based upon the conceptual model of the encapsulated landfill's water budget was pursued.

A water budget for an encapsulated site incorporates the flow of ground water into the site, the flow of ground water from the site, and changes in the

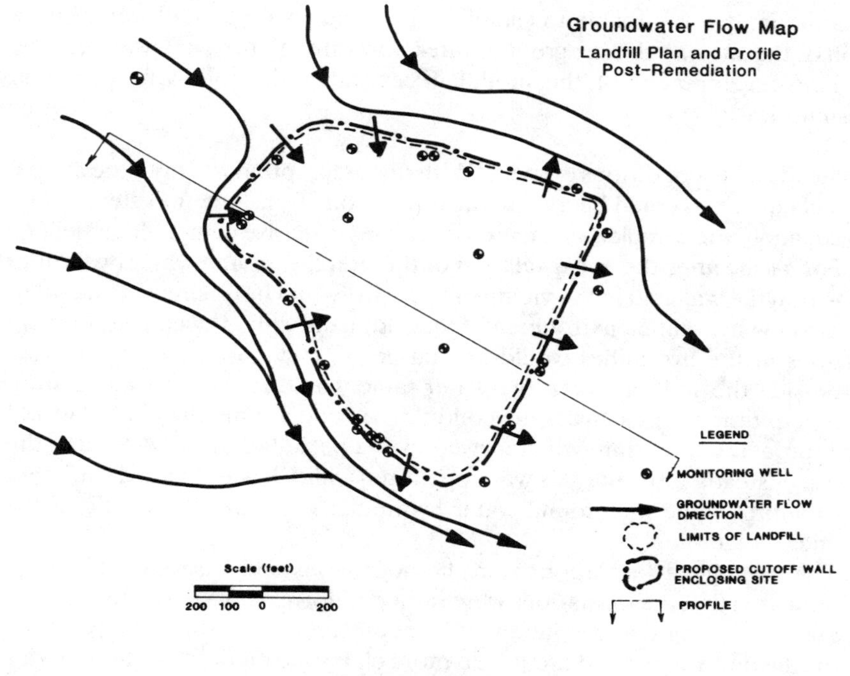

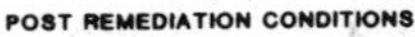

POST REMEDIATION CONDITIONS

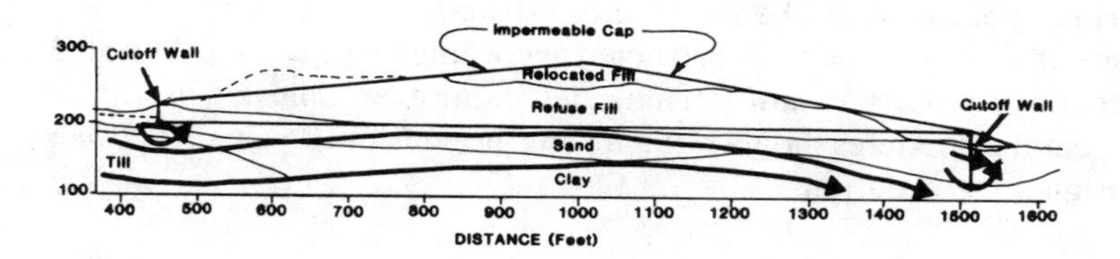

Figure 7-6. *(continued).*

volume of ground water stored within the site. Three avenues existed for ground water flow into or from the landfill site:

- Infiltration through the cap.
- Flow through the ground water cutoff wall.
- Ground water flow through the confining layer at the base.

The inflow and the outflow for each of the three avenues were evaluated separately, and the results of these evaluations were summed. With this technique, it was possible to establish the steady state ground water elevation of the site and the net ground water flow through the encapsulated site.

The development of the model necessitated the following principal assumptions:

- Steady state conditions existed. (With this assumption, steady state analytical equations could be used in the evaluation. Transient, non-steady state equations are complex, and their use requires extensive computer modeling.) For a time after the encapsulation of the site, the ground water hydraulics within the wall and in the vicinity of the site would be changing to adapt to the new hydraulic environment. Once adjustment had occurred, fluctuations in the hydraulics would be caused only by ground water changes outside the wall, as when the water table would react to recharge from precipitation. Thus, hydraulic conditions in and around the landfill would approach steady state when viewed over a period of years. Although the use of steady state analysis was not entirely valid, it was deemed sufficient to approximate the ground water hydraulics over the 30-year life of the encapsulation.
- The geological formations were homogeneous and isotropic. The data indicated that the formations were rather uniform, although all formations are, to varying degrees, nonhomogeneous and anisotropic. Accounting for the deviations would require a more elaborate model, but the superior accuracy was deemed superfluous for this study.
- Uniform materials would be used for construction of the encapsulation, and the cutoff wall would be constructed according to the design. Although perfect construction and uniform materials cannot be achieved, proficient construction procedures should result in only insignificant deviation between the design and execution.

The assumptions facilitated an adequate simulation of ground water hydraulics as they would prevail after construction of the encapsulation, yet they prevented the simulation from becoming overwhelmed by the vast complexities inherent in a more precise definition.

Precipitation was the principal source of ground water recharge in the vicinity of the landfill. To minimize the recharge, a cap with hydraulic conductivity less than 10^{-7} cm/s was designed to be placed atop the landfill. The cap would foster runoff and limit infiltration. To evaluate the potential ground water recharge through the cap, a water budget for the cap was examined. The rates of potential runoff and potential evapotranspiration were compared with the rate of precipitation to determine the recharge rate through the cap. The water budget indicated that the rate and depth of infiltration into the cap would be significantly less than the rooting depth of the vegetative layer: 0.4 inch compared with 6 inches (1 cm vs. 15 cm), respectively. It also indicated that the potential evapotranspiration rate of the vegetation would exceed the precipitation for a portion of the year. Therefore, no inflow into the landfill would occur.

The cutoff wall was designed as a barrier to ground water flow, and the material of construction had a hydraulic conductivity of less than 10^{-7} cm/sec. However, ground water flow, albeit minuscule, would occur. The flow to be expected was calculated using Darcy's equation for flow (a concept introduced in Chapter 3):

$$Q = K \frac{dh}{dl} A$$

In that equation, K = hydraulic conductivity; dh = difference in head (water table elevations) on either side of cutoff wall; dl = length of flow path through cutoff wall; A = vertical cross section of portion of cutoff wall affected (area = LH: L = length of cutoff wall, and H = height of portion of cutoff wall in contact with saturated sand). Inflow would be possible if the hydraulic head were higher outside the cutoff wall than inside it; outflow would be possible if the hydraulic head outside the wall were lower than that inside.

The cutoff wall was keyed into the low-permeability clay unit beneath the sand unit. Because ground water flowed through the clay, although in minute amounts, ground water would flow into and from the encapsulated mass. The rates of inflow and outflow through the clay were calculated. A flow net was constructed, and the following calculation was used:

$$Q = \frac{m\,K\,h}{n} \mathrm{L}$$

where m = number of flow tubes in the flow net; n = number of divisions of head in flow net; K = hydraulic conductivity; h = difference in head in clay between interior and exterior of cutoff wall; L = length of section beneath cutoff wall. Ground water could flow into the landfill through the confining layer where the hydraulic head in the clay outside the wall was higher than that inside it. Ground water outflow could occur where the hydraulic head of the clay within the limits of the wall was higher than that outside.

For a more accurate evaluation of the potential flow, the area of investigation for flow both through the wall and through the clay unit was divided into segments based upon the magnitude and direction of the head differences. The inflow and outflow rates were summed for each segment, and the result was an estimate of the total potential inflow and outflow. The calculations were made using an assumed constant hydraulic head within the limits of the cutoff wall.

The total inflow and outflow through the cutoff wall and the confining layer were integrated into the total ground water budget. Repetitive calculations of flow were made using different constant hydraulic heads within the wall in order to establish the equilibrium ground water elevation of the encapsulated site. When the calculated total inflow and outflow were equivalent, the site's ground water elevation was assumed to be at equilibrium. Inflow was projected to occur along the western portion of the landfill at equilibrium, and outflow was along its eastern portion. Table 7-6 presents the summation of the site water budget following encapsulation.

The summation represented the total ground water flow into and from the site. The net change in the site water budget reflected whether ground water was being added to or subtracted from storage. When the total inflow equaled the total outflow, the net change in storage was naught; under those circumstances, the ground water budget was in equilibrium, and a stable ground water elevation had been achieved.

The ground water budget for the landfill site attested that a ground water elevation within the limits of the cutoff wall of between 193 and 195 ft (58.8 and 59.4 m) would result in a balanced ground water flow. To be conservative,

TABLE 7-6. Ground water budget

	RATE OF INFLOW AT GROUND WATER ELEVATION:			
SOURCE OF INFLOW	193 ft (gal/day)	195 ft (gal/day)	58.8 m (L/d)	59.4 m (L/d)
Inflow through cutoff wall	842	723	3200	2747
+ Inflow through base	531	432	2018	1642
Total Inflow	1373	1155	5218	4389
	RATE OF OUTFLOW AT GROUND WATER ELEVATION:			
SOURCE OF OUTFLOW	193 ft (gal/day)	195 ft (gal/day)	58.8 m (L/d)	59.4 m (L/d)
Outflow through cutoff wall	− 899	−1071	−3416	−4070
+ Outflow through base	− 177	− 207	− 673	− 787
Total Outflow	−1076	−1278	−4089	−4857
NET GROUND WATER BUDGET	+297	−123	+1129	−467

the equilibrium ground water elevation was assumed to be 195 ft (59.4 m). Flow was calculated thus:

$$Q = K\,i\,A$$

where K = hydraulic conductivity (= 50 to 100 gal/day/ft^2 [2.4×10^{-3} to 4.7×10^{-3} cm/s]); i = hydraulic gradient (= 0.033 length/length); A = cross section (= 900 ft · 35 ft = 31,500 ft^2 [274 m · 11 m = 3014 m^2]).

The estimated outflow from the area within the limits of the cutoff wall at the landfill site, when equilibrium conditions existed, was projected to be 1200 gal/day (4560 L/d). Before remediation, the estimated ground water flow beneath the horizontal limits of the landfill was 52,000 to 104,000 gal/day (197,600 to 395,200 L/d). Therefore, the remedial construction was projected to result in a 98 to 99 percent reduction in the flow of ground water beneath the horizontal limits of the landfill. The remaining preconstruction flow would move around the encapsulated site and thus be diverted from the landfill. This reduced flow was projected to diminish significantly the rate of contaminant migration from the site. Therefore, the remedial design was adjudged capable of mitigating effectively any significant current and future releases or migration of hazardous wastes from the site. In addition, the ground water modeling effort identified the equilibrium ground water elevation within the site, and it thus facilitated the actual design of the elevation of the interface between the cutoff wall and the cap downgradient of the site. In sum, the ground water modeling effort provided the necessary technical basis to demonstrate that the proposed remediation would successfully attenuate current and future contaminant releases from the site.

ASSESSMENT AND SUMMARY

Ground water modeling is a tool that offers many significant benefits in the management of hazardous waste sites. It facilitates understanding of the processes affecting ground water flow and contaminant migration, and records the method of study for interested parties. The ground water model provides a systematic and well-defined method for evaluating conceptual models. As the means by which the hydrogeologist or engineer describes the actual physical system, the ground water model enables the organization and manipulation of large amounts of information. Ground water models, notably numerical models, can rapidly and systematically evaluate inhomogeneous, multidimensional, and multiphase hydrogeological systems, being several orders of magnitude better than techniques previously available to the hydrogeologist to interpret data and to test conceptual models. Because a ground water model is a controlled, systematic simulation, results can be duplicated, and individual variables can be altered to evaluate specific components of hydrogeological processes.

Ground water models, because of their organized and systematic properties, also afford a clear and detailed record of the modeling effort. The conceptual model is clearly documented in the input variables and in the conditions of the model. The theory by which the conceptual model is evaluated is documented in the model equations and the method of solution used. With this documentation, a ground water model and thus a conceptual model can be reviewed or even tested by other parties. Effective visual images and presentation aids can be developed from models, too. These images assist all pertinent parties, not the least of which is the public, in understanding the circumstances of contamination and the effects of remedial activities.

Finally, the financial and legal liabilities and the environmental risks inherent in hazardous waste investigations mandate that project designers make critical management decisions regarding the direction and design of inquiry. Information from modeling that accredits management decisions is valuable.

Ground water models, as do all tools, have limits that must be considered prior to and during their implementation. Their limitations are due to the theory and code of the model itself or the model's application. Limitations in the basic equations, in the numerical methods used to solve the equations, or in both must be addressed because the equations represent the physical, chemical, and biological processes upon which the model is based. The mathematical equations used in the model must correctly describe the processes that govern ground water flow and contaminant transport. If assumptions are used to reduce the equations for more manageable problem solving, these assumptions must correspond to known processes and site conditions. Incorrect equations and assumptions result in incorrect modeling results.[39] The computer code used to evaluate the mathematical model may have defects or mathematical errors that can significantly affect the results of the model. A model code that had not been adequately tested could clearly limit the accuracy and value of the model's results.

The application of a ground water model suffers if the data base for the model is limited or inadequate. The size and detail of the necessary data base generally correlate with the complexity of the ground water model employed. Complex, three-dimensional, numerical models demand a far more comprehensive data base than do simpler analytical models. The complex numerical model may be used with a limited data base only if simplifying assumptions are applied. Such simplification compromises the advantages of the complex model, is uneconomical, and can present the deceptive impression of a detailed understanding of the hazardous waste problem.

A final limitation of ground water models affects their ability to predict hydrogeological responses to existing or future stresses. Although ground water modeling has been conducted for about two decades, there has been little auditing of models' predictions to evaluate their accuracy. Some audits have found that short-term predictions of well-defined hydrogeological systems tend to be much more accurate than long-term predictions of complex systems.[40] As ground water modeling at hazardous waste sites often involves predictions ranging ten years or more, the results of such modeling must be carefully evaluated in light of this limitation.

Real aquifers do not fit the idealized assumptions of some model equations, and complete and correct mathematical equations for all processes do not yet exist. Nonetheless, models can approximate aquifer conditions with sufficient accuracy for many hazardous waste problems if they are employed with an awareness of their limitations. All ground water models are tools; hence, the selection and application of the appropriate ground water model should be based upon the goal desired and the data base available. A model will produce a result, regardless of whether the model accurately represents the true hydrogeological conditions. Therefore, all modeling should be undertaken in a scientific manner to assure that the results satisfactorily represent actual site conditions and provide information commensurate with the modeling goals.

NOTES

1. M. P. Anderson. Using models to simulate the movement of contaminants through groundwater flow systems. *Critical Reviews in Environmental Control.* 9(1979):97-156.

2. C. A. Appel and J. D. Bredehoeft. Status of groundwater modeling in the U.S. Geological Survey. USGS, Circular 737. 1979.

3. Y. Bachmat, et al. (Eds.). *Groundwater management: the use of numerical models.* American Geophysical Union, Water Resources Monograph Series-5. 1980.

4. C. R. Faust and J. W. Mercer. Ground-water modeling: recent developments. *Ground Water.* 18(1980):569-577.

5. C. R. Faust. The use of modeling in monitoring network design. *Proceedings* Nat. Symposium on Aquifer Restoration and Groundwater Monitoring, D. M. Nielsen (Ed.), pp. 156-162. 1982.

6. R. A. Freeze and J. A. Cherry. *Groundwater.* Englewood Cliffs, N.J: Prentice-Hall, 1979.

7. J. W. Mercer and C. R. Faust. Ground-water modeling: an overview. *Ground Water.* 18(1980):108-115.

8. J. W. Mercer and C. R. Faust. Ground-water modeling: mathematical models. *Ground Water.* 18(1980):212-227.

9. J. W. Mercer and C. R. Faust. Ground-water modeling: applications. *Ground Water.* 18(1980):486-497.

10. T. A. Prickett. Groundwater computer models—state of the art. *Ground Water.* U17 #2(1979):167.

11. T. A. Prickett. Modeling techniques for groundwater evaluation. *Advances in Hydroscience,* V. T. Chow. (Ed.). New York: Academic Press. 10(1975):1-143. (General state-of-the-art of models to 1973.)

12. H. F. Wang and M. P. Anderson. *Introduction to groundwater modeling.* San Francisco: W. H. Freeman & Co., 1982.

13. The International Ground Water Modeling Center, Holcomb Research Institute at Butler University in Indianapolis, Ind. maintains a listing of all documented and available ground water flow and transport models.

14. J. H. Lehr. Groundwater flow models simulating subsurface conditions. *Journal of Geological Education.* 11(1963):124-132.

15. W. C. Walton and T. A. Prickett. Hydrogeologic electric analog computers. *Proceedings* ASCE, Paper 3695, 89(HY6). 1963.

16. L. W. Gelhar, P. Y. Ko, H. H. Kwai, and J. L. Wilson. Stochastic modeling of groundwater systems. MIT, Ralph M. Parsons Laboratory Report No. 189. 1974. (Stochastic technique.)

17. International Ground Water Modeling Center. Mass transport models which are documented and available, Rep. GWM 82-02, Holcomb Research Institute, Butler University, Indianapolis, Ind. 1983.

18. K. Javandel, I. C. Doughty, and C. F. Tsang. *Groundwater transport: handbook of mathematical models.* American Geophysical Union, Water Resources Monograph 10. 1984.

19. L. F. Konikow and J. D. Bredehoeft. Computer model of two-dimensional solute transport and dispersion in ground water. USGS, *Techniques of Water-Resources Investigations,* Book 7, Ch. C2. 1978.

20. G. F. Pinder. A digital model for aquifer evaluation. USGS, *Techniques of Water-Resources Investigations,* Book 7, Ch. C1. 1970. (Finite difference.)

21. T. A. Prickett, T. G. Naymik, and C. G. Lonnquist, A random-walk solute transport model for selected ground water quality evaluations, Bull. 65, Illinois State Water Survey, Champaign. 1981.

22. T. A. Prickett and C. G. Lonnquist. Selected digital computer techniques for groundwater resources evaluation, Illinois State Water Survey Bull. 1971.

23. L. Smith and F. W. Schwartz. Mass transport, 1: A stochastic analysis of macroscopic dispersion. *Water Resources Research.* 16(1980):303-313.

24. L. Smith and F. W. Schwartz. Mass transport, 2: Analysis of uncertainty in prediction. *Water Resources Research.* 17(1981):351-369.

25. L. Smith, and R. A. Freeze. Stochastic analysis of steady state groundwater flow in a bounded domain, 2: Two-dimensional simulations. *Water Resources Research.* 15(1979):1543-1559.

26. P. C. Trescott, G. F. Pinder, and S. P. Larson. Finite-difference model for aquifer simulation in two dimensions with results of numerical experiments. USGS, *Techniques of Water-Resources Investigations,* Book 7, Ch. C1. 1976. (Finite difference.)

27. F. W. Bond, C. R. Cole, and D. Sanning. Evaluation of remedial action alternatives demonstration/application of groundwater modeling. *Proceedings* Nat.

Conf. Mgmt. Uncontrolled Hazardous Waste Sites. Silver Spring, Md.: Hazardous Material Control Research Institute, 1982.

28. A. B. Gureghian, D. S. Ward, and R. W. Cleary. A finite element model for the migration of leachate from a sanitary landfill in Long Island, New York—Part II: Application. *Water Resources Bulletin.* 17(1981):62-66.

29. L. F. Konikow. Modeling chloride movement in the alluvial aquifer at the Rocky Mountain Arsenal, Colorado. USGS Water-Supply Paper 2044. 1977.

30. T. G. Naymik. Modeling as a tool in monitoring well network design. *Proceedings* Second Nat. Symposium on Aquifer Restoration on Ground Water Monitoring, D. M. Neilsen (Ed.), NWWA, pp. 151-155. 1982.

31. T. G. Naymik and M. J. Barcelona. Characterization of a contaminant plume in groundwater, Meredosia, Ill. *Ground Water.* 19(1981):517.

32. J. F. Pickins and W. C. Lennox. Numerical simulation of waste movement in steady groundwater flow systems. *Water Resources Research.* 12(1976).

33. G. F. Pinder. A Galerkin-finite-element simulation of ground water contamination of Long Island, New York. *Water Resources Research.* 9(1973):1657-1669.

34. J. B. Robertson. Digital modeling of radioactive and chemical waste transport in the Snake River Plain Aquifer at the National Reactor Testing Station, Idaho. USGS open-file report. 1974.

35. S. G. Robson. Application of digital profile modeling techniques to ground water solute transport at Barstow, Calif. USGS Water Supply Paper 2050. 1978.

36. L. R. Silka and J. W. Mercer. Evaluation of remedial actions for groundwater contamination at Love Canal, New York. *Proceedings* Nat. Conf. Mgmt. Uncontrolled Hazardous Waste Sites. n.d.

37. J. L. Wilson and P. J. Miller. Two-dimensional plume in uniform ground-water flow. *ASCE Journal of the Hydraulics Division* 104, (1978)#HY4:503.

38. T. A. Prickett, T. G. Naymik, and C. G. Lonnquist, A random-walk solute transport model for selected ground water quality evaluations, Bull. 65. Illinois State Water Survey, Champaign. 1981.

39. F. W. Gillham. Applicability of solute transport models to problems of aquifer rehabilitation. *Proceedings* 2d Nat. Symposium on Aquifer Restoration and Ground Water Monitoring, NWWA. n.d.

40. L. F. Konikow. Predictive accuracy of a ground-water model-lesson from a postaudit. *Ground Water.* 24(1986):173-184.

PART II
REMEDIATION

Chapter 8
Developing the Feasibility Study

Uncontrolled hazardous waste sites are sometimes a source of jeopardy to human health and the environment. Diverse measures are available to mitigate or eliminate such dangers, but the circumstances at hazardous waste sites are complex with no two sites alike, and there is no single solution or set of answers to such a site. When confronted with an uncontrolled hazardous waste site, the engineering designer investigates the conditions as thoroughly as is practicable and seeks appropriate solutions. The remedial measures are chosen selectively and adapted to each specific site. The method used to find remedial measures is as complicated as the factors are at the site itself because the method considers all the aspects of the site and their prodigious interrelationships. The measures must be technically feasible; they also must be economical and in consonance with the expectations of the community of nearby residents. The goal of the solution is to mitigate the risks to human health and the environment.

When a solution to a site is sought, several alternative measures are posed. The examination and evaluation of these alternatives are often referred to as a feasibility study,[1] and this study follows the remedial investigation—terms broached in Chapter 1. Five essential steps are followed when performing a feasibility study:

1. Define the objectives of the remedial action.
2. Identify the technologies that can achieve the objectives.
3. A. Pose alternatives; the alternatives are composed of one technology or, more often, a set of technologies, and each alternative is capable of correcting the contamination problem on the site.
   B. Winnow the list of alternatives, and select only those that are superior based upon engineering, environmental, and economic criteria and that meet the expectations of neighboring communities.
4. Scrutinize each of the selected alternatives.
5. Record the method and results of the analysis.

This chapter was developed by Steven R. Garver, P. E., Douglas M. Crawford, and Therodore J. Jenczewski, Ph.D., of O'Brien & Gere Engineers, Inc.

All aspects of the feasibility study are controlled by government regulations. (See Chapter 1 for information on the Superfund Amendments and Reauthorization Act [SARA] and the National Contingency Plan [NCP]). Regulatory agencies and the U.S. EPA in consultation with state environmental and public health agencies generally must endorse the alternative ultimately implemented. However, the selected alternative also can result from a negotiated agreement or court settlement between the regulatory agencies and the owner or users of the waste site.

The response to an uncontrolled hazardous waste site addresses the three major components of risk:

- The contaminant sources, recognizing the matrices affected.
- The transport routes that accommodate the movement of contaminants.
- The receptors that are affected by the contaminants.

Sources of contamination include surficial or buried debris; liquid wastes; contained wastes; soils; structures, tanks, or other artifacts; sediments; and air, including air in soil pores. Routes available to transport contaminants, in addition to direct contact, include air for volatile or dispersible components, surface water, and ground water. The receptors that contaminants might reach include human populations, domestic or wild animals, and plant communities.

For the purposes of this book, the process employed in developing a feasibility study is best communicated through an example, the ABC Landfill, an abandoned industrial waste disposal site. The problems associated with ABC Landfill have been made simpler than those that would be encountered at an actual industrial landfill in order to illustrate better the sequence of tasks and the evaluation procedure employed in a feasibility study. In practice, hazardous waste sites often are multifarious, and a substantial effort is demanded to analyze and to develop a remedial program capable of meeting the public health and environmental objectives. The depth of the analysis, the volume of information to be assembled and evaluated, and the length of the feasibility study report are directly proportional to the complexity of the site.

Figure 8-1 depicts the uncontrolled hazardous waste site that serves as the prototype for this discussion. The ABC Landfill had been used by the ABC Casting Company for the disposal of general refuse from the plant, casting sands, and associated waste metal residues. The extent of the fill area was defined by analysis of aerial photographs and geophysical testing. Soil borings within the active fill area indicate that the refuse was placed above the ground surface; that the ground water table is approximately 5 ft (1.5 m) below the bottom of the fill material; that bedrock is approximately 50 ft (15 m) below grade. Lead concentrations in the fill material range between

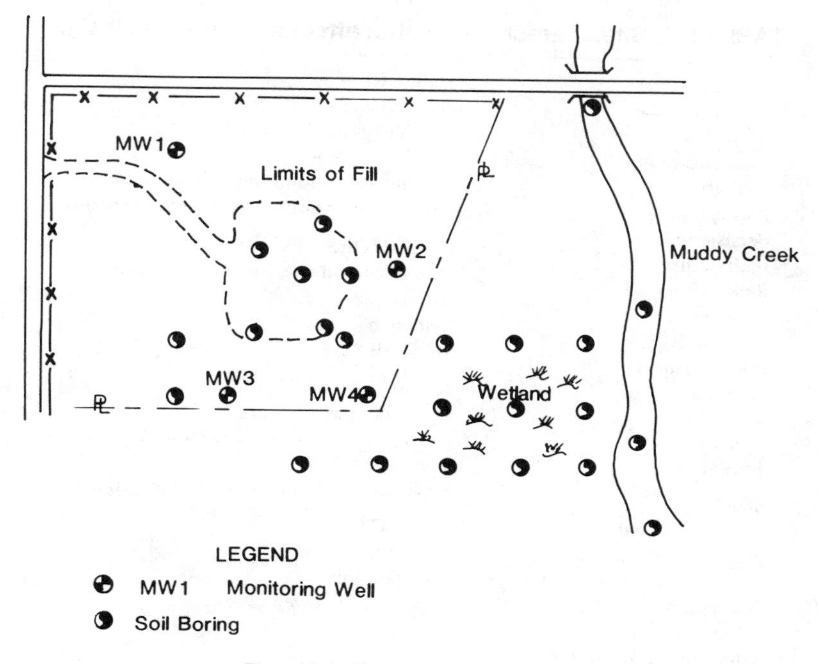

Figure 8-1. Plan view of ABC Landfill.

50 and 50,000 mg/kg dry weight. Extractable (EP Toxic) lead concentrations in the fill material exceed 5.0 mg/L; thus, the fill is a hazardous waste as defined by regulations pursuant to RCRA.[2] The concentrations of lead in the ground water sampled by the monitoring wells are below the state ground water quality standards. Surface erosion from the landfill has dispersed lead particles into the adjacent wetlands and into the sediments of Muddy Creek, but lead has not been detected in water samples taken from Muddy Creek either upstream or downstream of the ABC Landfill. The features that determine the selection of alternative corrective measures are summarized in Table 8-1.

DEFINE OBJECTIVES OF THE REMEDIAL ACTION

The design engineer must understand the site's characteristics in order to determine an approach to correct its problems. The circumstances of the site—a clear definition of the contaminant or contaminants of concern, the routes of migration, and the receptors of the contaminants—should already have been clearly defined through the remedial investigation, and the source documents describing the remedial investigation should be available to the designer performing the feasibility study. The remedial investigation also

TABLE 8-1. Site characteristics that affect the remedial design.

WASTE	SUBSURFACE
Quantity	Permeability
Composition	Depth to ground water
Concentration	Depth to bedrock
Toxicity	Ground water flow direction and rate
Biodegradability	Ground water discharge or recharge points
Persistence	RECEPTORS
Ignitability	On-site and off-site human populations
Reactivity	Animal life
Corrosivity	Critical habitats
Radioactivity	Wetlands
Infectiousness	ADJACENT LAND USES
Solubility	Upgradient waste sources
Volatility	Downgradient surface and ground water uses
Density	Residential, commercial, and industrial uses
Treatability	CLIMATE
SURFACE	Temperature
Source area and depth	Precipitation
Property boundaries	Prevailing wind velocities
Soil type and permeability	
Topography	
Vegetation	
Adjacent surface waters	

TABLE 8-2. Summary of problems, migration routes, and potential receptors at ABC Landfill.

SURFACE WATER
Lead-laden particles and soil particles have migrated and will continue to migrate via erosion and runoff from uncovered fill material, particularly during periods of heavy rain. Dispersed lead in wetlands is carried into Muddy Creek during rainfall. Lead in sediments of Muddy Creek can be carried downstream, especially during times of high stream flow.

GROUND WATER
Waste materials are not in direct contact with ground water during the highest ground water elevations observed. However, the potential exists for soluble lead residuals to leach into the water table. Existing monitoring wells do not show solubilization and ground water transport to be a current transport route.

AIR
No volatile organic compounds have been detected in air. However, the presence of putrescible refuse could release gases in sufficient quantity to disturb the integrity of a surface cap.

DIRECT CONTACT
Potential exists for human and animal life exposure to uncovered refuse and lead particles. In addition, humans or animal life could be exposed to the lead particles in the wetland.

should have defined the goals of the remedial action and have proffered generic response actions. The designer of the feasibility study examines the report of the remedial investigation thoroughly and from that information can prepare specific objectives that will govern the feasibility study.

Thus the objectives of the remediation effort are based on information on the conditions, problems, pathways, and receptors of the uncontrolled hazardous waste site. The statement of objectives clearly delineates the general and specific components of the site to be controlled. With each objective, criteria are identified that will be used to determine whether or not the objective is met because the remedial action eventually recommended must guarantee compliance with government standards. Table 8-2 summarizes the information prerequisite to the development of objectives.

Building on the information of Table 8-2, the designer establishes the purposes of the remedial effort and the standards to be used to measure achievement of those objectives. The data in Table 8-3 list performance criteria pertinent to the ABC Landfill.

TABLE 8-3. Objectives of the remedial actions and evaluation criteria.

PATH	OBJECTIVE	CRITERIA
SURFACE WATER	A. Mitigate transport of lead particles in landfill runoff.	A. [a]Maintain existing or lower lead concentrations in wetland and creek sediments.
	B. Reduce or eliminate contaminants in wetlands and creek sediments.	B. [a]Use concentrations previously documented as action level for lead in surface soils: 750 ppm.[b] (Note: A second option is to perform a site-specific risk assessment to determine an appropriate action level if no federal requirements exist.)
GROUND WATER	Mitigate lead from contaminating ground water concentrations that pose risk to public health.	Drinking water quality is based on National Interim Primary Drinking Water Regulations; maximum contaminant level for lead is 0.05 mg/L.
AIR	Limit releases of lead particles on site to levels that do not pose risks to human respiratory systems.	[a]
DIRECT CONTACT	Minimize potential for direct contact with contaminated fill, soils, and sediments.	[a]Use concentration previously documented as action level for lead in surface soils: 750 ppm.

a. No federal requirements exist for lead here.
b. Action level of 750 ppm derived from consent order.

IDENTIFY TECHNOLOGIES

In order to pose solutions that can be implemented on the site of ABC Landfill, the designer reviews all remedial responses that may be applicable to the site. This is the preliminary stage of the process that will result in selection of a remedial measure. Here, the designer is unencumbered by judgments of cost or effectiveness, and the purpose of this step is to insure that the universe of potential remedial actions is addressed—that no stone is left unturned. It is a time to let the imagination run wild and propose any approach, regardless of feasibility or history of prior use. The only criterion for selection of a technology is that it be appropriate to the known conditions and problems of the site. It is unnecessary at this time to define a complete solution to the problems, as that occurs later in the selection process, after the technologies have been evaluated on the basis of their feasibility.

A general listing of potential response actions and associated remedial technologies is a useful reference when the initial comprehensive list of technologies is developed. Table 8-4 lists a number of technologies and actions often considered for application at a hazardous waste site, whereas a more complete listing is contained in *Guidance on feasibility studies under CERCLA*. (Some of the specific technologies listed here are discussed in subsequent chapters of this book.)

Generally, the technologies employed to control transport routes are capping to control infiltration, volatilization, or suspension in air and surface water diversion or control of the upgradient ground water elevation. Receptors generally are protected by separating them from the contamination: water supplies can be treated or replaced; workers near the contamination wear protective clothing and safety equipment; persons can be denied entry to contaminated areas. Most hazardous waste sites exhibit more than

TABLE 8-4. Response actions and remedial technologies.

NO ACTION

SITE ACCESS CONTROL
Fencing; land use limits

SURFACE WATER CONTROL
Capping; grading; revegetation; collection systems.

GROUND WATER OR LEACHATE CONTROL
Capping; containment; pumping.

ON-SITE TREATMENT
Incineration; physical, biological, and chemical treatment.

CONTROL OF GAS MIGRATION
Capping; containment; collection.

REMOVAL OF MATERIALS
Excavation; grading; capping.

OFF-SITE TREATMENT
Physical, biological, and chemical

WATER OR SEWER LINE DECONTAMINATION
In situ cleaning; replacement; alternative supply.

one affected source area; for example, soil and ground water or ground water and air. The technologies address local or regional control, as necessary, and the geographical extent of the contaminant is matched by the scope of the control measure.

It is worthwhile to have other knowledgeable designers review the initial list of control technologies, as they may propose additional approaches that were overlooked. The listing is only expanded, not condensed, at this stage. The comprehensive list of alternative technologies is then screened to develop a manageable set of options, and each of these technologies is analyzed. The screening is based upon whether the technology is:

- Applicable to site conditions.
- Effective in nullifying the contaminants in the matrices of concern.
- Feasible.
- Reliable.
- Proved effective when used under similar circumstances.

Engineering judgment and experience play a major role as the technologies are critiqued. It is important that the rationale for the elimination of technologies be carefully explained and documented, as that record may be referenced in the future when the site's remediation is reviewed.

Table 8-5 presents the list of promising remedial technologies developed for the ABC Landfill.

TABLE 8-5. General response actions and associated remedial technologies potentially applicable at ABC Landfill.

GENERAL RESPONSE	APPLICABLE TECHNOLOGIES
No action	Monitoring, fencing, site use limitations
Containment	Capping, ground water barrier walls
Pumping	Ground water pumping, dredging
Collection	Subsurface drains, gas vents, gas collection
Diversion	Dikes and berms, grading
Complete removal	Excavation of waste material and sediments
Partial removal	Selective excavation of waste material and sediments
On-site treatment	Biological, chemical, or physical treatment
Off-site treatment	Treatment at a publicly owned facility
On-site disposal	Landfill
Off-site disposal	Landfill

Many technologies were dismissed because they did not suit the site conditions, were inadequate to the treatment task at hand, or were inapplicable. Those technologies are presented in Table 8-6, which lists both the specific technologies eliminated and the rationale for their removal.

POSE ALTERNATIVES; PERFORM PRELIMINARY SCREENING

The individual remedial technologies applicable to the site at hand are combined into alternative sets such that:

- Each set has the capacity to address the integrated contamination problems.
- Each set is appropriate to the site.

The designer then critiques these sets of remedial technologies and classifies them according to their environmental, engineering, and economic strengths. The superior alternatives are selected, and all others are dismissed.

The National Contingency Plan (NCP)[3] requires that the remedial action alternatives developed for control of hazardous waste sites include at least one alternative in each of the following categories:

1. No action.
2. Alternatives that do not attain applicable or relevant federal public health or environmental requirements but do reduce the likelihood of present or future threats from the hazardous substances.
3. Alternatives that attain applicable or relevant and appropriate federal public health or environmental requirements.
4. Alternatives that exceed applicable or relevant and appropriate federal public health or environmental requirements.
5. Alternatives for off-site treatment or disposal, as appropriate.

These five categories from the NCP will be referenced below as the example of the ABC Landfill is developed. Three of the categories (2, 3, and 4) characterize the remedial action in terms of its ability to achieve environmental standards.

As a first step in the critique, the designer evaluates the criteria used for selecting these alternatives. The following factors are weighed:

- *Environmental and public health effects:* The remedial alternatives must be able to meet the remedial response objectives; they must be capable of providing protection to the public health and welfare and not pose undue risk when implemented.
- *Engineering components:* The combination of remedial technologies that form the remedial alternative should be both feasible for the specific site

TABLE 8-6. General response actions and associated remedial technologies excluded from consideration for ABC Landfill.

GENERAL RESPONSE	ELIMINATED TECHNOLOGIES	CRITERIA
Containment	Bulkheads Gas barriers	No support from data.
Pumping	Liquid removal	Contaminated liquid (leachate) to be collected by other means.
Collection	Sedimentation basins	No support from data.
Diversion	Stream diversion ditches Trenches Terraces and benches Chutes and downpipes	Surface run-on and runoff better controlled by other technologies.
	Levees	Inapplicable to site.
	Seepage basins	No support from data.
Complete removal	Tanks Drums Liquid wastes	Inapplicable to site.
Partial removal	Tanks Drums Liquid wastes	Inapplicable to site.
On-site treatment	Incineration Solidification Land treatment	Waste characteristics not supportive of these technologies.
Off-site treatment	Incineration	Waste characteristics not supportive of these technologies.
In situ treatment	Permeable treatment beds Bioreclamation Neutralization Land farming	Random co-disposal of industrial waste mixtures not effectively treated by these technologies; site characteristics not supportive of these technologies.
Storage	Temporary structures	Inapplicable to site.
On-site disposal	Land application	No support from waste or site characteristics.
Off-site disposal	Surface impoundments Land application	No support from waste characteristics.
Alternative water supply	Cisterns Above-ground tanks Deeper/upgradient wells Municipal water system Relocation of intake structure Individual treatment devices	No support from data.
Relocation	Temporary/permanent relocation of residents	No support from data.

conditions and reliable in addressing the problems associated with the site.

- *Economic considerations:* The estimated capital, operating, and maintenance costs of the selected remedial technologies must be economical. Alternatives that provide the same level of remediation as other alternatives yet cost an order of magnitude more to implement should be eliminated from further consideration.
- *Community response:* It is desirable that proposed measures be acceptable to communities proximate to the site.

The most economical alternative is not necessarily the least expensive one; rather, it is the least costly of the alternatives that is also technically feasible and reliable, and that adequately protects the public health and welfare and the environment.

As the alternative sets of technologies are screened, it is desirable to categorize each technology employed in a remedial action in terms of its ability to meet each of the levels of protection defined by the five NCP categories, those criteria that characterize remedial actions' capacity to achieve public health or environmental standards. The sets of technologies are grouped into those that meet standards, those that exceed standards, and those that fail to meet standards but reduce jeopardy to public health and the environment. The ability to meet pertinent criteria is defined not only by the type of technology employed but also by the design criteria applied to the technology. For example, a 3-ft (1-m) clay cap may not be capable of achieving environmental standards, but a 6-ft (2-m) clay cap—the same technology—may meet standards. Therefore, the same technology may be employed in more than one remedial action classification.

The alternatives proposed for the ABC Landfill, for each criterion required by the NCP, are:*

1. *No action:* Monitor ground and surface water; fence perimeter of landfill; limit site use. (Although these activities in fact require intervention on the site, they represent minimal measures and are designed primarily to monitor the site conditions.)

2A. *Nonattainment:* Monitor ground and surface water; fence perimeter of landfill; limit site use; excavate no fill; excavate no sediments in wetland or creek bed; install low-permeability cover; collect leachate and treat on-site.

---

*Author's note: Between the time this chapter was written and the publication of the book, the types of alternatives specified in the NCP have been changed. However, the evaluation process remains largely as presented here.

2B. *Nonattainment:* Monitor ground and surface water; fence perimeter of landfill; limit site use; excavate no fill; excavate no sediments in wetland or creek bed; install impermeable cap; collect leachate and treat off-site.

2C. Non-attainment: Monitor ground and surface water; fence perimeter of landfill; limit site use; excavate all fill with lead concentration above action limit; excavate no sediments in wetland or creek bed; collect leachate and treat on-site; place waste in secure cell on-site; install low-permeability cover.

3A. *Attainment:* Monitor ground and surface water; fence perimeter of landfill; limit site use; excavate no fill; excavate all sediments in wetland and creek bed with lead ≥750 ppm and place on fill; collect leachate and treat off-site; construct ground water barrier wall; install impermeable cap on fill.

3B. *Attainment:* Monitor ground and surface water; fence perimeter of landfill; limit site use; excavate all fill with lead concentration above action limit; excavate all sediments in wetland and creek bed with lead ≥750 ppm; collect leachate and treat on-site; place waste in secure cell on-site; install low-permeability cover on remaining fill.

4A. *Exceeding:* Monitor ground and surface water; fence perimeter of landfill; limit site use; excavate no fill; excavate all sediments in wetland and creek bed with detectable levels of lead and place on fill; collect leachate and treat off-site; construct ground water barrier wall; install impermeable cap on fill.

4B. *Exceeding:* Monitor ground and surface water; fence perimeter of landfill; limit site use; excavate all fill with lead concentration above action limit; excavate all sediments in wetland and creek bed with lead concentration above action limit; collect leachate and treat on-site; place waste in secure cell on-site; install low-permeability cover on remaining fill.

5A. *Off-site:* Monitor ground and surface water; fence perimeter of landfill; limit site use; excavate all fill with lead concentration above action limit; excavate no sediments in wetland or creek bed; install low-permeability cover on remaining fill; collect leachate and treat on-site; place waste in secure cell off-site.

5B. *Off-site:* Monitor ground and surface water; fence perimeter of landfill; limit site use; excavate all fill with lead concentration above action limit; excavate all sediments in wetland and creek bed with lead ≥750 ppm; install low-permeability cover on remaining fill; collect leachate and treat on-site; place waste in secure cell off-site.

5C. *Off-site:* Monitor ground and surface water; fence perimeter of landfill; limit site use; excavate all fill with lead concentration above action limit; excavate all sediments in wetland and creek bed with lead con-

CATEGORY	Ground & Surface Water Monitoring	Fencing	Site Use Limitations	FILL EXCAVATION		WETLAND/CREEK EXCAVATION			Low Permeability Cover	Impermeable Cap	LEACHATE TREATMENT		SECURE CELL		Ground Water Barrier Wall
				None	Pb < Detectable	None	Pb> 750 ppm	Pb < Detectable			On-Site	Off-Site	On-Site	Off-Site	
1	X	X	X												
2A	X	X	X	X		X			X		X				
2B	X	X	X	X		X				X		X			
2C	X	X	X		X	X			X		X		X		
3A	X	X	X	X			X			X		X			X
3B	X	X	X		X		X		X		X		X		
4A	X	X	X	X				X		X		X			X
4B	X	X	X		X			X	X		X		X		
5A	X	X	X		X	X			X		X			X	
5B	X	X	X		X		X		X		X			X	
5C	X	X	X		X			X	X		X			X	

Figure 8-2. Remedial alternatives developed for ABC Landfill.

centration above action limit; collect leachate and treat on-site; place waste in secure cell off-site; install low-permeability cover on remaining fill.

The low-permeability cover would be compacted soil 1.5 ft (45 cm) thick with permeability $\geq 10^{-5}$ cm/s overlain by 0.5 ft (15 cm) of soil supporting vegetation. An impermeable cap is designed according to regulations pursuant to RCRA.[4] Such a cap incorporates a double-liner system, a lateral drainage layer, and a layer of soil to support vegetation. Permeability is $\geq 10^{-7}$ cm/s. The secure cell would also be constructed according to the requirements of RCRA.[5] The component technologies in each alternative can be arrayed in a chart for ease of comprehension. See Figure 8-2 for an example.

The designer elucidates the technologies used in each remedial action alternative in a written report; the explanation includes a summary of the applicability of the technology and, perhaps, a statement describing its prior applications. One purpose of this report is to document the rationale for having eliminated other remedial alternatives from further consideration. Each of the factors used to screen alternatives—environmental, engineering, economic, public response—is addressed. The discussion may be supported by a review of the technical literature.

SCRUTINIZE SELECTED ALTERNATIVES

The remedial alternatives that fulfill the screening criteria are subjected to a detailed analysis. The nontechnical issues are examined in greater depth, and the costs are estimated more accurately than was the case in the preliminary screening of the alternatives. The following aspects of each alternative are assessed:

- Technical merit.
- Institutional capability.
- Public health and environmental benefit.
- Economy.
- Community response.

Technical Analysis. All aspects of the implementation and performance of the remedial alternative are judged. Technical factors generally can be grouped into two areas:

- *Site factors:* Elements of the locale that influence the ability of the remedial action to achieve the objectives of the remedial program. (See Table 8-1.)
- *Engineering factors:* Components of the alternatives relating to the construction and operation of the facilities that implement the remedial action. Their probable performance is reviewed in light of the intended lifetime of the remedial program. A number of these components are listed in Table 8-7.

TABLE 8-7. Technical analysis of remedial alternatives: engineering factors considered.

- Demonstrated ability to meet the contaminant level and exposure objectives
- Significant site-specific construction problems
- Time period for construction and start-up
- Operation and maintenance requirements
- Safety considerations
- Time required to achieve remedial objectives

The technical evaluation is recorded and used as support, both for comparing the technical merits of the alternative actions and for emphasizing the major areas of technical uncertainty. A technical comparison of remedial alternatives for the ABC Landfill follows. (Refer to Figure 8-2 for the components of each alternative.)

Category 1. Remedial alternative 1 is the no-action alternative. The monitoring program would identify changes in site conditions as they occur. The fencing would limit direct contact by humans with the wastes on the site but would probably not limit direct contact by animal life. Site use limitations in the form of restrictive covenants in the deed or zoning controls or both would impede current and future use of the property.

The no-action alternative would be straightforward and quick to implement. Each technology is reliable. The monitoring program provides excellent means to identify changes in site conditions. A data base would be developed and used to identify whether future site remediation could be warranted. The fence would require simple but minor repairs. Once established, the site use limitations would be somewhat reliable, as they would establish legal barriers to future use.

With regard to safety, the monitoring program would identify any long-term and short-term threats to receptors in the environs and to the environment. As the monitoring program and fence are implemented, the workers would potentially suffer additional direct contact with the contaminated wastes, but even these exposure levels would be managed because standard safety procedures would be employed.

Category 2. The low-permeability cover specified for Alternatives 2A and 2C would effectively reduce percolation of water by enhancing evapotranspiration and controlling runoff. Therefore, the lead released from the landfill would be reduced. This cover would also obviate direct contact of receptors with the contaminated wastes in the landfill. Such covers are constructed using standard construction procedures and materials; regular observation and maintenance of the cap would sustain its integrity.

The impermeable cap would permit less infiltration and, therefore, generate less leachate than the low-permeability cover. Standard construction methods and materials would be employed. This cap also would require regular observation and maintenance.

A leachate collection system would be included with each alternative of Category 2. An on-site treatment system would be employed with Alternatives 2A and 2C because of the larger volume of leachate collected with the low-permeability cover than with the impermeable cap. A suitable treatment facility off-site would be found for Alternative 2B. It would be more efficient to treat the larger volumes of leachate associated with the low-permeability cover on the site with a package treatment plant than at a facility off the premises. Similarly, the relatively small volumes of leachate associated with the impermeable cap would be more economically treated at an off-site treatment facility.

With Alternative 2C, workers would be exposed to additional risk of direct contact with the contaminated materials during excavation of the landfill. In addition, there are risks associated with additional construction activities. Risks due to the excavation of materials in the fill other than the site's hazardous components would also be manifest. Standard construction requirements and maintenance procedures would apply to the secure cell.

Category 3. Removal of wetland and creek sediments having lead concentrations above 750 ppm would effectively remove all contaminated material for the purposes of this action level. With Alternative 3A, the excavated materials would be placed atop the landfill, and they and the landfill would be shielded by the impermeable cap. The ground water barrier wall would complete encapsulation of the wastes. Leachate would be collected and treated at a facility on-site. In combination, these technologies would effectively isolate lead-bearing materials in excess of the action level and insure that the ground water would not become contaminated. All these technologies are reliable and can be implemented using standard construction procedures and materials. The safety of the construction workers would be protected through standard practices.

With Alternative 3B, all technologies are reliable and feasible, requiring only standard construction techniques and materials. Risks of direct contact by workers would be encountered during excavation and transportation to the landfill.

Category 4. Alternatives 4A and 4B respond to lead concentrations in the soil below the action level and maintain ground water lead concentrations below the maximum contaminant level for lead. Alternative 4A is similar to Alternative 3A, except that all wetland or creek sediments with concentrations above the action level (750 ppm) would be removed and placed on the landfill. Alternative 4B is similar to Alternative 3B, but all materials with lead concentrations above background would be excavated and placed in a secure cell on site. These alternatives are evaluated on the same basis as the alternatives of Category 3.

Category 5. Alternatives 5A, 5B, and 5C each specify disposal of contaminated materials in an off-site secure cell. The technologies to be implemented are all reliable, especially the associated operational and maintenance activities.

All construction practices and materials are standard. Workers would risk direct contact during excavation of the landfill materials, and the public would be at risk during transport of the materials off-site. Alternative 5A could not attain the applicable or relevant and appropriate federal requirements, but Alternative 5B could. Alternative 5C would exceed the requirements.

Institutional Analysis. In this book, *institutional* refers to government and private organizations and to government regulations. The objective of a remedial action is to reach applicable or relevant and appropriate government standards with regard to the permissible concentrations of contaminants in the environment. Federal government standards have predominance over state or municipal standards. However, the designer should understand and record in the feasibility study report how state and municipal standards differ from the federal standards. A remedial action that meets or exceeds federal standards is selected for implementation, unless unusual circumstances are involved. In the event that no applicable or relevant and appropriate federal requirements nor any state or municipal standards control the contaminants found on the site, a risk assessment specific to the site can be performed. The risk assessment would identify the appropriate level of remediation of the contaminants and mitigate jeopardy to human health and the environment. (Risk assessment is addressed in detail in Chapter 4.)

The pertinent federal requirements are found through study of the regulations and guidance policies promulgated by government agencies, notably the U.S. EPA, and the relevant state environmental control and public health agencies. Table 8-8 lists the principal sources of these requirements.

It is critical that the designer consult with the government agencies

TABLE 8-8. Principal federal legislative acts and policies applicable to uncontrolled hazardous waste sites.

Comprehensive Environmental Response, Compensation, and Liability Act (CERCLA) PL 96-510, 42 U.S.C. 9601 Superfund Amendments and Reauthorization Act of 1986 (SARA) PL 99-499
Resource Conservation and Recovery Act (RCRA) PL 94-580, as amended, 41 U.S.C. 6901
Safe Drinking Water Act (SDWA) PL 93-523 as amended, 42 U.S.C. 300f *et seq.*
Clean Water Act (CWA) PL 92-500, as amended, 33 U.S.C. 1251 *et seq.*
Ground Water Protection Strategy (GWPS) U.S. EPA Report WH 550, November 1980.
Clean Air Act (CAA) PL 90-148 as amended, 42 U.S.C. 7401 *et seq.*

responsible for implementing environmental regulations to be certain that all relevant requirements are acknowledged. These agencies, too, can provide information on the local geology and soil conditions, the transport and handling of toxic and hazardous substances, human health and safety considerations, and the protection of biological, historical, and recreational resources. Table 8-9 lists regulatory agencies frequently consulted in completing a feasibility study.

Through the institutional analysis, the designer reviews all applicable standards and determines how those standards will be realized on the uncontrolled hazardous waste site at hand. Within the documentation of the feasibility study, relevant federal, state, and municipal regulations and guidelines affecting the response actions are summarized. A listing of the government agencies contacted during the investigation and a brief description of the results of the communication are also included. The summary institutional analysis performed for the ABC Landfill follows.

Category 1. The no-action alternative cannot be expected to comply with pertinent regulations. Monitoring would be conducted, so the reports of results would be submitted to the appropriate regulatory agencies.

Category 2. The on-site treatment system would have to comply with the permit restrictions of the National Pollutant Discharge Elimination System (NPDES, or SPDES if the state controls the program).[6] Periodic monitoring reports would be submitted to the regulatory agencies. Alternative 2C attains the requirements for the fill but not for the sediments in the wetland

TABLE 8-9. Principal federal agencies involved in feasibility studies.

Agency
Environmental Protection Agency
Federal Emergency Management Agency
Department of Health & Human Services
Army Corps of Engineers
Department of the Interior
Geological Survey
Bureau of Land Management
Fish & Wildlife Service
Department of Transportation
Coast Guard

and creek; Alternatives 2A and 2C attain none of the requirements. All alternatives, however, reduce the risks associated with the wastes.

Category 3. Alternatives 3A and 3B attain the requirements for lead. The impermeable cap and the ground water barrier wall in Alternative 3A would encapsulate the lead wastes and the lead-contaminated sediments as required by RCRA. The on-site secure cell in Alternative 3B similarly would contain all wastes and sediments contaminated beyond action levels. The annual monitoring reports would be submitted to the regulatory agencies.

Category 4. Alternatives 4A and 4B would exceed the requirements. Other institutional issues are similar to those presented for Category 3 alternatives.

Category 5. All of these alternatives specify off-site disposal of contaminated materials. Alternative 5A corresponds to Alternative 2A, except that the landfill wastes above the detection limit would be excavated and disposed of in a secure cell off-site. This alternative would attain the pertinent requirements for the wastes deposited in the landfill but not for the contaminated sediments. Alternative 5B attains the requirements for lead and is comparable to Alternative 3B, except for the off-site disposal of excavated materials. Alternative 5C exceeds the requirements for lead in all media and is similar to Alternative 4B. A NPDES permit would be required for the on-site treatment system specified for each alternative. The annual monitoring reports would be submitted to the regulatory agencies.

Public Health and Environmental Analysis. The potential effect of implementation of the remedial alternatives on public health is assessed and recorded. The effects during both construction and operation are evaluated. In order to provide a baseline against which to judge the consequences of each alternative, the health effects of the no-action alternative are articulated. If the contaminants have migrated from the site, the level of exposure to humans and the environment is estimated. Table 8-10 lists the data included in the record.

TABLE 8-10. Background data vital to the evaluation of public health and environmental effects.

Site contaminants	Concentration and volume of hazardous substances remaining on-site
Levels of contamination	Populations at risk
Extent of contamination	Mechanisms of contact
Extent of migration	
Estimated levels of exposure	

Public Health Analysis. The standards defined in the government regulations are utilized to compare the various levels of effectiveness of both the remedial alternatives and the no-action alternative. For each alternative, the chemicals that are controlled and those not effectively controlled are listed; the contaminant levels anticipated on the site and the points of contact along migration routes are cataloged; and the time required to reduce the concentrations of the contaminants to acceptable levels both on the site and at receptor locations is indicated. Once those data are analyzed, the effectiveness of each alternative in safeguarding the public health is documented.

Environmental Analysis. In addition to their effects on public health, the remedial alternatives will alter the environment. These effects are assessed to evaluate potential adverse consequences of the construction and implementation of the remedial alternatives and to determine if the alternatives afford satisfactory protection of the environment. The assessment begins with a review of the repercussions of the no-action alternative. The current situation and that projected in the future, absent any remedial action, are described. The concentrations of contaminants currently found on and off the site and the concentrations projected in the future for each additional alternative are compared with the criteria developed by the institutional analysis, including tolerable concentration levels in surface water, ground water, air, soils, and sediments. The short- and long-term beneficial effects of implementing each of the remedial alternatives are also assessed, and these effects are compared both with the judgmental criteria and with the no-action alternative. Beneficial effects include achieving low levels of contaminants in the media of interest; improvements in environmentally sensitive lands such as wetlands and forests; decreases in the destruction of sensitive ecosystems; and improvements in resources such as farmland, fisheries, and recreational lands. Adverse effects of implementing the various alternatives are also estimated unless the screening analysis indicated no evident adverse effects. In any event, adverse effects of the no-action alternative are always projected. Potential adverse effects include new airborne emissions; contaminated discharge to surface or ground water; and depreciation of a sensitive or unique environmental, biological, or human resource.

Assessment of ABC Landfill. For the ABC Landfill, the public health and environmental analyses were combined. They are related below.

The no-action alternative, Category 1, would reduce the risk of direct contact of contaminated materials at the landfill site by humans because the area would be fenced. Otherwise, the no-action alternative provides no beneficial effects, and the adverse effects manifest from the site would simply continue.

The landfill caps specified in Alternatives 2A, 2B, and 2C, Category 2, would effectively prevent runoff of lead-contaminated particles from the landfill, would prevent direct contact of the landfill materials, and would reduce air dispersion from the landfill. The leachate collection systems would also prevent lead from migrating into the ground water system. However, the risks due to lead concentrations in the wetland and creek would not be addressed. Risks associated with transport of contaminants in the air and surface water and direct contact by receptors in the environs would result because the contaminated fill would be excavated.

Alternatives 3A and 3B, developed for Category 3, would address the risks associated with all areas contaminated by lead. Each remedial objective would be considered, and all criteria would be met. However, excavation of the landfill wastes and the contaminated sediments in the wetlands and creek would create the risk of air and surface water transport of contaminated materials into the environs, direct contact during excavation by workers, and spills during excavation.

The consequences of implementing Alternatives 4A and 4B of Category 4 are similar to those for Alternatives 3A and 3B. The risks associated with all contaminated areas would be addressed, but the excavation would increase the potential for direct contact by workers and for distribution of the contaminated material into the environs.

The risks inherent in Alternatives 5A, 5B, and 5C, developed for Category 5, relate to distribution of the contaminant due to excavation, transportation, and disposal of contaminated materials. Alternative 5A would not reduce the risks associated with the contaminated sediments of the wetland and creek.

Economic Analysis. Information about the relative costs of the various remedial alternatives is essential in the selection process. The cost estimate for each alternative includes all the costs associated with the remedial action: construction, operation, and termination. Each remedial action, it is assumed, has a determinate lifetime, the time during which the alternative must function in order to achieve the remedial objectives. Over that period, funds are expended to operate and maintain the facilities of the alternative.

The economies of remedial alternatives are compared using present worth analysis, a method that estimates the initial capital costs and the annual operating costs for each year of the lifetime of the project in terms of the current cost of money. The cash flow of each year is discounted to a present worth to furnish an estimate of the present value of the total expenditure required for each remedial action. This analytical method is especially useful in comparing an alternative with a high initial capital cost and low operating costs to an alternative with a lower capital cost but higher operating costs. The present worth method is fully explained in standard economics texts.[7] Typical cost items are listed in Table 8-11.

TABLE 8-11. Items included in capital and operating costs for remedial action alternatives.

CAPITAL COST ITEMS	OPERATING COST ITEMS
Site preparation	Operating labor
Buildings	Operating supervision
Equipment	Maintenance labor
Utilities	Maintenance materials
Construction labor	Raw materials
Indirect costs	Fuel and utilities
• Engineering design fees	Administrative overhead
• Legal fees	Taxes
• Start-up expenditures	Insurance
• Contractor fees	Working capital
Contingency expenditures	Capital expenditures to replace worn items

The capital costs for each remedial alternative are determined using standard engineering cost estimating procedures; they include all direct and indirect costs. Capital costs are determined using traditional sources of information: published sources such as *Means Cost Estimating Guide* and *Engineering News Record* and unpublished data such as quotations from equipment vendors and service suppliers, past estimates for similar projects, and local labor costs. Operating costs include all expenditures associated with continuance of the remedial action from the day operation commences through attainment of the objectives and any necessary removal or demolition of the remedial facilities. Annual expenditures required to operate and maintain the facilities are estimated over the lifetime of the installation in order to calculate the cash flow and thus to furnish figures for the present worth analysis. Cost estimates are prepared with as much accuracy as possible, and items with great cost uncertainty should be noted in the written record. Ambiguity in the figures for capital and operating costs is often due to a lack of design data, data that should have been developed as fully as possible during the earlier remedial investigation.

The present worth analysis compares the diverse remedial alternatives on an equivalent basis. Sensitivity analysis is used to reveal uncertainties in the major variables that influence the economics of the alternatives—the discount rate, fuel and utility costs, time period for construction, and cost of labor. The resulting cost estimates for all the alternatives are tabulated in sufficient detail to permit future modifications of the estimate. In addition to the cost

TABLE 8-12. Summary of remedial action costs for ABC Landfill.

REMEDIAL ALT.	CAPITAL COST	30-YEAR PRESENT WORTH OPERATION/MAINT.	TOTAL PRESENT WORTH
1	$ 80,000	$ 90,000	$ 170,000
2A	1,000,000	400,000	1,400,000
2B	3,000,000	450,000	3,450,000
2C	4,000,000	475,000	4,475,000
3A	4,250,000	550,000	4,800,000
3B	4,250,000	500,000	4,470,000
4A	5,000,000	800,000	5,800,000
4B	5,000,000	750,000	5,750,000
5A	4,000,000	500,000	4,500,000
5B	4,500,000	550,000	5,050,000
5C	5,250,000	800,000	6,050,000

figures, the basis of design for the facilities and the accompanying assumptions are clearly stated; significant cost differences among the remedial alternatives should be noted. The total capital costs, the operating costs, the present worth analysis, and the sensitivity analysis for the alternatives are summarized in a table and included in the documentation of the feasibility study. Table 8-12 is a completed example of the economic analysis for the prototype.

Community Response. Government regulations require a program of public education and information when intervention is planned at an uncontrolled hazardous waste site. This program is instituted as part of the remedial investigation and continued through the analysis of remedial alternatives. Interested members of the public are provided with the full record of the designer's evaluation of the remedial alternatives and allowed a three-week period to comment. A public notice and fact sheet summarizing the remedial response alternatives are issued two weeks prior to the beginning of the comment period. The public's comments on the evaluation process and the response to those comments are included with the final report describing the feasibility study, submitted by the designer to the regulatory agencies.

TABLE 8-13. Outline for report of feasibility study.

SECTION	CONTENT
	Executive Summary
1	Introduction
2	Screening of Remedial Action Technologies
3	Screening of Remedial Alternatives
4	Detailed Analysis of Remedial Alternatives
5	Summary of Alternatives
6	Community Response Summary
	References
	Appendixes

RECORD THE FEASIBILITY STUDY

The regulatory agencies, usually the U.S. EPA with the participation of the state environmental control and public health agencies, and the site's owners or users, or both, choose the remedial alternative befitting the conditions at hand. To assist the parties in making their decision, the designer submits the records of all aspects of the development of the remedial alternatives and their critiques. All the figures, tables, and charts developed as the analyses were performed are incorporated into the report of the feasibility study. The regulatory agencies use this report when selecting the appropriate alternative. The format for the report recommended by the U.S. EPA is shown in Table 8-13.[8]

The executive summary is an overview of the complete study. It provides a brief description of the site background and problems, the objectives and criteria for the remedial action, the remedial options considered, and a summary comparing the remedial alternatives with the no-action option. This summary is presented in tabular form and includes a brief description of each alternative, the results of capital and present worth cost estimates, and a list of the public health, environmental, technical, and community concerns raised during the process. Such summary information is presented in Table 8-14.

The introduction of the report gives a complete description of the site background and history, and presents the objectives of the remedial effort. Section 2 of the report, "Screening of Remedial Action Technologies," includes a table listing all the technologies considered and the criteria used to winnow them. For each technology dismissed, the reason why it was

TABLE 8-14. Summary of remedial alternatives for ABC Landfill.

	COSTS ($1000)		ISSUES				
ALTERNATIVE	CAPITAL	PRESENT WORTH	PUBLIC HEALTH	ENVIRONMENT	TECHNICAL	COMMUNITY	OTHER
1. No action	80	170	Lead exposure: ingestion; inhalation.	Lead exposure: surface water; wildlife; ground water; soils.	—	Unacceptable	—
2A, 2B. Cap landfill; collect/treat leachate	1,000 3,000	1,400 3,450	Removes direct contact route for landfill, migration of lead via runoff. Risks remain for lead in wetland/creek.	Lead in wetland and creek.	—	Unacceptable	Nonattainment alternative.
2C. Excavate fill; on-site secure cell; cap other fill; collect/treat leachate.	4,000	4,475	Removes risks assoc. with landfill. Risks remain for lead in wetland and creek.	Lead in wetland and creek.	Excavation in landfill.	Unacceptable	Nonattainment alternative.
3A. Excavate landfill, wetland, creek; on-site secure cell; cap landfill; collect/treat leachate.	4,250	4,750	Removes risks assoc. with landfill and wetland/creek sediments.	Removes high lead concentrations; protects ground water.	—	Acceptable	Fulfills criteria.
3B. Excavate landfill, wetland, creek; on-site secure cell; cap; collect/treat leachate.	4,250	4,750	Removes risks assoc. with landfill and wetland/creek sediments.	Removes high lead concentrations; protects ground water.	Excavation in landfill.	Acceptable	Least expensive attainment alternative.

4A. Excavate detectable lead in wetland, creek; place on fill; cap; build ground water barrier wall; collect/treat leachate.	5,000	5,800	Removes risks assoc. with landfill and wetland/creek sediments.	Removes all lead concentrations; protects ground water.	—	Acceptable	Exceeds criteria.
4B. Excavate detectable lead; place in on-site secure cell; cap other fill; collect/treat leachate.	5,000	5,750	Removes risks assoc. with landfill and wetland/creek sediments.	Removes all lead concentrations; protects ground water.	Excavation in landfill.	Acceptable	Least expensive alternative that exceeds criteria.
5A. Excavate detectable lead in landfill; dispose in off-site secure cell; cap other fill; collect/treat leachate.	4,000	4,500	Removes risks assoc. with landfill. Risks remain for lead in wetland and creek.	Removes lead in wetland and creek; protects ground water.	Excavation in landfill.	Unacceptable	Off-site alternative; nonattainment alternative.
5B. Excavate detectable lead in landfill; lead over action level in wetland and creek disposed in off-site secure cell; cap other fill; collect/treat leachate.	4,500	5,050	Removes risks assoc. with landfill and wetland/creek sediments.	Removes high lead concentrations; protects ground water.	Excavation in landfill.	Acceptable	Off-site alternative; attains criteria.
5C. Excavate detectable lead; dispose in off-site secure cell; cap other fill; collect/treat leachate.	5,250	6,050	Removes risks assoc. with landfill and wetland/creek sediments.	Removes all lead concentrations; protects ground water.	Excavation in landfill.	Acceptable	Most expensive alternative; exceeds criteria.

ill-suited is specified. This table was developed when technologies were screened during the evaluation procedure. Complete descriptions of the remedial alternatives are contained in section 3 of the report. The descriptions include preliminary flow sheets, material and energy balances, and illustrations of the site arrangement. The chart that arrays the remedial alternatives, which was developed when the alternatives were evaluated, is included. All the analyses developed when the alternatives were critiqued are inserted into section 4, "Detailed Evaluation of Remedial Alternatives." These are the studies of the technical feasibility, the institutional, public health, and environmental evaluations, and the cost estimates. Section 5, "Summary of Alternatives," generally includes the tables and figures used in the executive summary, together with additional detail concerning comparisons of the alternatives. If requested by the U.S. EPA, a single remedial action is recommended, and the rationale for that choice is stated. The final section of the report recounts rejoinders from the public who reviewed the remedial alternatives.

SELECTION OF THE REMEDIAL ALTERNATIVE

The NCP, as outlined in CERCLA and SARA, requires that the remedial alternative selected for implementation: be economical; mitigate and minimize threats to public health and the environment; and provide adequate protection to public health, welfare, and the environment. SARA favors the use of long-term remedial actions that include treatment, recycling, reuse, and other means to reduce permanently the volume, toxicity, or mobility of the contaminants at a site; remedial alternatives that call for off-site transport and disposal of contaminants without treatment are discounted. The original CERCLA had no such biases.

Therefore, the selected alternative should attain or exceed the pertinent regulatory standards that apply to the site and should realize sustained effectiveness. An alternative that does not meet the standards may be selected in the following situations:

- The selected alternative is not a final remedy, but is part of a more comprehensive remedy.
- The remediation is financed by the federal Superfund, the urgency for action at another site prevails, and an alternative that approaches the standards can be financed.
- No alternative that attains or exceeds the standards is feasible.
- All alternatives that attain or exceed the requirements will result in significant adverse environmental effects.

- The remediation is financed by the Superfund, but money is unavailable; there is strong public interest in the remediation; and litigation would likely not result in the desired remedy.

Whenever any of these situations occurs, the selected remedial alternative is that which most closely approaches the level required by federal public health and environmental standards.

SUMMARY

A feasibility study is utilized to develop and evaluate remediation scenarios applicable to a specific uncontrolled hazardous waste site. The goal of the feasibility study is to integrate a set of technologies that, in concert, are intrinsically capable of comprehensively eliminating the site's risks to the degree required by pertinent government standards; this set of technologies must also be economical and acceptable to the communities in the environs of the site. The method employed by the designers of the feasibility study is systematic and guided by government regulations, yet such a design demands multifarious creativity.

The following specific steps are implemented to develop a feasibility study:

1. Develop remediation objectives that suit the contamination problems of the specific site.
2. Select remedial technologies befitting the problems of the site and practicable to implement.
3. Combine and group the selected technologies into sets; these sets are alternative remedial actions. Government regulations and guidelines outline the global constitution of the action alternatives and require also an evaluation of suspending remedial work, thus taking no action. Each alternative remedial action is designed to address all contaminants on the site and the routes of contaminant migration. Each alternative must achieve remediation to the levels defined by government standards or, in the absence of standards, to the levels determined by a discrete risk assessment.
4. Evaluate the remedial alternatives with respect to their technical, institutional, public health, environmental, and economic properties and in accord with their acceptance by the communities in the site's environs.

The progression of the feasibility study is documented in a report, the official record of the designer's work, and that report becomes public information. All technologies, alternatives, and evaluation criteria are recorded. The report furnishes the basis on which the remedial alternative is selected. The alternative to be implemented is ordinarily endorsed by the regulatory agency, though customarily the regulatory agency consults with the site owner or user. Alternatively, the chosen remedy may derive from a settlement reached in a court of law.

BIBLIOGRAPHY

Rishel, H. L., T. M. Boston, and C. J. Schmidt. *Costs of remedial response actions at uncontrolled hazardous waste sites; Pollution technology review No. 105.* Park Ridge, N.J.: Noyes Publications, 1984.

Rogoshewski, P., H. Bryson, and K. Wagner. *Remedial action technology for waste disposal sites; Pollution technology review no. 101.* Park Ridge, N.J.: Noyes Publications.

U.S. EPA. *Compendium of costs of remedial technologies at hazardous waste sites.* (Office of Research and Development and Office of Solid Waste and Emergency Response.)

———. *Remedial action costing procedures manual.* (Office of Research and Development and Office of Emergency and Remedial Response.) Sept. 1985.

———. *Guidance on feasibility studies under CERCLA.* Apr. 1985.

———. *Handbook for remedial action at waste disposal sites.* (Office of Emergency and Remedial Response and Office of Research and Development.) June 1982. EPA-625/6-82-006.

NOTES

1. The approach outlined here and the terms employed are taken from: U.S. EPA. *Guidance on feasibility studies under CERCLA.* Prepared for the Hazardous Waste Engineering Research Laboratory (Office of Research and Development) and Offices of Emergency and Remedial Response, and Waste Programs Enforcement (Office of Solid Waste and Emergency Response). Apr. 1985. Modifications here are consistent with the National Oil and Hazardous Substances Pollution Contingency Plan promulgated 20 Nov. 85 (40 CFR 300) and the Superfund Amendments and Reauthorization Act of 1986 (SARA, PL 99-499). For specific and detailed information about feasibility studies, see the source document and 40 CFR 300. (SARA was introduced in Chapter 1.)
2. See 40 CFR 261.24.
3. 40 CFR 300. 20 Nov. 85.
4. 40 CFR 264, Subpart N.
5. See 40 CFR 264, Subpart N.
6. See 40 CFR 122.
7. E. L. Grant. *Principles of engineering economy,* 7th Ed. New York: John Wiley & Sons, Inc., 1982.
8. *Guidance on feasibility studies under CERCLA.*

Chapter 9
The Recourse of Closure On-Site

In many cases, the most economical, environmentally sound, and publicly acceptable solution to remediation of a hazardous waste site is securing the wastes on the premises where they were first disposed of. The nature of the waste, the hydrogeological conditions, the types of soils available at the site, and the risks of moving wastes off-site are factors that are evaluated to determine the feasibility of on-site closure. In-place containment techniques are applied to isolate the hazardous waste from the environment. Sound engineering solutions should evaluate the economy, feasibility, and risks associated with the remediation alternative of on-site closure, and these risks should be compared with the alternate solution of off-site disposal.

Contaminants from a hazardous waste site can be transported in the ground water, in the surface water, and through the air. The designer working on remediation of a hazardous waste site assesses each of these pathways before proposing a method to contain the wastes. Data specific to the site, obtained through the hydrogeological investigation and laboratory analysis, provide the necessary information for the assessment. If the designer concludes that the wastes may be contained in place, specific technologies are applied to impede the transport of contaminants from the site of interment.

Of the pathways, the most important to consider is ground water because when wastes are buried, contaminants may be transported to and with the ground water underlying the site. Gravity and the relative permeability of the soil influence the movement of dissolved materials in the vadose zone. Once the contaminant contacts the ground water table, the transport mechanics change, and the ground water flow regime and the properties of the soil affect transport. Contaminants may travel in the saturated zone distinct from the ground water as a separate liquid phase; alternatively, they may be dissolved and travel in the ground water. Either way, a plume of contamination is created. A broad range of alternatives is available to retard the movement of this plume, both by limiting the recharge of ground water and by intercepting the plume.

Contaminants are able to travel in surface waters either adhered to surface

---

This chapter was developed by Richard D. Jones, P.E., David G. VanArnam, P.E., and Dharma R. Iyer, Ph.D. of O'Brien & Gere Engineers, Inc.

soil, especially as the contaminated soil is eroded, or dissolved in the aqueous phase, then carried with the surface water. By either means, the contaminants may enter a stream or body of water, or they may percolate into the ground water and mobilize a plume of contamination. Methods that abate the transport of contaminants in surface water focus on preventing contact between the water and contaminants at the surface.

Contaminants also may move through the air as a gas after volatilization; alternatively, they may adhere to soil particles or be in the form of small particles carried by the wind. Factors that influence the rate of volatilization include the nature of the contaminant (coefficient of transfer to the atmosphere), contaminant concentration, air temperature, wind velocity, and area of exposure. Wind shear is a principal agent of the erosion of fine particles on the surface, the likelihood of such erosion being especially high under arid conditions and where contaminants have a strong affinity for exposed soils. Methods that control the transport of contaminants in the air focus on collecting and treating gases and separating the contaminated particles from exposure to the wind.

Three techniques that block transport routes and permit on-site containment of wastes are discussed in this chapter:

- *In-place containment:* If the substrata of a hazardous waste site are sufficiently impermeable to the flow of ground water, many wastes can be contained in place. By controlling the flow of ground water through the wastes, leaching of contaminants is virtually stopped. Likewise, controlling surface water atop the landfill prevents erosion of soils containing contaminants and also prevents water from percolating through the landfill. When surface water is controlled, the hydraulic gradient is diminished, and, thus, the potential for transport of the contaminants in the ground water is reduced.
- *Secure burial cell:* Certain site conditions, such as the presence of permeable substrata and native soils or shallow depth to ground water, render in-place containment impracticable. An alternative on-site disposal option is construction of a new containment cell to hold the wastes. Moving the wastes off-site or treating them on-site may be more costly and less environmentally sound than constructing an on-site secure cell. Such cells are regulated by the multifarious requirements of RCRA[1] and the Toxic Substances Control Act,[2] as well as state and local laws. Among the requirements are foundations, liner systems, leak detection systems, leachate collection and treatment systems, proper excavation and placement procedures, capping, and long-term monitoring.
- *Neutralization:* The nature of the wastes at a hazardous waste site may make using in-place containment or on-site secure cells infeasible because

the wastes may be incompatible with liners, or they may be too mobile and easily leached from the site. Chemical fixation and stabilization processes are available that can immobilize hazardous wastes and make them nontoxic and noncorrosive. After the wastes are stabilized, on-site disposal in a secure cell may be suitable. In addition, off-site disposal may also be suitable because the risks associated with transporting the materials are substantially reduced.

After the waste is contained, the facility is closed, and post-closure maintenance and monitoring are initiated. Monitoring systems, often including leak detection systems, surface water monitoring, and ground water monitoring wells, are activated. The containment facility is maintained on a regular schedule for 30 years or more.

CONTAINING WASTES IN PLACE

This section presents techniques that, when implemented, result in the remediation of hazardous wastes on site without excavation. In certain circumstances, the most acceptable and economical solution to remedying a hazardous waste site is not to move the wastes but to leave them in place. The movement of ground water and surface water through the waste and the potential for erosion are controlled; thus, migration of the wastes from the site is prevented. Ground water can be controlled by barrier systems, leachate collection systems, and ground water collection systems. Surface water is controlled by diverting runoff around the landfill, grading the landfill to control drainage, and capping the landfill to prevent percolation through the site or volatilization from it.

Control of Ground Water.

Subsurface Containment Barriers. Subsurface barriers are constructed to control the horizontal flow of ground water, generally being constructed upgradient of the plume of contamination to divert uncontaminated ground water around the wastes. Because of interactions between the material of the cutoff wall and most contaminants, such walls cannot be used downgradient of the plume to impede the migration of contaminated ground water. Full, circumferential closure of a site also can be effected. The containment barrier usually is joined to an impermeable confining layer beneath the site. Such a stratum may be clay, silty clay, or a rock formation. The addition of a cap effectively encapsulates the contamination. Barrier walls typically are constructed as cutoff walls or deep granular drainage trenches.

The *soil-bentonite wall* is probably the most widely used containment

TABLE 9-1. Design and specification issues: soil–bentonite cutoff wall.

Permeability testing to verify effectiveness of barrier wall	Required hydration time and resultant properties of slurry
Methods of mixing bentonite slurry and backfill to assure homogeneous backfill	Acceptable sources of soil to be mixed as backfill
Measures to check and control water-bentonite viscosity and density	Allowable methods of placing backfill
Source of mixing water for bentonite-water slurry	Laboratory and field testing, including viscosity and permeability

technique.[3] It is generally applicable where there is a defined and acceptable confining layer, and where the tests of chemical compatibility between the soil-bentonite mixture and the wastes to be contained support its use. When correctly constructed, such a wall provides a complete and effective ground water barrier. The designer must determine the stability of the trench and the composition of the bentonite-water slurry required to hold open the excavation. Important, too, is examination of the compatibility of the leachate, the soil material, and the water used in the slurry with the soil-bentonite mixture. Compatibility is apparent if the soil-bentonite mixture hydrates or swells in the presence of the leachate. The wall may furnish inadequate containment if shrinking occurs. Specific design considerations are displayed in Table 9-1.

Construction of the soil-bentonite slurry trench cutoff wall begins with excavation of native materials. The trench is filled with a water-bentonite slurry. Trench depths of 35 to 50 ft (11 to 15 m) may be attained using an extended boom backhoe; a clamshell is used to dig trenches 100 ft (30 m) or deeper. During excavation, the backfill material is carefully placed into the trench, where it displaces the bentonite-water slurry.[4] The wall requires adequate inspection by resident engineering and geotechnical experts during construction. It is necessary that the methods of both construction and field testing be documented. Excavated soils not utilized in the backfill are disposed of according to applicable regulations or design considerations.

A number of factors may cause failure of soil-bentonite slurry walls, including:[5]

- Nonhomogeneous backfill material.
- Spalling of material from the excavation wall.
- Inadequate excavation of the material into which the wall is to be keyed.
- Encapsulation of pockets of the bentonite-water slurry in the cutoff wall.

A cement-bentonite cutoff wall is constructed similarly to a soil-bentonite wall. However, instead of a soil-bentonite backfill material, the designer utilizes a bentonite-cement backfill. This material hardens in 7 to 30 days to form an impermeable barrier similar to a stiff clay.[6] The admixture should be tested to insure compatibility with the leachate. A cement-bentonite wall is generally more costly than a soil-bentonite wall because of the need to import materials; however, it may be the most appropriate remedy where suitable native soil is not readily available.

The *vibrating beam* technique for installing a barrier wall has been employed in several cases to control ground water migration. A pile driver vibrates an H-pile beam to a specified depth, and a slurry may be used to lubricate the beam's penetration of the ground. As the beam is removed, slurry is injected through nozzles at the end of the beam into the cavity created by the beam's removal. This process is repeated to construct segments of the wall. Either a bituminous-asphalt mixture or a cement-bentonite mixture is used as the slurry. The bituminous blend is generally a mixture of asphalt, cement, sand, and water.[7]

Drainage Trenches. When the water table is near the ground surface, a trench may be used to intercept contaminated ground water. Trenches may be used either as the primary remedy or as a supplement to other actions. Drainage trenches are installed to extend at least 3 to 4 ft (90 to 120 cm) below the water table in order to withdraw ground water. These trenches require long-term maintenance, as do recovery wells. However, a trench can be used in strata of low permeability where wells would not constitute a continuous hydraulic barrier.

Water intercepted in a drainage trench either flows to a collections sump where it is pumped from the trench, or it is pumped directly from the trench. The depth of the trench is limited by the excavation equipment available and the ability of the soil to support the walls of the trench without caving in. Open trenches deeper than 8 ft (2.5 m) are generally impractical.

A perforated pipe may be laid in the trench, which may be backfilled with gravel or crushed rock. A filled trench may be constructed deeper than an open excavation. This approach produces a permeable bed; the ground water infiltrates the gravel packing, enters the pipe through the perforations, and is conveyed by gravity to a collection sump. The contaminated water is pumped from the collection sump.

Interception drainage trenches are sometimes used in conjunction with low-permeability barrier walls. The interior of the trench's downstream wall is lined with an impermeable material and thus creates a downgradient barrier. The contaminated ground water is collected before it reaches the wall to prevent further migration.

Surface Water Control. At a hazardous waste site, the flow of surface water runoff is controlled. Surface water would otherwise infiltrate, produce leachate, and contaminate the ground water system. When surface water is controlled, the migration of contaminants is attenuated, and the transportation of contaminated sediments that would be carried by the surface water is reduced. Control measures include stormwater drainage facilities; site grading, covering, and vegetation; and impermeable caps or cover systems. Stormwater drainage measures are a permanent element of in-place containment facilities. In design, the goal is to prevent any stormwater from contacting the sealed and contaminated material. Typical components of stormwater drainage facilities incorporate the slope, the landfill cap, stormwater ditches to convey runoff, diversion structures, and swales.

Stormwater drainage facilities should, at a minimum, be designed for the maximum intensity storm that could occur in an average ten-year period. More stringent design standards may be dictated by federal, state, or local requirements, depending upon the nature of the site and the specific remedial action.

To determine the correct size of the drainage channels, the volume of runoff from the drainage basins is computed using any of a number of stormwater engineering design methods. Factors considered in the design are the shape and size of the channel, the velocity of the flow as determined by the slope or pitch of the channel, the liner material for the channel, and the route of the channel with respect to the remainder of the design of the landfill.

During construction, measures are implemented to ameliorate damage from surface water runoff and to control the transport of contaminants. The construction management techniques used include limiting the exposed work area, covering the working faces daily and compacting the cover soil to seal the surface, placing a temporary covering, and advancing the work in a manner that minimizes the potential for erosion. Temporary silt control dams constructed of hay bales or geotextile fabrics and temporary impoundment areas may also be necessary to divert runoff, to control run-on water, and to control erosion. Also during construction, it may be necessary to collect and treat any surface water that contacts the contaminated substances in the exposed, working face. Depending upon the nature of the substances and the length of the contact, the treatment mode may be simple settling, or it may require chemical precipitation, filtration, or activated carbon treatment.

Cover Systems. The surface of a hazardous waste landfill is a critical component in remediation. It is graded, covered, and planted in order to control runoff water; so the potential for surface infiltration and erosion is reduced. Depending upon the site and materials stored, a system employing an impermeable cap may be necessary; if a cap is unnecessary, simply grading the site may be effective. Grading may be used as either a temporary or a permanent measure and may be sufficient if facilities are already in place to collect leachate or to control ground water. In some instances, landfills that contain hazardous substances may require a synthetic impermeable cap in conjunction with other barriers.

The cover system for a landfill is selected according to the requirements specific to the site. Generally, it is composed of a number of layers of natural or synthetic material. Cover systems range in complexity from simple single-layer graded native soils to multiple-layer systems composed of natural and synthetic materials.

The exact type of cover material for simple grading often depends on the availability of native soils that can be compacted and can support vegetation. Well-designed grading and planting promote the sheet-flow of runoff. The transpiration of the vegetation also removes a volume of water that otherwise would remain on the surface or percolate into the ground atop the impermeable cap. Erosion is thereby minimized, and, because the velocity of the water is even and minimal, the transport of particles is low.

An impermeable cap minimizes the infiltration of precipitation or surface water. Such caps are composed of impermeable soils, soil admixtures, or impermeable synthetic liners. If native clays are available at or near the site, they may be economical to use. In addition to permeability, other important engineering properties are considered. Table 9-2 lists some of the standards often found in government regulations.

The natural soil used must restrict the movement of water and, typically, must have a hydraulic conductivity of less than 10^{-7} cm/s. Also, the soil generally should be free of stones and other objects that could puncture the overlying flexible membrane. As the soil is recompacted, its natural texture is homogenized and becomes less variable, so that pathways for water or

TABLE 9-2. Typical engineering properties specified for soils of impermeable caps.

Property	Specification
Permeability	$\leq 10^{-7}$ cm/s
Compaction density	90-95% modified Proctor density
Lift thickness	6-12 inches (15 to 30 cm)
Liquid limit	≥ 30
Plasticity index	≥ 15
Water content	$\pm 3\%$ of optimal moisture content
Slopes: Maximum	4 horizontal to 1 vertical
Minimum	2 to 5%

contaminant migration are minimized, and a uniformly low permeability is insured. If suitable soils are unavailable within a reasonable distance of the site, the permeability of native soils may be reduced and their stability enhanced by adding bentonite, cement fly ash, or other materials. Whenever an admixture is used, its fabrication in the field is monitored to insure that the proportion of materials is maintained consistently; were there a deviation, the engineering properties of the admixture would be uneven and potentially unreliable.

Synthetic membrane liners are generally made from polymers, rubber compounds, thermal plastics, or combinations of these materials. They are available in varying thicknesses.

The physical qualities of the liner material are the primary consideration, including tensile properties, resistance to puncture and tearing, resistance to degradation from sunlight, and strength at various temperatures. All liners are subject to degradation from the environment. Over time, sunlight attacks certain polymeric liner materials and weakens the exposed portions. Temperature stresses, caused by both heat and cold, shrink and crack the liner and adversely affect its performance. Liner materials susceptible to these effects can be protected by limiting the time the liner is exposed to the elements during construction. Because the liner does not contact the wastes, chemical compatibility is not a concern.

Liners are subject to numerous loads and stresses, both during and after installation. In all cases, it is prudent to design the liner system and to control the liner's installation to diminish physical harm. In addition, high strength and extra thickness improve the liner's resistance to breaching. The American Society for Testing and Materials has developed test procedures for measuring the tensile properties of flexible membrane liners, and these procedures are commonly cited in specifications for liner materials.[8]

The weakest part of a synthetic liner system is its seams—and no liner system is better than its weakest part. Because of the areal size of most landfills, liners often must be joined in the field from several prefabricated rolls or sheets; during joining, workers may make errors that subsequently lead to liner failure. The techniques used to join liners include: chemical adhesion; solvent welding; and dielectric, thermal, or extrusion-welded seam formation. In every case, accurate timing and the proper application of pressure are critical.

As the liner is installed, field testing and quality assurance are imperative. Such testing and quality assurance apply to both the manufacture and the installation of the materials. The liner system should be installed by a reputable and experienced contractor in order to warrant that it will perform according to specifications when in place. It is sometimes desirable to require that the contractor installing the liner have a representative of the liner's manufacturer at the site during the installation. Generally, the design specifies field testing of the seams with vacuum or pressure techniques.

Certain landfills require several different liners to control multiple adverse environmental effects. One liner can protect against ground water migration, another against surface water migration, and yet another against volatilization. A number of designs for liner systems have been proposed,[9] and several have been applied successfully. Figure 9-1 shows typical cover systems.

Leachate Collection and Handling. Leachate can be produced by infiltration and percolation of rain water through a landfill prior to closure or by occasional contact between the landfill materials and elevated ground water.[10]

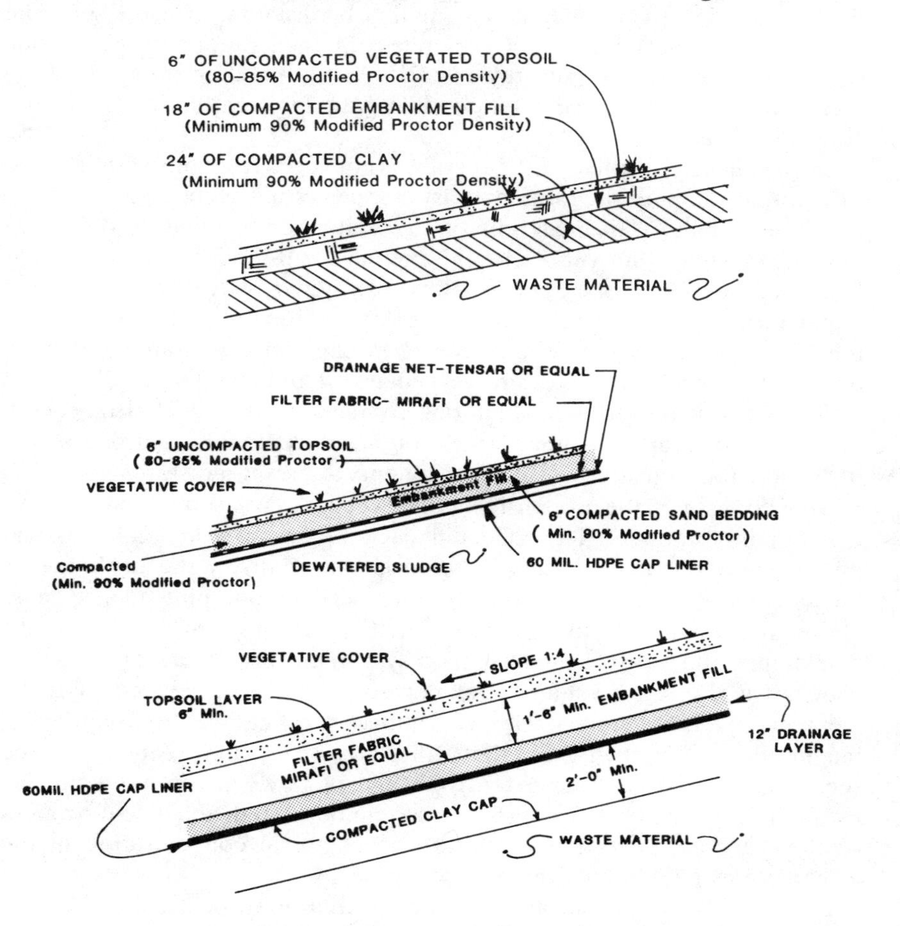

Figure 9-1. Typical landfill covery system.

When wastes are interred, leachate is frequently collected to prevent its mingling with the ground water system. The object of a leachate collection system at a waste disposal site is both to control drainage and to remove the leachate that is produced.

Leachate typically is collected either with a permeable downgradient interception trench or with a network of interception pipes and trenches, whichever fits the site's conditions. A leachate interception trench generally consists of highly permeable aggregate placed at the appropriate depth. A geotextile as a wrap around the leachate collection pipe or as a lining in the interception trench, or as both, often is necessary to prevent soil from accumulating and clogging the pipe.[11] In many instances, the effectiveness of the interception trench may be augmented by installing "fingers" into the downgradient edge of the landfill. Perforated pipe, generally polyvinyl chloride (PVC) or high density polyethylene (HDPE), depending upon chemical compatibility, is placed in the trench with a granular backfill to intercept and convey the leachate.

Once collected, the leachate generally is conveyed from the waste deposit to a tank or sump on-site for temporary storage, or it is pumped directly to a treatment facility. The materials of the pumps and piping used for this purpose must be compatible with the chemical substances in the particular leachate. Therefore, they are usually constructed of the same synthetic materials used in the manufacture of the selected flexible membrane liner. If the collected leachate is to be conveyed through an uncontaminated area, double-wall pipe may be desirable because it reduces the possibility of exfiltration. A leak-detection monitoring mechanism may be included between the walls. However, such monitoring is often subject to false alarms because of the inherent sensitivity of the instruments to errant moisture.

Depending upon the propensity of the leachate to volatilize, the vented air may require treatment. The collected leachate may be stored in a tank and pumped periodically to a truck for removal. Alternatively, the leachate may be collected in a tank or basin station for periodic pumping, placed in an above-ground storage tank, or transmitted directly to a treatment facility. The pumps and pipes of the collection system are sized to accommodate a wide range of leachate volume. Peak flows occur while the cell is open to the elements as it is being filled with waste; these flows can be several orders of magnitude greater than those that will be conveyed and treated once the secure cell is closed. The extensive range of flows affects not only the hydraulic capacity of the conveyance system but also the chemical strength of the leachate. During peak flows, the chemical concentration in the leachate may be diluted significantly.

Leachate is treated in a variety of ways, similar to those described for the treatment of ground water in Chapters 10 and 11. However, treatment of leachate is complicated by inconsistencies: a widely ranging volume and variability in both the flow and the concentration of the chemicals in the leachate. Equalization lagoons or tanks often are used to mitigate these variables.

Control of Gases. When hazardous wastes are buried, a gas may be produced. This is especially an issue when industrial or chemical wastes are buried together with biodegradable, usually municipal, wastes. These biodegradable wastes produce methane and carbon dioxide gases, and certain hazardous wastes can volatilize. The designer of the containment facility must control volatilization and the production of other gases; otherwise, the integrity of the cap could be compromised through chemical attack or pressure induced by the gas produced. The factors considered are the volume of gas to be yielded and its chemical composition.

Volatilization is usually controlled by placing a gas interception network over the waste material; the gas thus is collected, and, if necessary, it can be treated. This network, similar to a leachate collection system, consists of coarse, permeable gravel in a layer or in a network of trenches. Pipes may be inserted into the collection trenches. The gas venting system generally consists of risers through the cover system at collection points—the high elevations—of the gas collection network. The design of this system can be based on natural or mechanical ventilation, and the combustibility of gases, particularly methane, is considered in designing the ventilation system. In some cases, the gases require treatment before they are released to the atmosphere. If combustible gases are present, the gas may be flared. Treatment also can be effected by: carbon cartridge filters to remove volatile compounds and odors; chemical precipitation; stripping; and scrubbing.

ON-SITE SECURE LAND BURIAL

When in-place containment of hazardous wastes is infeasible, it may still be possible to contain the wastes on-site by constructing a new on-site secure burial cell. New on-site secure cells require careful preparation. The base is designed to be structurally sound and impermeable; liners are specified to be physically strong to prevent tearing and puncture, and to be compatible with the myriad chemical components of the wastes; monitoring systems are installed to detect and signal any leaks; leachate and gas systems are installed to collect wastes for treatment; impermeable caps are constructed to prevent surface water infiltration; and monitoring and sampling systems are installed to satisfy long-term closure requirements. Development of a secure on-site land burial cell is particularly attractive when commercial off-site facilities are unavailable or are beyond the limits of economical transport.

Extensive government regulations apply to the development of an on-site secure land burial cell, as various state and federal guidance criteria have been established that define the minimum requirements for solid waste facilities.

A secure cell basically consists of:

- A liner system, to inhibit the migration of leachate from the waste into the ground water or the surrounding soil.
- A cap system, to minimize the percolation of incident precipitation, to promote surface water runoff, and to control volatilization of the waste.

Some of the elements of on-site secure land burial are common to the on-site remedial alternative discussed earlier in this chapter. The common elements include:

- Provision for the control of surface and ground water contact with the waste, both during and after construction.
- Control of waste volatilization.
- Final capping.

The control of gases was addressed in the preceding section of this chapter and is not repeated here.

On-site secure land burial cells differ from in-place containment principally in the means by which the wastes are isolated from contact with ground water. Additionally, on-site secure land burial requires the removal of the waste material from its original location and placement into the prepared cell. The handling of the hazardous waste material introduces additional elements into the remedial program that merit consideration; not the least of these elements is the health and safety of workers during the excavation and movement of the waste. Another issue is the selection and utilization of equipment to excavate and haul the waste material.

The secure land burial cell should be constructed on a suitable and adequately prepared base. Ideally, the base materials should enhance the overall containment system by having low permeability, generally 10^{-7} cm/s or less and providing adequate separation between the deposited waste materials and ground water. Most federal and state regulations governing the construction of secure land burial cells specify minimum separation requirements, generally 5 ft (1.5 m) or more. In addition, the base must be structurally capable of supporting the filled cell. Soils having a low bearing capacity influence the height to which the secure cell can be constructed. The ultimate volume of waste that the secure cell is capable of holding is thereby limited. For this reason, it is important to understand well the stratigraphic foundation of the proposed secure land burial cell and to understand the soil properties of the proposed site. Both factors limit the ultimate size to which the cell can be constructed safely.

The properties of the soil are learned through the geotechnical testing and evaluation techniques commonly used in routine construction. With planning, this information can be obtained in conjunction with the initial site hydrogeological investigation. A typical geotechnical investigatory program includes soil sampling, classification of soil types by grain size using mechanical sieve and hydrometric analysis, and determination of Atterberg liquid and plastic limits. Consolidation tests and bearing capacity tests also provide valuable information related to the structural properties of the soil. Samples used for these tests should be representative of the overall site conditions and are collected in accord with standard sampling practices.[12]

As a case in point, an evaluation was performed at a site that had effectively no off-site disposal locations available for the waste material. Therefore, the client was limited to construction of an on-site land burial facility. Because the construction cost of such a cell is almost directly proportional to its area, it was desirable that the cell be constructed to the greatest height practical in order to achieve the lowest unit disposal cost. The geotechnical evaluation performed as part of the preliminary design for this project concluded that the underlying soils were compressible and had a low bearing capacity so that the soils limited the height of the cell. Given these circumstances, the client was influenced to reevaluate other remedial alternatives such as incineration, an alternative that had previously been considered more expensive than the secure cell.

As the soil used for the liner is compacted, the soils immediately beneath the cell are consolidated; thus, the potential for future settlement of the base and its potential failure as a liner are diminished. In order to achieve proper compaction, the soil is temporarily removed from the area where the base is to be constructed. It is then replaced in multiple compacted lifts with conventional earth-moving equipment. Irregularities in the soil or stones or other protrusions could pierce or rip any synthetic materials used in the liner construction. Loads applied to the liner from the heavy equipment used for construction of the cell and the subsequent placement of the waste materials could cause a puncture or tear.

Once the structural and hydraulic properties of the secure cell base are determined, a liner system is designed. The liner system generally consists of:

- A subliner of natural, recompacted soils or a secondary synthetic membrane, or both.
- A leak detection layer for the primary liner.
- A synthetic primary membrane.
- A drainage layer for leachate collection.

These layers are illustrated in Figure 9-2. The subliner system provides a second line of defense against the migration of contaminants in case the primary containment membrane develops a leak. Most federal and state guidelines condone composite subliner systems comprised of recompacted, low-permeability soils, generally 2 to 3 ft (60 to 90 cm) thick, overlain by a flexible synthetic membrane.

The primary concern in the selection of a material for the liner is the chemical compatibility of the liner material and the leachate to be generated by the wastes buried in the secure cell. Typical wastes, such as organic solvents, often affect the chemical bonds of the synthetic materials of the liner, and, over time, the liner can deteriorate. A substantial body of data has been compiled describing the compatibility of specific chemicals with various liner materials. Such information is relevant during design when candidate liner materials initially are screened. On the other hand, the long-term effects of chemicals on lining materials generally are less well defined, and available data are based on limited field data. If the secure cell will receive a variety of waste products, it is difficult to identify a suitable liner material because of unknown but synergistic effects of the chemicals that will form in the leachate. Therefore, testing is conducted in the laboratory to determine the compatibility of candidate liner materials and actual samples of the leachate. If it is impossible to collect the leachate, the chemicals that will be buried are synthesized into a representative sample of the expected leachate. Lengthy test procedures have been developed for this purpose.[13] In most

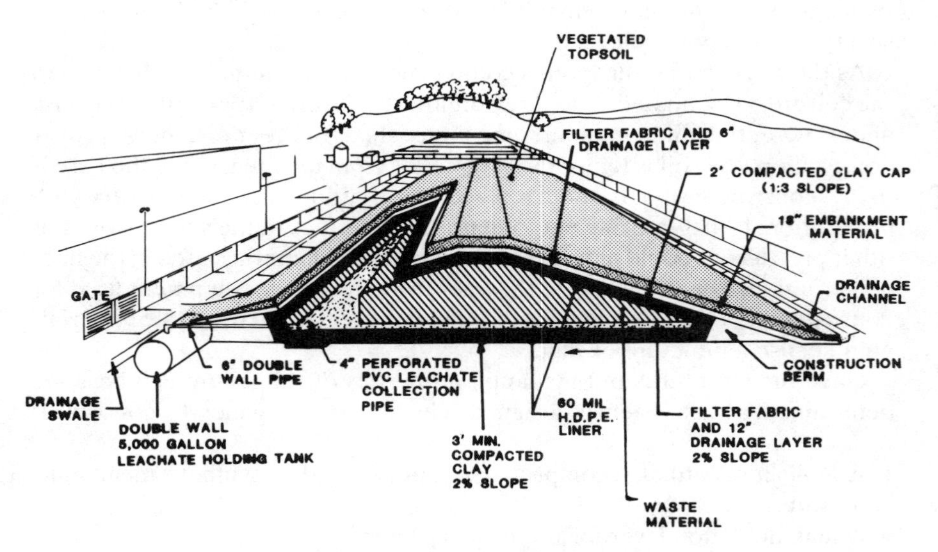

Figure 9-2. Typical design of secure cell liner system.

applications, a material resistant to the widest range of chemicals is used. Extrathick liners frequently are specified to counter the uncertainties of chemical compatibility.

A leak detection system generally is installed between the primary and secondary synthetic membranes. Thus, any leaks that might occur in the primary membrane can be discovered. Lysimeters, devices used to collect soil moisture, sometimes are used in leak detection systems. Most leak detection systems are composed of a permeable layer of material placed between the liners and sloped to a sump. Any seepage occurring between the liners drains through the permeable layer and is collected and monitored. The material in the drainage layer may be sand or gravel, or, if volume or height restrictions are a concern for the design of the secure cell, one or more layers of synthetic geo-nets. Regardless of the material used, the drainage layer normally is wrapped with filter fabric to prevent clogging by soil fines. Leak detection can be accomplished either by physical measurement of the collected moisture or by automated, current-activated alarm systems.

In addition to the bearing capacity of the underlying soils, it is important to consider the suitability of native soils for use in the construction of the berms and dikes that are to be part of the secure cell. The materials used for berms and dikes should have properties that allow their construction at the maximum slope considering an appropriate factor of safety. Therefore, their design depends upon the particular properties of the soils used. For each berm or dike design considered, slope stability calculations are performed. Various computer programs based on accepted methods of analysis can be employed.[14] These calculations are required by regulatory agencies to support the design of the secure cell and for the approval of environmental permits. Economic considerations nearly dictate that berms and dikes use materials readily available to the site of the work. However, if the materials at hand are unsuitable for use in the construction of the secure cell, the design must include provisions for bringing suitable materials to the site or for the use of reinforcing materials to enhance the stability of the slope. Special geotextiles and synthetic geo-grids have been used for this purpose. Gabions, stepped embankments, and other construction techniques have also been applied to reinforce slope stability.

The way that leachate is best handled depends upon the length of time the secure cell will be open to the environment and the probable volume of precipitation expected during that time. The volume of water can be estimated by calculating a water balance based on the design of the cap and by accounting for the anticipated incident precipitation during construction of the cell. A number of procedures for predicting leachate quantities have been published; a widely used procedure is the U.S. EPA Water Budget method.[15] The treatment given the leachate is dependent upon the range of volume, flow, and concentration of the contaminants. Lagoons or tanks can be used to reduce the effects of inconsistencies on the treatment process. (See also Chapter 10 for treatment technologies.)

Once the secure cell is prepared, excavation and relocation of the waste material commence. As with any remedial option that necessitates direct exposure to contaminated waste materials, safety is an essential consideration. Field personnel often must wear protective clothing, hard hats, goggles, and, in some cases, respirators. (See Chapter 6 for details.) The use of safety equipment is mandatory, and workers require time to don it at the start of each work shift and for cleanup or decontamination after work. The designer factors in such time when scheduling the productivity of each work crew.

As the plan for placement of the waste is developed, the scheduling is constrained by the time allowed for the total remedial program. The schedule, moreover, dictates the number of pieces of equipment and the requisite capacity of the equipment to be employed in the work. Hazardous wastes usually are excavated with conventional earth-moving equipment: backhoes and loaders generally are used for precise excavation and excavation at depth. Motorized excavating pans and bulldozers typically are used for shallow and areally extensive waste removal. In equipment selection, a serious consideration is the amount of dust that each piece will generate.

During excavation, too, the safety of persons off-site is considered. Intense rain could wash away contaminated material; high winds could blow contaminated dust toward receptors in the environs of the site. The phase of the project when waste is excavated is programmed so that the active area exposed at any one time is limited. Through well-planned sequencing and staging of the waste removal, the potential for volatilization of waste material and the exposure of waste to wind and surface runoff can be diminished. Thus, the potential for contaminants to migrate from the site is reduced. Monitoring programs and contingency plans are developed to address any migration that does occur. Components of these plans include procedures to notify emergency response teams and government officials and to evacuate residents, if such action is essential. During one experience when contaminated soil was excavated and relocated, a public road had to be closed temporarily. In addition, an extensive dust control program was implemented. In such situations, the project team necessarily works closely with local health and safety officials.

The waste material is placed and compacted in the secure cell. The techniques used are the same ones that would be employed for the construction of any embankment. The material is spread and graded with a bulldozer, and it is compressed with a roller or another type of compacter. The waste is placed in rather thin lifts, of 6 to 12 inches (15 to 30 cm), to achieve uniform compaction. Density tests normally are performed on each completed lift to ensure stable construction.

As the waste is transported from the site of removal to the new secure cell, small quantities of contaminated materials could become spread along the haul roads. For this reason, haul roads on the site of the excavation are strictly delimited, and transport vehicles are confined to those routes only. Once the hauling is complete, the roads are inspected for residual contamination, and, if necessary, cleanup measures are implemented. Contaminated material from the haul roads is placed into the secure cell.

After all the wastes are deposited in the secure cell, it is sealed with an impermeable cap similar to that described for in-place containment facilities above. In order to enclose the containment completely, the synthetic membrane liner included in the cap is joined to the liner of the cell bottom.

CHEMICAL FIXATION AND STABILIZATION

Some hazardous wastes do not readily lend themselves to containment in place, to deposition in new secure land burial cells, or to removal from the site unless they first are stabilized. A number of chemical techniques are available to stabilize waste, both organic and inorganic, which may then be buried. This section of the chapter generally describes some of these chemical processing techniques. To be practicably implemented, the processing of hazardous waste should be relatively easy and economical. Neither the toxic material nor the chemical agents used for solidification should be reactive or easily degradable.

Chemical fixation or stabilization renders hazardous wastes nontoxic. Chemical agents fix or solidify the waste, and the toxic components within the solidified material are immobilized. The pioneering work with solidification processes began with radioactive wastes. Recently, solidification has become an increasingly attractive alternative to the remediation of improperly discarded hazardous wastes. The toxic components are bound into a form immune to leaching. The binding is achieved by fixing the components within the solid matrix of the chemical agents or by encapsulating the components within a solid form of the chemical agents, and the net effect is to reduce or eliminate the surface area available for the release of the toxic components into the environment.

The disposition of hazardous wastes through chemical fixation and stabilization processes depends on the physical and chemical characteristics of the wastes and the characteristics of the resultant material. Of particular importance is the matrix containing the toxic components. Not all hazardous wastes are suitable for fixation or solidification because, depending upon the chemical characteristics of the waste, the solidified material may be unqualified for safe disposal. For solidification to be applicable as a remediation alternative, the following qualities are necessary:

- The toxic components in the waste are in a form immune to leaching.
- The process will result in improved waste handling.
- The material is not reactive or degradable.
- The material is structurally stable.

Regulations require that solidified hazardous wastes be tested for reactivity, EP toxicity, ignitability, and corrosivity before disposal.[16] The solidified wastes are usually disposed of in sanitary or secure landfills. The landfill selected for disposal may have liners or caps. Consequently, in order to support the cap of the secure cell, the solidified wastes must maintain strong structural integrity over the life of the disposal site. In addition, the solidified material should be stable under varying environmental conditions.

A variety of chemical fixation or stabilization processes are available, the selection of a particular process being mainly dependent upon the physical and chemical characteristics of the waste. It is essential to test the compatibility of the waste with the candidate fixation processes. Other considerations are pretreatment requirements, ease of operation, specifications of the finished product, cost of processing, and the method of disposal. Chemical fixation or stabilization processes are classified on the basis of the chemical agents used to immobilize the hazardous or toxic components or the physical aspects of the processing. Table 9-3 lists typical techniques.

Except for encapsulation, the processes listed consist of creating a homogeneous mixture of the hazardous waste materials and the chemical agents. That mixture then solidifies into a monolithic block of waste with high structural integrity. The toxic materials are thus contained within the solidified matrix.

Cement-based processes for solidification normally use common construction materials such as portland cement in conjunction with other additives to form a rigid concrete-like mass incorporating the toxic components within the matrix. Commonly used additives include fly ash, clay, vermiculite, and sodium silicate in addition to several proprietary additives.

TABLE 9-3. Fixation processes.

Cement-based processes	Thermo-plastic based processes
Lime-based processes	Self-cementing techniques
Silicate-based techniques	Surface encapsulation
Organic polymer-based techniques	

Solidification with cement and other additives is best suited for hazardous waste containing heavy metals, where the solidified material is not subjected to wide variations in pH. Ideally, wastes contain moisture in the range of 30 to 50 percent by weight. Wastes in the form of slurries with a greater water content are dewatered to minimize the proportion of chemical agents required for solidification.

Metals are incorporated into the solidified matrix as hydroxide or carbonate compounds, or directly into the cement matrix as ions. Certain materials such as asbestos, latex, metal filings, and plastics tend to enhance the properties of the solidification mixture because they increase the strength and stability of the mixture and decrease setting and curing times. Certain inorganic and organic compounds contained in waste cause problems in the setting and curing of the waste-cement mixture, so that additives are required to overcome their adverse effects. Organic materials such as silt, clay, coal, and lignite may delay the setting and curing of portland cement for several days. Sodium salts of arsenate, borate, phosphate, iodate, and sulfide can both prolong the curing process and weaken the structural integrity of the treated waste product. Salts of copper, lead, magnesium, tin, and zinc may have the same effects.

The proportion of cement and other additives required is usually equal to or greater than the amount of waste. Therefore, the total quantity of waste that must be disposed of is at least doubled. Because of high transportation and disposal costs, cement usually is used in conjunction with other solidification processes where the result is less volume. A major advantage of the cement-based process is the relatively low cost of cement and additives. Also, equipment for processing is readily available and easy to operate.

Lime with fine-grained siliceous materials and water also is used to form a pozzolanic concrete-like solidified mass to fix hazardous waste. Waste products such as fly ash and cement-kiln dust commonly are used as additives. In general, lime-based and cement-based techniques are applicable to similar hazardous wastes and also have similar setting and curing problems.

As a result of the problems associated with cement- and lime-based processes, silicate-based solidification techniques have been developed. A siliceous material is used to bind toxic components; lime, cement, gypsum, or another suitable additive bolsters setting and curing. Other additives are used for specific purposes. Clays bind water and specific anions or cations; emulsifiers and surfactants incorporate immiscible organic liquids; proprietary adsorbents also are used as additives. Silicate-based solidification is applicable to most divalent cations and certain organic solvents. Current research is concentrating on the use of newer siliceous compounds for the solidification of hazardous materials, particularly those containing difficult components such as organic solvents. Materials such as sodium borate, calcium sulfate, potassium dichromate, carbohydrates, and oil and grease can interfere with the formation of bonds with the solidified material. Silicate-based processes generally are proprietary in nature because of the very selective mix of additives developed for specific wastes. These processes can be inexpensive if the chemicals used are commonly available, but they may be costly otherwise.

The polymerization of an organic monomer in the presence of a catalyst and the hazardous waste is a technique used to trap the constituents within the polymeric mass. The most commonly used monomer is urea formaldehyde; others include vinyl ester-styrene and polyester. The polymeric material does not combine chemically with the waste but merely forms a spongy mass that traps the solid particles while permitting some liquid to escape. Organic polymer-based techniques are advantageous because they require low volumes of additives, the polymer may be applied to both dry and wet sludges, and the waste-polymer mixture has a low density. A major disadvantage is that the solidified product must be placed in a container because of the loose matrix and the possibility of biodegradation of certain organic polymers. In addition, certain processes are acidic and may cause metals that are soluble at low pH values to escape with the water.

Hazardous wastes may be fixed within plastic materials such as asphalt, bitumen, paraffin, and polyethelene, bitumen being the most commonly used thermoplastic material. Dried hazardous waste material is mixed uniformly with a heated thermoplastic and then cooled. The resulting solidified material is rigid but plastic, and it is generally more resistant to water or biodegradation than the untreated waste.

Some wastes, notably those high in sulfate and sulfite, are amenable to self-cementing techniques. Gas cleaning or desulfurization sludges containing large amounts of calcium sulfate can be self-cemented. This process is rather similar to cement-based techniques because 8 to 10 percent of the waste, by weight, is calcined to form cement-like products—calcium sulfate or calcium sulfite—under strictly controlled conditions. Pretreatment may be necessary; the sludge may require dewatering, or fly ash may have to be added to obtain an optimum moisture content prior to processing. The major advantage of self-cementing is that the main constituent of the solidified products is inorganic salts. Thus, the product is stable, inflammable, and nonbiodegradable. Setting and curing times are less than those of the lime-based process. The process is costly, however, because it is energy-intensive, and because skilled labor and specialized equipment are required.

Hazardous waste or a solidified form of the waste may be micro-encapsulated within an impermeable jacket. Most of the techniques available are applicable only to inorganic wastes, particularly those containing heavy metals. Organic compounds are, as a class, the least amenable to solidification, although methods to solidify them are under development. Anions of sulfates and chlorides, common in hazardous wastes, are usually difficult to bind into insoluble compounds; therefore, the anions are precipitated by means of additional chemicals.

In sum, a variety of chemical stabilization techniques are available to assist in the disposal of hazardous wastes. Among the additives are cement, lime, silicates, organic polymers, and thermoplastics. In addition, there are self-cementing techniques and surface encapsulation methods. Whenever chemical fixation is selected for implementation, chemical compatibility must be assured between the waste and the stabilization additive and between the resultant product and the environment. When wastes must be stabilized before disposal, handling costs and risks are initially higher than they would be if the wastes were buried immediately.

POST-CLOSURE MAINTENANCE AND MONITORING

After the wastes are in place and construction has ended, waste burial facilities—both in-place containments and new secure cells—are maintained and monitored. The long-term management of on-site hazardous waste facilities includes routine inspection and maintenance of the landfill, any associated facilities required for the treatment of the leachate generated over the post-closure period, surface water controls, and ground water monitoring facilities.

The flow of ground water and leachate are audited and controlled for, usually, a minimum of 30 years. A program of ground water monitoring is central to the post-closure program of auditing the integrity of the facility. Permanent monitoring wells are installed for this purpose. A minimum of one upgradient well and three downgradient wells normally are installed, but more wells frequently are required in order to characterize adequately any changes in ground water quality. The minimum information collected and maintained through the post-closure period includes ground water quality parameters and indicators of ground water contamination. The specific parameters are determined by the characteristics of the wastes deposited in the site.

The post-closure maintenance and monitoring methods described here are typical. However, many valid and effective variations are practiced.

Routine Inspection and Maintenance. The landfill and area immediately adjacent should be inspected on a routine schedule, usually monthly, and the inspection and maintenance activities should be carried out by trained personnel. The inspector should observe the condition of the cap and its vegetative cover. No trees, shrubs, brush, or deep-rooting weeds should be allowed on the cap. If observations indicate that deep-rooting vegetation has become established, a vegetation control program should be initiated. Monthly inspections also scan for any signs of erosion, insect damage, and disease or thinning of the vegetative cover. Such conditions should be corrected immediately upon identification. It is important to keep the vegetation as dense and uniform as possible, both to impede erosion and to optimize transpiration. The grasses on the cap should be mowed periodically to maintain a maximum height of, typically, 3 to 4 inches (8 to 10 cm). Taller grass can inhibit erosion and increase evapotranspiration.

Roads leading to the facility must stay in good repair in order to support the maintenance activities. During the post-closure period, access by unauthorized persons and vehicles should be barred. Fencing, gates, and signs installed as part of the site security plan should be routinely inspected and maintained in order to insure their integrity.

Leachate generated during or after the landfill is closed and collected by the leachate collection system typically is conveyed to on-site storage or to an aqueous treatment facility. The design of a leachate collection system should consider the long-term operation of the facility and include adequate access for maintenance purposes. Manholes for clean-out points should be placed at critical locations such as pump stations, at the locations of key valves, and at regular and frequent intervals along the conveyance lines. The valves on the network of pipes should be designed to permit periodic pressure testing of the line segments to search for leaks and to allow the operator to isolate segments should a leak occur. If lines are installed outside a liner system, a mechanism for leak detection such as current-activated alarm systems or shallow trench lysimeters should be incorporated. Double-wall pipe also can be used effectively to provide secondary containment to guard against leaking pipes.

Maintenance of the leachate collection system typically includes visual inspection of any sumps and pumps as well as instrumentation and control systems. All floats and switches should be operated periodically to insure their ability to function. Visual inspection of leachate discharge lines usually is made at the clean-outs and at the discharge manholes as well as through sumps and riser access casings. Pumps and valve seats should be checked periodically for deterioration. Replacement parts should be stocked and readily available. Backup systems and failure alarms are necessary to assure a rapid response to malfunctions and to preserve the integrity of the containment. If a leak is detected, the appropriate valves should be closed to isolate the leaking line, and the line segment in question can be pressure-tested using the line clean-outs. If the lines are plugged, they can be cleared with flushing or by applying pressure at the clean-outs.

Drainage facilities, including all surface water diversion ditches, should be inspected routinely and after major storms in order to identify any accumulation of debris or any obstruction that may have occurred. Any accumulated materials should be removed and properly disposed of immediately. Any needed repairs to the ditches also should be implemented as soon as discovered.

To maintain their integrity, monitoring systems require inspection on a routine basis, and they should be maintained as necessary. Maintenance activities include replacement of cracked protective casings in ground water monitoring wells, cleaning or replacement of water quality monitoring probes, and calibration of flow measurement devices.

Post-closure Monitoring. A monitoring program should be developed that is fitted to the particular circumstances of the site. A statement of goals is articulated first in a written monitoring plan; in that plan, the objectives of the monitoring program are devised. These monitoring objectives depend on the contaminants present and the potential routes these contaminants could travel to jeopardize receptors. The monitoring program should specify the parameters to be monitored and the frequency of analysis. It also itemizes the tasks and schedule of the site inspector. In its most global form, the monitoring program can encompass ground water, surface water, air, sediment or soil samples, and leak detection. The results of the monitoring program are assessed periodically, typically annually, to insure that they are providing the information necessary to observe the integrity of the site.

Ground water sampling and analysis usually are required on a quarterly basis. Usually, composite leachate samples are collected from the facility sumps or storage tanks and analyzed in accord with the predetermined monitoring program. The sampling and analysis should be performed with all the controls associated with proper field methods: sampling and analytical protocols are specified; equipment is maintained, cleaned, and decontaminated as necessary; a field log book documents the sampling and sample handling; the chain of custody is recorded; and quality control/quality assurance measures are implemented.

SUMMARY

On-site containment of hazardous wastes often is the most justifiable remedial alternative. A properly designed, constructed, and maintained containment can minimize risks to the environment and to public health. It is frequently the most economical alternative because the wastes are not moved, or they are moved a relatively short distance. Because of these advantages, on-site closure of a hazardous waste site is often the most acceptable solution.

Three on-site techniques are discussed in this chapter. In in-place containment, the wastes are not moved, and the transport routes through which the wastes could escape into the environment are effectively blocked. In the secure burial cell technique, the wastes are excavated and sealed in a new secure burial cell constructed adjacent to the original waste deposit. In neutralization, various chemical treatments are used to render the wastes nonhazardous; the wastes may then be sealed in place or put into an on-site secure cell.

In order to determine which of the three methods is suitable for a site, an economic assessment, an environmental assessment, and a risk assessment are performed. Facilities associated with the on-site containment require long-term maintenance. Caps and leachate collection and treatment facilities are given special attention to insure that they are functioning as designed. Ground water wells around the site are monitored to make sure that no leachate is escaping from the containment.

The design and construction measures together with the long-term maintenance and monitoring procedures are effective remedies for a hazardous waste site. During each phase of the remediation work—from assessment to design to construction to post-closure activities—care must be exercised to insure maximum protection to the environment and to public health.

NOTES

1. 41 U.S.C. 6901.
2. 15 U.S.C. 2601.
3. See refs. 9 through 14 in *Proceedings* of national conference on management of uncontrolled hazardous waste sites, 1982. Silver Spring, Md.: 1982 Hazardous Materials Control Research Institute, 1982. p. 182. See also pp. 175-202.
4. U.S. EPA et al. Hazardous waste management conference. 1982. Fig. 4, p. 179.
5. U.S. EPA et al. Hazardous waste management conference. 1985. p. 370.
6. U.S. EPA et al. Hazardous waste management conference. 1982. p. 180.
7. U.S. EPA et al. Hazardous waste management conference. 1982. p. 181.
8. ASTM D1004. ASTM D3083. ASTM D638.
9. U.S. EPA *Corporate design guidance.*
10. U.S. EPA et al. Hazardous waste management conference. 1984. p. 114.
11. U.S. EPA, MERL. *Field assessment of site closure, Boone County,* Ky. Aug. 1983. p. 21. PB83-251629.
12. *Annual Book of ASTM Standards.* 1984. Section 4, "Construction Volume 04.08, Soil and Rock Building Stores."
13. U.S. EPA. Chemical compatibility testing. Method 9090.
14. S. G. Wright. SSTABI—a general computer program for slope stability analyses, Personal Report No. GE-74-1, Department of Civil Engineering, University of Texas. Aug. 1974. See too Geocomp Corp. GEOSLOPE. Concord, Mass.
15. P. R. Schroeder, J. M. Morgan, and A. C. Gibson. "The Hydrologic Evaluation of Landfill Performance (HELP) Model EPA/530-SW-84-009 and EPA/530-SW-84-010 (2 vols.). June 1984.
16. 40 CFR 261.20 and 261.24. Rev. 1 July 1982.

Chapter 10
The Disposition of Ground Water

A review of potential corrective actions for an unsecured hazardous waste site may reveal that the most economical and practicable remedy for contaminated ground water is to collect and treat it. In many instances, collection and treatment constitute the only remedy that effectively removes contaminants from the ground water system. Collection and treatment are often the chosen alternative when containment methods are infeasible for economical or technical reasons, as, for example, when constructing a containment wall to the depth required would be impracticable, or when the contaminants of concern are incompatible with the materials available for use in a containment wall. Many of the considerations pertinent to the collection of ground water are the same as the factors important in the initial investigation, and the information presented about monitoring wells in Chapter 3 is pertinent to the discussion here.

Ground water collection provides several benefits leading to the remediation of a site:

- Removal of the highest contaminant concentrations from the ground water at the source.
- Prevention of downgradient migration of the contaminant plume.
- Prevention of off-site migration of the contaminant plume.

These benefits are manifested both through a separate collection system and when a collection system is combined with a containment structure.

Ground water can be treated in a variety of ways. In order to remove contaminants to levels that satisfy regulatory requirements, techniques capable of separating the compound or compounds of concern from the water are employed. The physical and chemical properties of each compound are evaluated to determine the appropriate treatment technology.

Implementation of ground water treatment utilizes the traditional engineering design approach to a construction project. Through the logical

---

This chapter was developed by John C. Tomik, J. Andrew Irwin, P.E., and Robert C. Ganley, P.E. of O'Brien & Gere Engineers, Inc.

progression of conceptual, preliminary, and final design, a remediation project evolves from problem definition to implementation of a technically sound and economically feasible solution.

COLLECTION PRACTICES

Several techniques are available to remove ground water from subsurface geological formations, with the collection system selected determined by the hydrogeological conditions of the site. The important factors are:

- Depth to ground water.
- Depth, nature, and concentration of the contaminants.
- Hydraulic conductivity of the aquifer.
- Anticipated yield of ground water.
- Thickness of the aquifer.
- Distribution of grain size in the aquifer.
- Transmissivity and storage capacity of the aquifer.

Information about these factors is developed through the hydrogeological site assessment. (See Chapter 3.)

Well Construction. In order to collect contaminated ground water, one or more wells may be installed. The variability of hydrogeological factors makes it impossible to standardize the design of a collection well. However, the American Water Works Association has published standards for deep wells and provides descriptions and illustrations of the most common types of well construction.[1] See Figure 10-1 (pp. 258-261) for examples.

To provide data about the hydrogeological conditions of a site with the detail necessary to design a well system, a pilot hole is drilled at each proposed well location. In the drilling of the pilot hole, these data are collected: grain size of soils; ground water levels; chemical characteristics of ground water, from analysis of collected samples; and thickness of the aquifer. The development of data from the pilot holes minimizes the risk of constructing an inadequate well system.

Wells constructed in consolidated aquifers generally are composed of a casing extending into the bedrock formation and an open borehole extending below the casing to a depth where a suitable well yield is obtained. In an unconsolidated formation, a well is composed of a well casing extending above a well screen to provide an intake area where water may enter the well from the aquifer. Depending upon the geological conditions of the well site, an artificially graded gravel material may be required in a zone immediately around the well screen to optimize well efficiency.

The minimum diameter of a well casing is based on both the anticipated yield of the well and the open area of the well screen required to maintain the maximum hydraulic efficiency of the well. The diameter of the well itself is designed to furnish sufficient clearance for the installation and operation of a pump with the necessary capacity. Table 10-1 (p. 262) provides basic data on the relationship between casings and yields.[2]

The well casing is most frequently constructed of a steel that conforms to standards of the American Petroleum Institute[3] or the American Society for Testing Materials.[4] A properly designed well screen has two properties. It provides a large enough intake area to allow water to flow into the well at a low velocity, and it provides a sufficiently small intake area to prevent sand from entering the well with the water. Several types of well screens are available, including perforated casing, slotted pipe, louver-type screen, and case-type wire-wound screen. The wire-wound screens are the most efficient. They consist of a continuous winding of round- or wedge-shaped wire on a cage of vertical rods. The slot size of the screen is fabricated to conform to the aquifer materials and can range from 0.003 to 0.250 inch (0.08 to 6.35 mm).

Selecting the slot size of the well screen is the critical factor in well design. The selection is based upon analysis of the grain size of soil samples from the aquifer. If the grain size analysis indicates a uniformity coefficient of 5 or less, the slot size should retain 40 to 50 percent of the aquifer material. (The *uniformity coefficient* is computed by dividing the 40 percent retained size of sediment by the 90 percent retained size.[5]) If the uniformity coefficient is greater than 5, the slot size should retain 30 to 50 percent of the aquifer material.[6] The fine material that passes through the screen is removed during well development; the coarse material is held outside the screen. Therefore, a natural gravel pack develops around the well screen.

A gravel-packed well is one in which an artificially graded gravel envelope is placed around the well screen to prevent fine sand from entering the well and to provide an annular zone of high permeability. Although a gravel-pack well is more costly than a naturally developed well, site conditions may require the gravel-pack well in order to optimize well efficiency. For example, a gravel-packed well is required under the following conditions: fine, uniform sand where the well screen slot size selected for a naturally developed well is less than 0.010 inch (0.25 mm); aquifers consisting of thin alternating layers of fine, medium, and coarse sand; and highly weathered or fractured bedrock formations.[7] Gravel-pack material consists of clean, sorted, well-rounded grains of siliceous material. The specification of a gravel pack

(continued on p. 262)

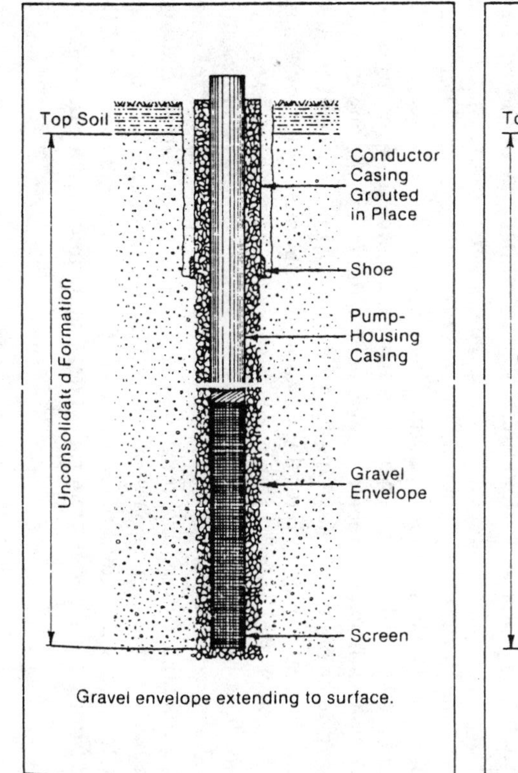

Gravel envelope extending to surface.

Gravel-packed well with conductor casing grouted in place and gravel envelope extending to surface.

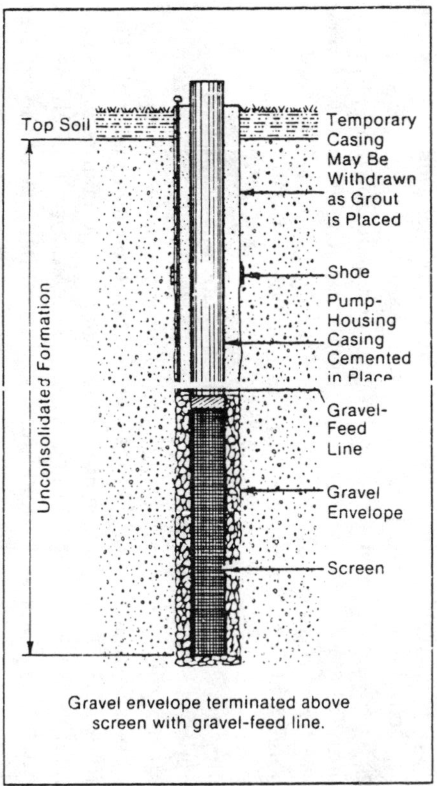

Gravel envelope terminated above screen with gravel-feed line.

Gravel-packed well with well casing cemented in place and gravel envelope terminated above the top of the screen.

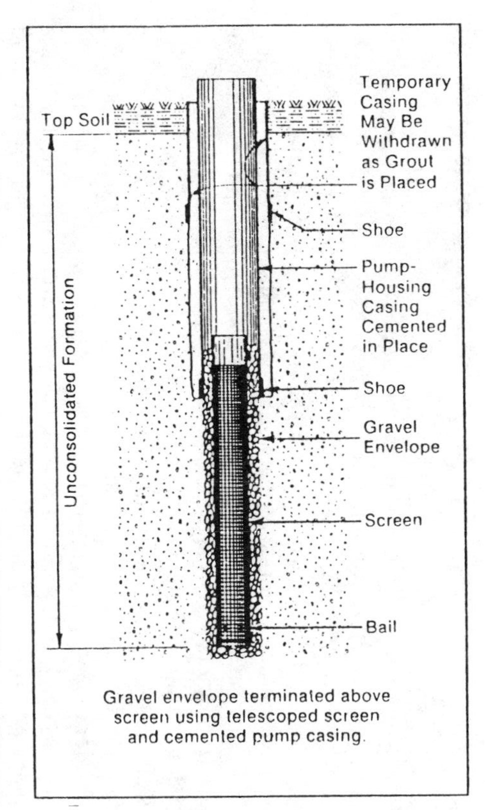

Gravel envelope terminated above screen using telescoped screen and cemented pump casing.

Gravel-packed well with telescoped screen, well casing cemented in place, and gravel envelope terminated above the top of the screen.

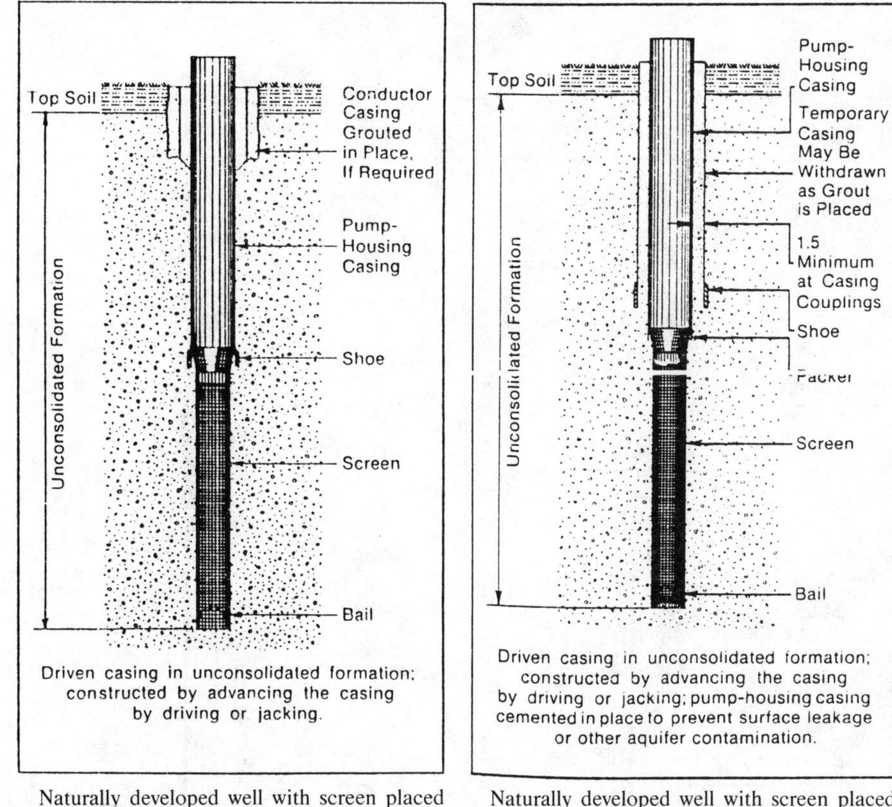

Naturally developed well with screen placed across aquifer and well casing driven or jacked into place.

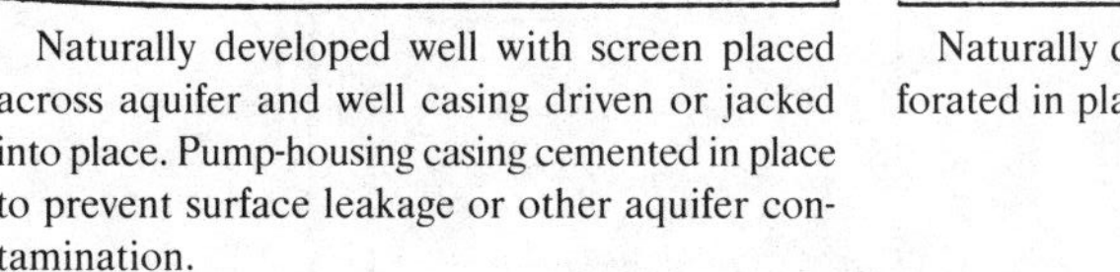
Naturally developed well with screen placed across aquifer and well casing driven or jacked into place. Pump-housing casing cemented in place to prevent surface leakage or other aquifer contamination.

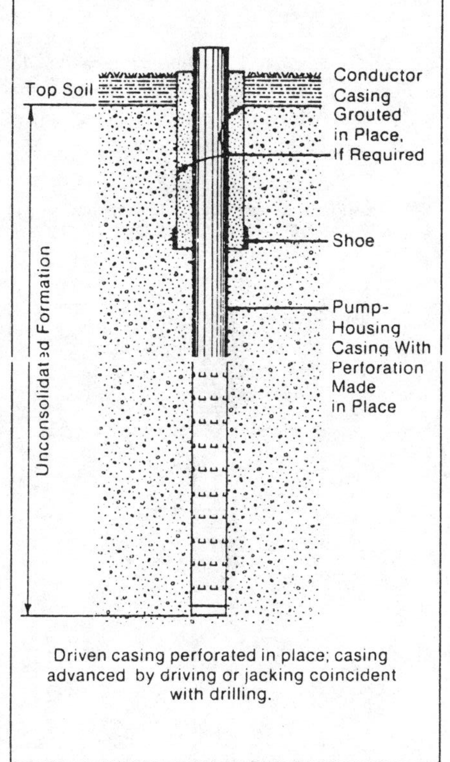

Naturally developed well with well casing perforated in place.

Figure 10-1. Common types of well construction: consolidated and unconsolidated formations. Reprinted from AWWA Standard for Water Wells, by permission. Copyright 1984 American Water Works Association

Overburden
Consolidated Formation
Creviced Rock Formation
Water-Bearing Rock
Temporary Casing May Be Withdrawn as Grout is Placed
Pump-Housing Casing Cemented in Place
Gravel Envelope
Screen

Well with grouted pump-housing easing, screen and gravel packed; used when formation yields water with turbidity unless filtered through gravel pack; conductor casing installed through overburden.

Gravel-packed well completed in consolidated rock and well casing cemented in place.

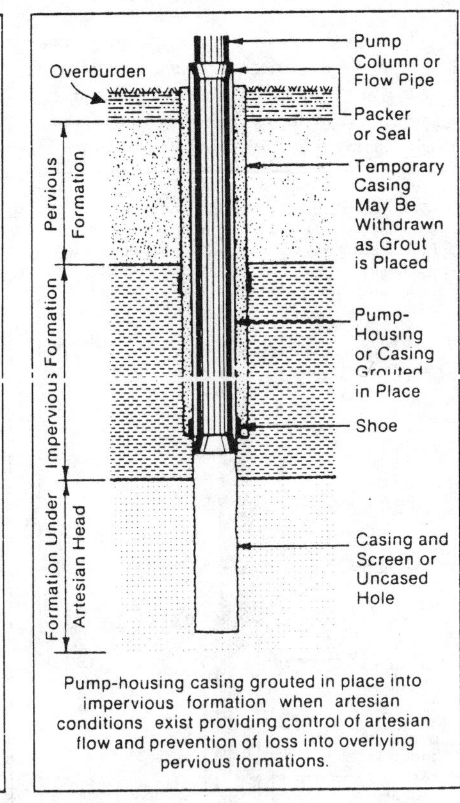

Pump-housing casing grouted in place into impervious formation when artesian conditions exist providing control of artesian flow and prevention of loss into overlying pervious formations.

Well completion where piezometric level is above ground elevation.

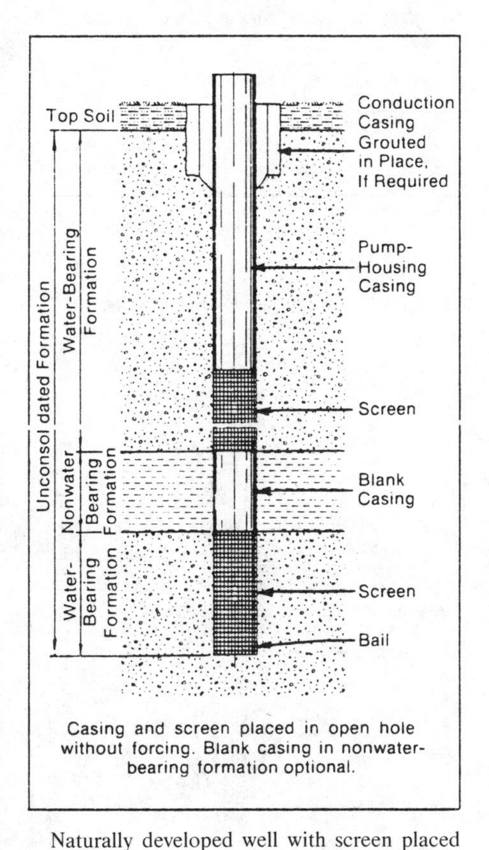

Casing and screen placed in open hole without forcing. Blank casing in nonwater-bearing formation optional.

Naturally developed well with screen placed across aquifer and well casing set in place in an open hole larger than the casing and screen without forcing.

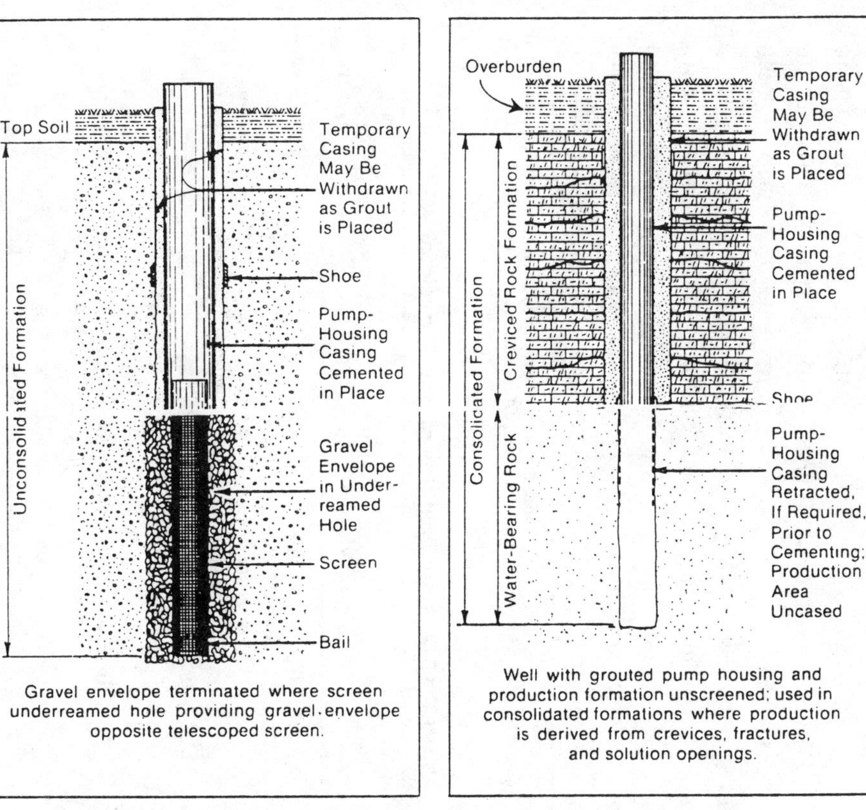

Gravel-packed well with bore hole underreamed for screen and well casing cemented in place.

Well with open hole completion in consolidated rock and well casing cemented in place.

Figure 10-1. *(continued).*

TABLE 10-1. Recommended casing sizes for specific well yields.

ANTICIPATED WELL YIELD		NOMINAL SIZE OF PUMP BOWLS		OPTIMUM SIZE OF WELL CASING*		SMALLEST SIZE OF WELL CASING*	
gpm	m^3/d	in.	mm	in.	mm	in.	mm
<100	<545	4	102	6 ID	152 ID	5 ID	127 ID
75-175	409-954	5	127	8 ID	203 ID	6 ID	152 ID
150-350	818-1,910	6	152	10 ID	254 ID	8 ID	203 ID
300-700	1,640-3,820	8	203	12 ID	305 ID	10 ID	254 ID
500-1000	2,730-5,450	10	254	14 ID	356 ID	12 ID	305 ID
800-1800	4,360-9,810	12	305	16 OD	406 OD	14 OD	356 OD
1200-3000	6,540-16,400	14	356	20 OD	508 OD	16 OD	406 OD
2000-3800	10,900-20,700	16	406	24 OD	610 OD	20 OD	508 OD
3000-6000	16,400-32,700	20	508	30 OD	762 OD	24 OD	610 OD

*The size of the well casing is based on the outer diameter of the bowls for vertical turbine pumps and on the diameter of either the pump bowls or the motor for submersible pumps.

TABLE 10-2. Specifications of gravel pack for wells.

UNIFORMITY COEFFICIENT (U_C) OF AQUIFER	GRAVEL PACK CRITERIA	SCREEN SLOT SIZE
<2.5	(a) U_C between 1 and 2.5 with the 50% size not greater than 6 times the 50% size of the aquifer. (b) If (a) is unavailable, U_C between 2.5 and 5 with 50% size not greater than 9 times the 50% size of aquifer.	$\leq$10% passing size of gravel pack
2.5-5	(a) U_C between 1 and 2.5 with the 50% size not greater than 9 times the 50% size of the formation. (b) If (a) is unavailable, U_C between 2.5 and 5 with 50% size not greater than 12 times the 50% size of aquifer.	$\leq$10% passing size of gravel pack
>5	(a) Multiply the 30% passing size of the aquifer by 6 and 9 and locate the points on the grain-size distribution graph on the same horizontal line. (b) Through these points, draw two parallel lines representing materials with $U_C \leq 2.5$. (c) Select gravel pack material that falls between the two lines.	$\geq$10% passing size of gravel pack

material depends on the grain size of the aquifer's materials; the criteria sometimes used are summarized in Table 10-2.[8] The thickness of the gravel around the well screen ranges between 3 and 6 inches (8 and 15 cm).

Installation of the gravel pack generally involves:

1. Installation of a temporary and oversized well casing.
2. Installation of the well, casing, and screen.
3. Placement of the gravel pack through tremie pipes.
4. Removal of the temporary casing.

At a minimum, the gravel-pack material extends 2 ft (61 cm) above the top of the well screen. Once the gravel-pack material is selected, the slot size of the well screen may be chosen. The screen should retain 90 percent or more of the gravel pack.

After a well is installed, it is developed to optimize efficiency and to preclude fine materials from entering. Development of a well is a mechanical process: a flow of water is drawn into the well and flushed into the formation to remove the fine materials from the formation immediately surrounding the well screen. Development is continued until the movement of fines from the formation lessens and the well reaches its desired yield. Development methods vary, depending upon the aquifer materials, type of well construction, and well installation procedure. A number of methods are presented below.[9]

- *Mechanical surging:* This method involves the up-and-down movement of a surge block inside the well casing, generally a leather or rubber disk sandwiched between wooden disks that have a slightly smaller diameter than the well screen. The movement of the surge block creates a piston-like action; it moves the fine materials into the well and forces a surge of water into the formation. This method is commonly employed when the cable tool drilling technique is used for well installation. It minimizes the amount of water removed from the well, an important consideration if the water requires treatment.
- *Air surging:* An air compressor is connected to an air lift system consisting of an air line inside a discharge pipe; the air lift system is installed within the well. Development is initiated by extending the air line below the discharge line and abruptly surging air at pressures of 100 to 150 lb/in^2 (70,300 to 105,500 kg/m^2) into the well screen. The air line is then raised inside the discharge pipe, and the well is pumped by conventional air lift until the water runs free of sand. This method of development is especially useful in gravel-packed wells.

- *Overpumping:* The simplest method of well development is to pump the well at a yield exceeding its anticipated pumping rate. Pumping is commenced at a low discharge rate and increased in stages until the rate exceeds the anticipated flow rate. This method usually stops fines from entering the well, but it often cannot maintain the optimum well efficiency. Another drawback to this method is that a large volume of water is removed from the well—water that may have to be treated before it is discharged.

Function of Well Systems. When ground water is collected as part of the remediation of a hazardous waste site, the well system is designed such that:

- The total discharge of the wells provides a cone of influence that completely intercepts the width of the contaminated plume.
- The well system reverses the hydraulic gradient and the downgradient plume boundary to prevent additional migration.
- Discharge rates of the well system are sufficient to reverse the hydraulic gradient but not so high that drawdowns reach the intake of the pump.

A major consideration in designing a well system is to predict the quantity of ground water to be pumped from a single well or from the entire system. This quantity is estimated by using data obtained from the hydrogeological investigation as values in equilibrium equations applied to the water table under, it is assumed, confined conditions; see Figure 10-2.[10] Equilibrium equations assume ideal aquifer conditions: a homogeneous and isotropic aquifer, horizontal and uniform flow, velocity proportional to the tangent of the hydraulic gradient, and steady radial flow. The equations generally furnish a reasonable prediction of a well's performance.

The most accurate way to determine the radius of influence of a well is to conduct an aquifer performance test: a withdrawal well is pumped at a consistent yield, and water levels are measured in nearby observation wells. However, hydrogeological investigations conducted to determine the extent of contamination generally do not involve the installation of large-diameter wells capable of performing aquifer performance tests, and it is often necessary to develop preliminary estimates of the radius of influence from the characteristics of the aquifer. These estimates are checked following the installation of a recovery well by means of an aquifer performance test.

Equations are available to estimate the radius of influence for a well under various aquifer conditions. Equations applicable to areas where the hydraulic gradient is relatively flat are invalid for areas with a hydraulic gradient greater than 0.01 length/length. Other equations are available for wells

MODEL	BASIC EQUATION	US UNITS*	METRIC UNITS**
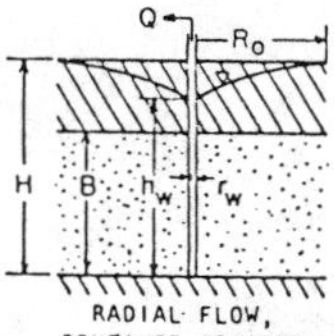 RADIAL FLOW, CONFINED AQUIFER	$Q=\frac{2\pi KB(H-h_w)}{\ln R_o/r_w}$ K = PERMEABILITY	$Q=\frac{KB(H-h_w)}{229 \ln R_o/r_w}$	$Q=\frac{KB(H-h_w)}{2653 \ln R_o/r_w}$
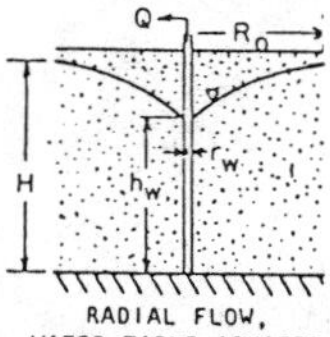 RADIAL FLOW, WATER TABLE AQUIFER	$Q=\frac{\pi K(H^2-h_w^2)}{\ln R_o/r_w}$ K = PERMEABILITY	$Q=\frac{K(H^2-h_w^2)}{458 \ln R_o/r_w}$	$Q=\frac{K(H^2-h_w^2)}{5305 \ln R_o/r_w}$
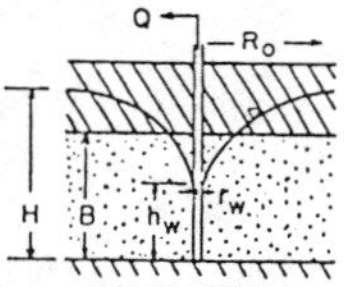 RADIAL FLOW, MIXED AQUIFER	$Q=\frac{\pi K(2BH-B^2-h_w^2)}{\ln R_o/r_w}$ K = PERMEABILITY	$Q=\frac{K(2BH-B^2-h_w^2)}{458 \ln R_o/r_w}$	$Q=\frac{K(2BH-B^2-h_w^2)}{5305 \ln R_o/r_w}$
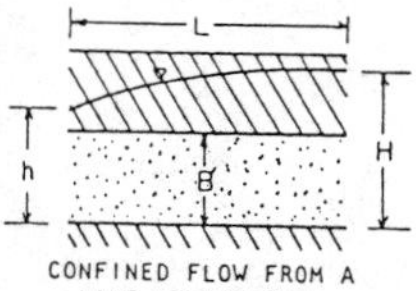 CONFINED FLOW FROM A LINE SOURCE TO A DRAINAGE TRENCH	$\frac{Q}{X}=\frac{KB(H-h)}{L}$ X = UNIT LENGTH OF TRENCH. K = PERMEABILITY	$\frac{Q}{X}=\frac{KB(H-h)}{1440\ L}$ FOR FLOW FROM 2 SIDES, USE TWICE THE INDICATED VALUE.	$\frac{Q}{X}=\frac{KB(H-h)}{16{,}667\ L}$
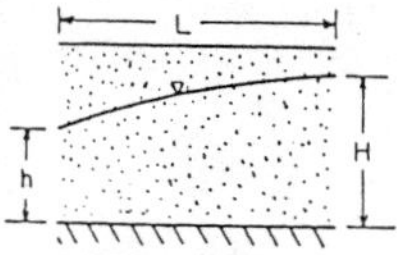 WATER TABLE FLOW FROM A LINE SOURCE TO A DRAINAGE TRENCH	$\frac{Q}{X}=\frac{K(H^2-h^2)}{2L}$ X = UNIT LENGTH OF TRENCH. K = PERMEABILITY	$\frac{Q}{X}=\frac{K(H^2-h^2)}{2880\ L}$ FOR FLOW FROM 2 SIDES, USE TWICE THE INDICATED VALUE.	$\frac{Q}{X}=\frac{K(H^2-h^2)}{33{,}333\ L}$
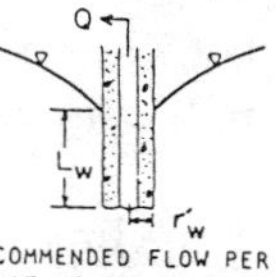 RECOMMENDED FLOW PER UNIT LENGTH OF WET BOREHOLE (SICHART)	$Q = 2\pi r'_w C\sqrt{K}$ C = EMPIRICAL COEFFICIENT	$Q = .035\ L_w r'_w \sqrt{K}$ r'_w In Inches L_w In feet	$Q = 0.0247\ L_w r'_w \sqrt{K}$ r'_w In mm L_w In m

*Except where noted: Q in gpm; H, B, R_o, r_w In feet; K in gpd/ft^2
**Except where noted: Q in M^3/m n; H, B, R_o, r_w in meters; K in μ/sec

Figure 10-2. Summary of ground water flow equations. Reprinted from *Construction Dewatering: A Guide to Theory & Practice,* by J. Patrick Powers, P.E. Copyright 1981 by John Wiley & Sons, Inc. Reprinted with Permission from the publisher.

pumped from a confined aquifer with a sloping hydraulic gradient. The equations are:[11]

Low hydraulic gradient:

$$R_0 = r_S + \left[\frac{T_t}{C}\right]^{1/2}$$

where R_0 = radius of influence
r_S = well radius
T = transmissivity
t = time since pumping started
C = constant, depending on the units

$$R_0 = 3(H - h)\sqrt{K}$$

where R_0 = radius of influent in feet
$H - h$ = water level drawdown in feet
K = hydraulic conductivity (μ/s)

High hydraulic gradient:

$$Y_L = \frac{Q}{2Kbi} \qquad X_L = \frac{Q}{2\pi Kbi}$$

where Y_L = radius of inflow perpendicular to direction of flow
X_L = radius of inflow downgradient to well
Q = discharge rate
i = natural piezometrical slope
K = hydraulic conductivity

(The equations applicable to wells with significant hydraulic gradients may also be applied to unconfined aquifers by replacing b with the uniform saturated aquifer thickness [h_0], assuming the drawdown is small compared to aquifer thickness.[12])

Based on the aquifer coefficients estimated from the preliminary hydrogeological investigation, the configuration of the contaminant plume, and the equations that calculate the wells' area of influence, a preliminary design for a recovery well system can be established; this design will include the flow rate and the required number and spacing of wells. At least one recovery well is installed, and an aquifer performance test is conducted in order to confirm the radius of the cone of influence. The ground water itself is collected with a well point system, gravity drains, or artificial recharge.

Well Point Systems. A well point system may provide the most economical method of ground water collection where the water table is within 10 ft (3 m) and the contamination within 30 ft (9 m) of the surface. A well point system consists of several individual well points spaced at 2- to 6-ft (1- to 2-m) intervals along a specified alignment. A *well point* itself is a 1½- to 3½-inch-diameter (4- to 9-cm-diameter) well screen or perforated pipe, between 18 and 40 inches (46 to 102 cm) in length, with a conical steel drive point at the bottom. Well points may be installed either with deep well methods or by driving, jetting, or augering methods. The individual well points are attached to a riser pipe (1 to 3 inches [3 to 8 cm] in diameter) and connected to a header pipe (6 to 8 inches [15 to 20 cm] in diameter). The header pipe is connected to a centrifugal pump at midpoint. At the top of each riser pipe is a valve to control the withdrawal of water. Because the yield of different well points may vary, the valve is adjusted so that the total drawdown is not large enough to expose the top of the screen and thus pull air into the system. The pump can provide 20 to 25 ft (6 to 8 m) of suction, but friction losses reduce the effective suction to 15 to 18 ft (about 5 m). If added water lift is required, larger-diameter well points, furnished with jet ejector or submersible pumps, are necessary. Submersible pumps can lift water from depths of 100 ft (30 m) or more, but both the capital and the operation and maintenance costs are high.

In silts and clays and other materials with low permeability, ground water is not easily drained because of the soils' high specific retention, but ground water flow can be assisted with a vacuum well point system. Such a system includes a screen extending above the water table with a sand pack around the screen and an impermeable seal above the sand pack.

Gravity Drain Systems. Ground water collection often can be achieved by means of a trench where circumstances are conducive to gravity flow. The trench functions as an infinite number of extraction wells and creates a continuous zone of depression along its length. A gravity drain system generally is used where the ground water table is within 10 ft (3 m) and the contamination is confirmed within 30 ft (9 m) of the land's surface. Installation of a gravity drain is simple. A trench is excavated perpendicular to the flow of ground water to a depth below the water table; a perforated pipe is placed in the trench, and the remainder of the furrow is backfilled with gravel. The ground water is collected in a main collector pipe, the header, and flows to a sump; it is then pumped to the surface for treatment or discharge.

In order to design a gravity drainage system, two factors are important:

- The relationship between depth and flow.
- The upgradient and downgradient influence of the trench.

The upgradient influence is determined thus:[13]

$$D_U = 4/3\, H \tan\theta$$

In that equation:

D_U = effective distance of drawdown upgradient, in meters
H = saturated thickness of water-bearing strata not affected by drainage, in meters
θ = angle between initial water table or ground surface and horizontal plane

The downgradient influence is determined thus:

$$D_d = \frac{K \tan\theta}{g}(h_1 - h_2 + D_2)$$

In that equation:

D_d = distance downgradient from the drain where the water is lowered to desired depth
K = hydraulic conductivity
g = drainage coefficient (meters/day)
h_1 = effective depth of the drain, in meters
h_2 = desired depth to water table after drainage, in meters
D_2 = distance from the ground surface to water table before drainage at the distance D_d downgradient from the drain, in meters

Artificial Recharge. To optimize the effectiveness of collection systems, the ground water can be recharged artificially. Artificial recharge accomplishes the following: provides a method for the discharge of treated ground water, increases the hydraulic gradient to expedite movement of ground water toward a collection system, creates a hydraulic barrier to restrict the migration of a contaminant plume, and provides a method for introducing solutions into the ground water to enhance the mobility or treatment of a contaminant plume. Recharge basins and injection wells are the standard means of artificial recharge.

Ground water most often is recharged through recharge basins, which require certain conditions: highly permeable surface soil with no layers of low permeability in the unsaturated zone; a water table relatively deep to keep the recharge mound beneath the base of the basin; a shallow, unconfined, and highly permeable aquifer to allow lateral flow of recharge water; and recharge water relatively free of suspended solids. The following equations describe the theory of horizontal ground water flow and are used in the design of recharge basins:[14]

$$H_{x,y,t} - H = \frac{V_{at}}{4f}\{F[(\overline{W}/2 + x)\,n, (L/2 + y)n]$$
$$+ F[\overline{W}/2 + x)\,n, (L/2 - y)n]$$
$$+ F[\overline{W}/2 - x)\,n, (L/2 + y)n]$$
$$+ F[\overline{W}/2 + x)\,n, (L/2 - y)n]\}$$

where:

$h_{x,y,t}$ = height of water table above impermeable layer of x, y coordinates from center of basin and time t
H = original height of water table above impermeable layer
V_a = arrival rate at water table of water from infiltration basin
t = time since start of recharge basin
f = fillable porosity $(1 > f > 0)$
L = length of recharge basin (in y direction)
W = width of recharge basin (in x direction)
n = $(4t\ T/f)^{-½}$

$$F(\alpha, \beta) = \int_0^1 erf(\alpha\tau^{-½}) \cdot erf(\beta\tau^{-½})dt^*$$

Recharge basins generally require considerable maintenance because the bottoms and banks of the basin become clogged with fine materials. The clogging may be caused by settling of suspended sediments, growth of bacteria and algae, chemical precipitation, or air trapped within the pore spaces. Pretreatment of the water can be minimized by means of filtration, sedimentation tanks, chlorination, sequestration, or the use of surfactants.

Wells also can be used to achieve artificial recharge. Recharge wells are used when conditions do not favor recharge basins, such as when the aquifer is confined, when fine-grained layers of low permeability are present within the vadose zone, when the water table is close to the land surface, or when there is insufficient land available for a recharge basin. Ground water recharge wells are constructed in a manner similar to that of deep wells and well points, but specific construction stipulations also apply. Cable tool drilling or air rotary drilling is preferred over mud rotary drilling because the mud might obstruct the aquifer, and a fast grout seal is necessary when the well is pressurized to prevent water from migrating along the casing either to other aquifers or to the surface.

A recharge well is developed to maximum efficiency initially, because it

*See tables by Hantush.[14]

cannot develop itself as pumping wells can.[15] Occasional redevelopment is necessary to increase efficiency and to prevent obstruction of the aquifer from suspended solids, bacteria, or the precipitation of calcium carbonate. A downspout and a discharge point below the water level are necessary to prevent cascading water, and aeration is needed to minimize screen clogging due to precipitation of iron or agitation of aquifer materials outside the screen. To preempt clogging, the water reinjected may require pretreatment.

Petroleum hydrocarbons have unique physical and chemical properties; therefore, special collection and treatment procedures and equipment are used to remedy ground water contaminated due to a petroleum leak or spill. Petroleum itself is insoluble and floats atop the water table. However, petroleum products sometimes contain toxic substances that can dissolve in the ground water. The ground water recovery system selected for petroleum hydrocarbons is specific to the conditions of the site, and routine ground water collection systems are modified in order to achieve collection of the immiscible petroleum product from the surface of the water—the free product. Conditions that influence the migration and recovery of petroleum hydrocarbons are listed in Table 10-3.

The surface layer of product atop the ground water can be pumped from the earth to an oil-water separator; alternatively, the product can be removed with an oil-water separator floating on the water table. The most common method of petroleum hydrocarbon recovery is use of recovery wells. Wells recover product rapidly because the resulting cone of depression expedites product migration toward the wells. The product is removed from the ground water surface at the wells.

To modify a recovery well in order to recover petroleum hydrocarbons, the well screen is lengthened and extended above the water table to allow the free product to enter the well. The diameter of the well casing and screen generally is enlarged too. The specific size of the well depends on the

TABLE 10-3. Factors governing movement of petroleum hydrocarbon in ground water.

SOIL CHARACTERISTICS	PROPERTIES OF PRODUCT	GROUND WATER CONDITIONS
Porosity	Volume of loss	Depth
Permeability	Viscosity	Gradient of flow
Texture	Solubility	Velocity of flow
	Chemical composition	Anticipated well yield
	Product thickness	

pumping equipment selected; for example, single-pump systems generally require a minimum diameter of 6 inches (15 cm), but a dual-pump system requires a minimum diameter of 10 inches (25 cm). Different configurations of these pumping systems are available to suit the hydrogeological conditions at hand, the type and size of the spill, and the type of equipment available. Configurations include single-pump systems utilizing one recovery well, single-pump systems using multiple wells, two-pump systems using one recovery well, and two-pump systems using two recovery wells. See Figure 10-3.

Single-pump systems often are employed if the ground water is in a formation with low permeability. The pump is installed below the water table, and the pumping level is maintained close to the pump intake to recover the floating product. With a two-pump system, one pump—the water table depression pump—is installed at the bottom of the well to pump ground water and develop a cone of influence in the water table. It thus hastens the migration of product toward the recovery well. The second pump—the product pump—is installed at the oil-water interface to separate and remove the product as it accumulates in the recovery well. Level controls are installed within the well to cycle the product pump and to stop the water table depression pump if the product approaches the pump intake.

Water table depression pumps are similar in construction to pumps installed in water wells. They may be submersible, centrifugal, air lift, or suction pumps. Because petroleum hydrocarbons are corrosive, the pump and discharge lines are constructed of chemically resistant materials. Plastic impellers, bowls, and rubber bearings are replaced with stainless steel, Teflon™, or Viton™. The pumping system also is specified to be explosion-proof for safety because pure hydrocarbons may be pumped.

The factors that determine the most suitable product pump are the conditions of the site, the long-term objectives of the recovery system (for example, the final allowable product thickness), and the limitations of the equipment. Several manufacturers furnish product pumps. Some are similar to a submersible pump and are capable of pumping at depths greater than 100 ft (30 m); they can pump a product only if it is several inches thick. Another pump can recover product to less than 0.25 inch (0.6 cm) but is limited to a suction lift of 10 ft (3 m) and requires a minimum casing of 24 inches (61 cm) diameter. One manufacturer utilizes an air lift method of product collection that permits the complete recovery of product; however, an above-ground oil-water separator is required. Another product pump bails the product from the water surface and automatically separates the oil from the water.

Single-pump systems are less expensive than two-pump systems, but they

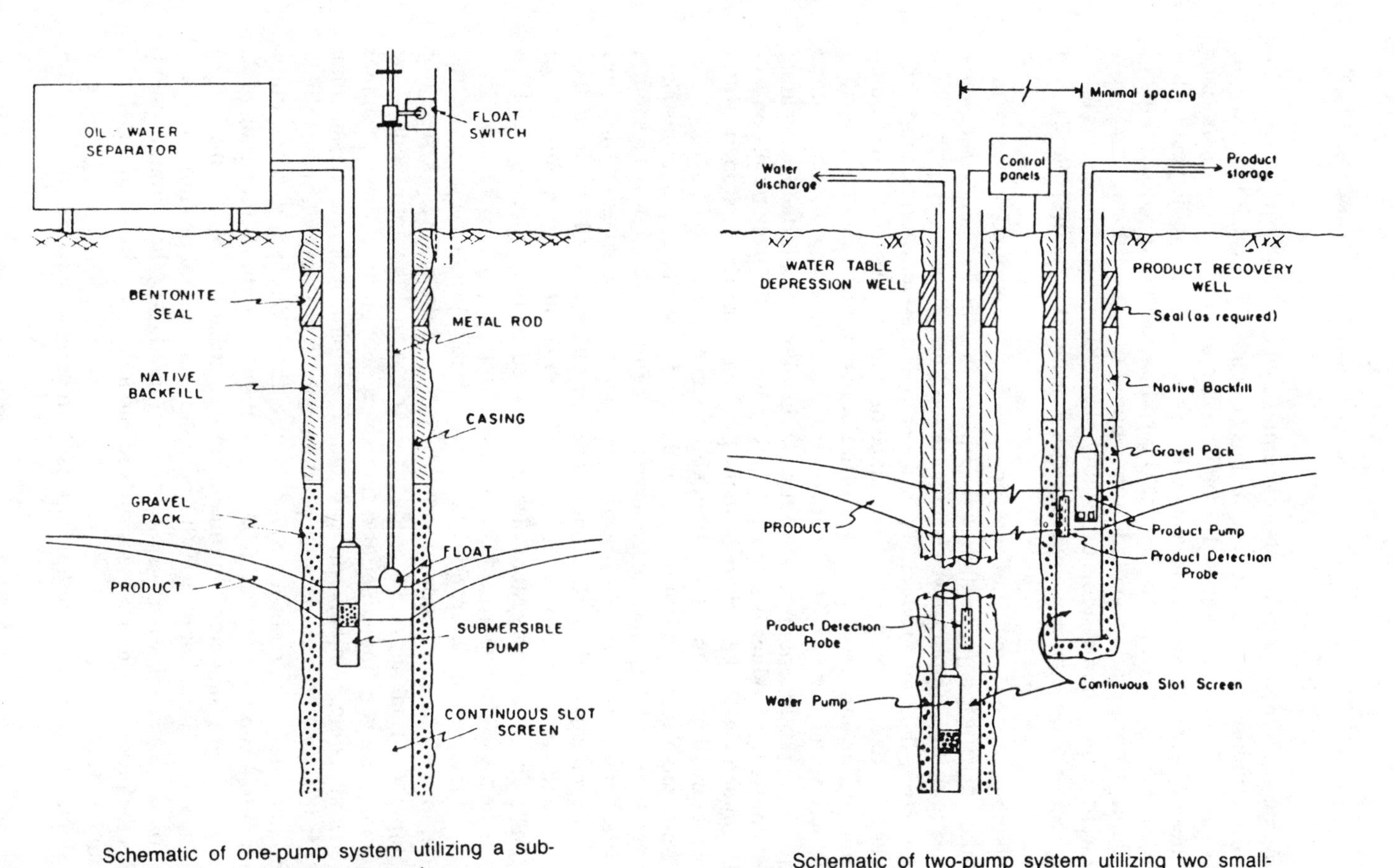

Schematic of one-pump system utilizing a submersible pump and float controls

Schematic of two-pump system utilizing two small-diameter wells

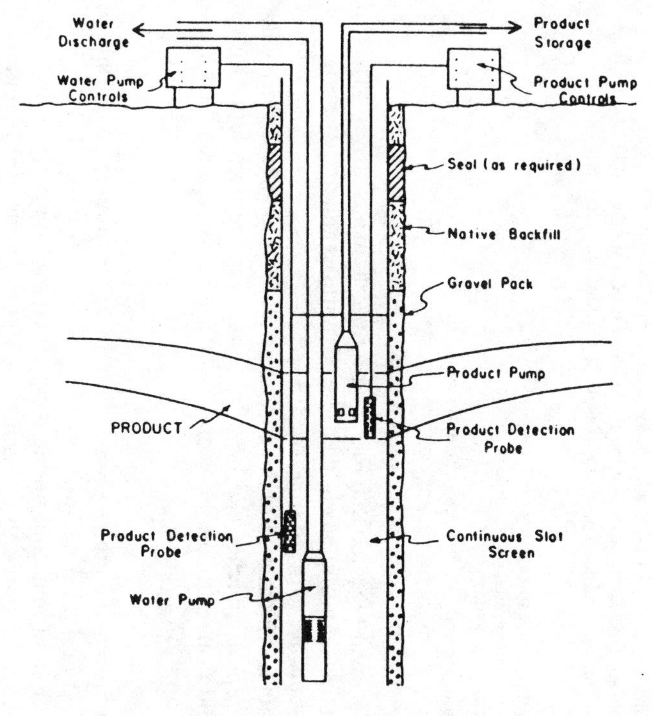

Schematic of two-pump system utilizing one recovery well

Figure 10-3. Recovery well pump systems. Adapted from "Underground Oil Recovery" by S.B. Blake and R.W. Lewis, Proceedings of the second national symposium on aquifier restoration and ground water monitoring, copyright 1982 by the National Water Well Association. Reprinted with Permission from the publisher.

TABLE 10-4. Considerations in use of one- or two-pump systems.

ONE-PUMP SYSTEM

Disadvantages:
Above-ground oil-water separators are required to recover product.
System efficiency is low because pumps cycle on and off.
Agitation of product and water may emulsify product, making separation difficult.

TWO-PUMP SYSTEM

Disadvantages (in contrast to single-pump system):
Larger-diameter, more expensive casing and screen are required.
Electronic controls are costly, require maintenance, and must be inspected at least weekly.
The product pump may be unable to pump thin layers of product.

Advantages:
Above-ground oil-water separators may not be needed.
The product may be recovered to a reusable grade.
Oil and water are not emulsified.
The system is automatic and can operate continuously.

have several disadvantages, so that the two-pump system has proved to be the more efficient and economical choice. The advantages and disadvantages of the pump systems are arrayed in Table 10-4.

If the product loss is small, and the ground water is within 8 ft (2.5 m) of the surface, interceptor trenches can be used to recover the petroleum hydrocarbons. In addition to the methods employed for gravity drains, the following measures apply:[16]

- The trench bisects the entire plume; therefore, the plume can be only of limited extent and must be accessible.
- Skimming or pumping equipment is operated continuously; otherwise, the product accumulates and migrates around the ends of the trench.
- An impermeable barrier is installed on the downgradient side of the trench such that it stops the migration of product but allows the ground water to pass below.

Case Study: Collection of Mineral Spirits Leakage. Upon the sale of a manufacturing facility, its premises were assessed to ascertain whether any environmental liability was present. The facility stored odorless mineral spirits in two underground 3000-gal (11.4-m^3) tanks, and Freon was stored in an above-ground tank. The site had been used as a warehouse, and the underground tanks previously had stored diesel fuel. Volatile organic compounds and a petroleum substance were found in three test wells. The objectives of a detailed hydrogeological investigation were:

- To determine the type of petroleum materials released.
- To define the source and extent of the product within the soils.
- To develop a remedial program to recover the product.

The land was flat and covered predominantly by manufacturing buildings and paved parking areas. Surface water runoff flowed to the north and northeast, entering two small intermittent streams that flowed along the northern and eastern boundaries of the site. The geological features of the site included fine to coarse sand with some gravel and silt extending from the ground surface to a depth of 20 ft (6 m). This material was found over a fine sand and silt deposit that was at least 6 ft (2 m) thick. Ground water was encountered at an average depth of 8 ft (2 m) and flowed northeast under a gentle hydraulic gradient of 0.001 length/length.

The site investigation included two steps:

1. Determination of hydrogeological characteristics and the presence and type of any hazardous materials or oil in the subsurface.
2. Installation of additional ground water monitoring wells to identify the source and extent of any materials detected during Step 1. Data from these wells provided information useful to the development of the remediation plan.

During Step 1, three soil borings were completed and developed into monitoring wells. The samples collected were tested in the field with a photoionization detector (HNU) to assess qualitatively the volatile organic content of the soils. This field screening revealed that the highest concentrations of volatile organics were present at the water table; the implication was that the source of the product was a slow leak rather than a sudden spill.

In the monitoring wells, a layer of free petroleum was discovered atop the ground water. To determine whether this product was the mineral spirits currently stored in the tanks or the diesel fuel formerly stored, samples of the product were taken from two wells, and a sample of the stored mineral spirits was taken from a tank. All three samples were sent to a laboratory where they were analyzed by a gas chromatograph with a flame ionization detector. The analysis demonstrated an identical match between the sample taken from the storage tank and the samples removed from the wells and detected no other petroleum products in the ground water.

A supplemental analysis was performed to establish whether any soluble components of the mineral spirits had entered the ground water. A sample of water from one of the monitoring wells was analyzed for all priority pollutants, but none was detectable by the instruments. Therefore, investigators concluded that the product plume was having no significant effect on the quality of the ground water.

During the Step 2 investigation, the objective was to define the horizontal and vertical extent of the free product pool. To develop data for the assessment, nine additional monitoring wells were installed. The ground water elevation was measured at all well points, and these data were used to define both the configuration of the ground water table and the direction of ground water flow. The monitoring wells were also checked for the occurrence of the product and the product's thickness.

A map, shown in Figure 10-4, was developed with the data collected; it indicated that the free product pool was essentially within the premises of the manufacturing facility and contained within a 60-ft (18-m) radius of Well 8. The maximum thickness was recorded at 0.58 ft (18 cm) at Well 2. The symmetrical configuration of the plume was attributed to the relatively flat hydraulic gradient. The information about product thickness as well as the preliminary screening of the data indicated that the source of the free product could be leakage from either the underground tanks or related pipelines.

Site Remediation. Because the source of the mineral spirits could have been either the underground tanks or the pipelines, both the tanks and the lines were emptied and removed from service. Steps then were taken to remove the free product from the subsurface. The characteristics of the site were opportune for development of a product recovery well:

- The high permeability of the subsurface sand and gravel materials promoted rapid flow of ground water and product toward a recovery well.
- The symmetrical configuration and small radius of the free product pool coupled with the rather flat hydraulic gradient of the water table required the use of only one recovery well to contain the entire free product pool effectively.

As is always the case, the site investigators were conscious that the product recovered had to be disposed of acceptably. Potential disposal points were the sanitary sewer system, the stormwater sewer, and the ground water system, each of which was evaluated. The recovered product could not be sent to the sanitary sewer because the municipal sewerage system was already overloaded. The stormwater sewer drained into a stream that subsequently flowed into a municipal water supply reservoir; so that option, too, was barred. The only option, then, was ground water recharge. Not only was ground water recharge the most suitable option, but it also delivered auxiliary benefits. A correctly designed ground water recharge system would enhance the recovery system because it would create a ground water mound that would increase the hydraulic gradient and, therefore, increase the flow

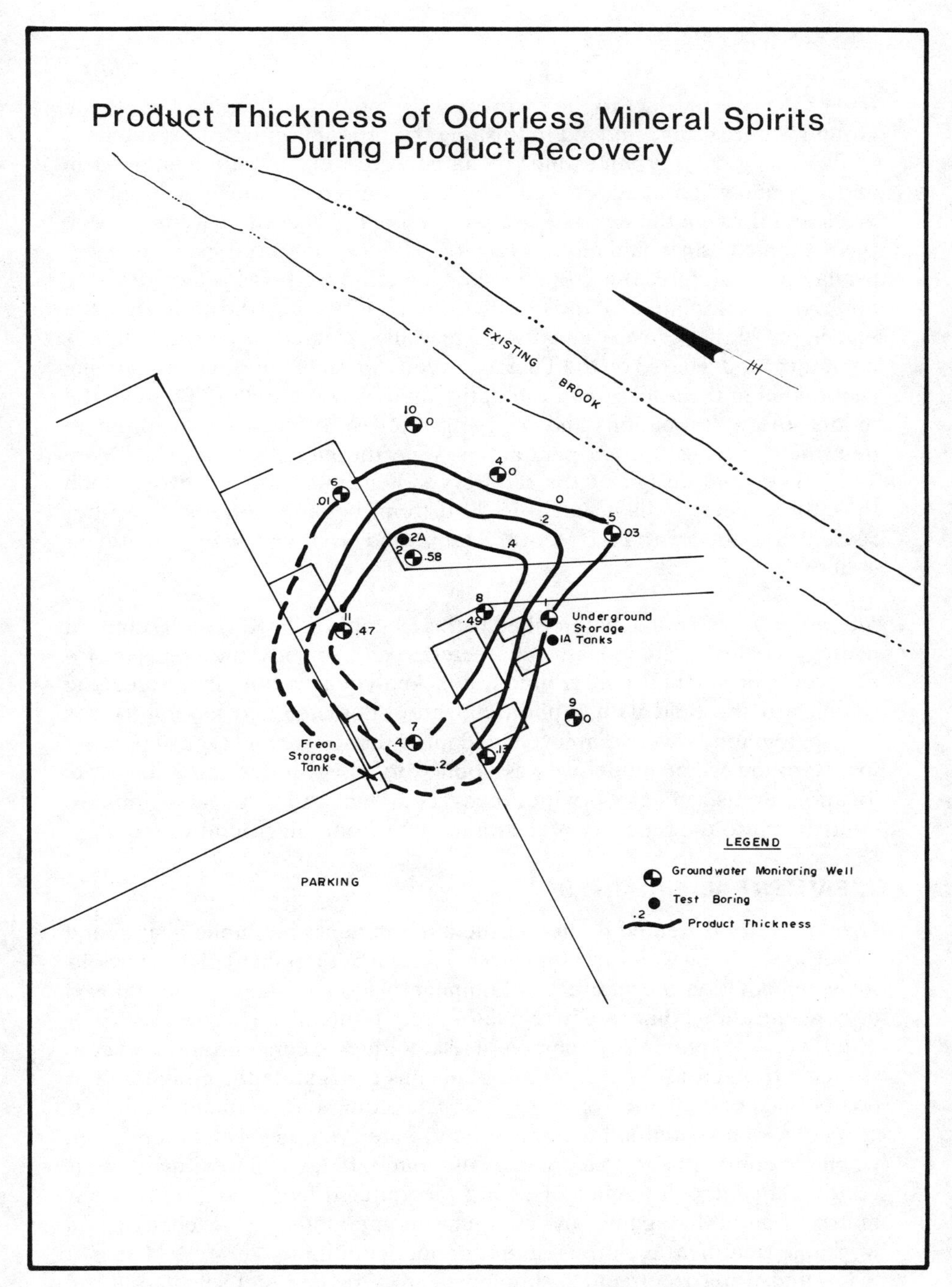

Figure 10-4. Product thickness of odorless mineral spirits.

rate of the free product to the recovery well; moreover, it would elevate the ground water surface and aid in flushing the product in the vadose zone.

Two methods of ground water recharge were weighed: a recharge basin and a ground water injection well. The size required of a recharge basin was evaluated through the use of a computer model.[17] According to the model, the recharge basin would require 1200 ft^2 (110 m^2)—far more space than was available. Therefore, the ground water injection well was selected as the appropriate alternative for the site. To optimize efficiency, the design and the placement of the recovery and discharge wells were studied painstakingly. A computer model based on the Theis equations for recharging and discharging wells aided in the selection of the optimum loci of the wells.[18] Through the model, the well locations and pumping rates were fine-tuned in order to determine the best design parameters. (See the model's results in Figure 10-5.) The final design of the recovery system incorporated two 24-inch (61-cm)-diameter wells installed to 20 ft (6 m) below ground level, a filter scavenger unit, a product recovery tank, and a ground water treatment system.

Summary. The problem described in this example utilized several standard hydrogeological techniques to characterize and remedy the leakage of a petroleum product into the ground water. Analytical techniques—screening samples in the field with a photoionization detector and laboratory gas chromatography—were employed to identify the source and type of product loss. Ground water modeling was applied to assist in designing the type, location, and size of the product recovery system. All of these techniques contributed to the recovery of a contaminant from the ground water.

TREATMENT ALTERNATIVES

Ground water is treated to protect the environment and public health, and sometimes ground water treatment is a necessary component of a hazardous waste remediation alternative. Contaminants in ground water can be removed to concentrations that satisfy regulatory requirements for water quality or criteria for acceptable environmental risk. Reducing concentrations of constituents from ground water requires a means to separate the compound or compounds of concern from the water. One common treatment method is stripping volatile compounds from ground water, whether with air or steam, when the contaminant is an organic compound; biological treatment also is used. Inorganic compounds often are precipitated from the ground water matrix. For both organic and inorganic compounds, ion exchange is a workable treatment technique under certain conditions.

Ground water treatment technologies are screened and selected at the

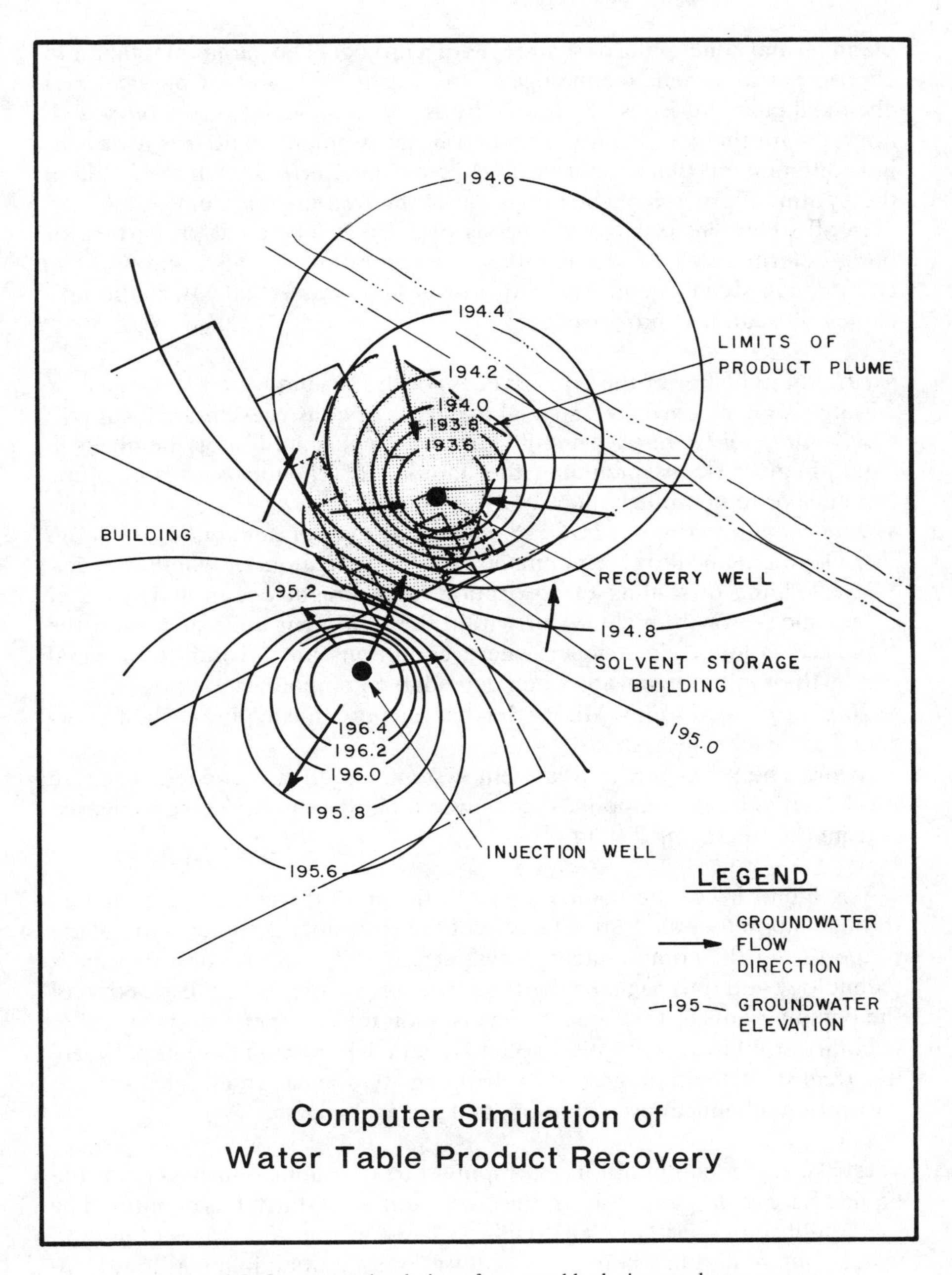

Figure 10-5. Computer simulation of water table during product recovery.

planning and conceptual design stages of a project. The paramount factor in choosing a treatment technology is the waste itself and its physical and chemical characteristics. Additionally, as the treatment mode is proposed, not only are the compounds regulated by government standards reviewed, but contaminants that are not regulated are characterized, as they may affect the treatment process or the operation of the treatment system.

In all treatment processes, various physical or chemical properties, or both, determine whether separation, concentration, or destruction of the contaminants can occur. Properties of ground water that affect the efficiency of treatment processes are:

- *pH:* An indicator of the corrosiveness of the ground water. If the ground water is highly corrosive, special materials of construction are justified.
- *Alkalinity and acidity:* An indicator of the consumption of chemicals in treatment systems, including the addition of chemicals to precipitate metals or to neutralize the water.
- *Total organic carbon (TOC), biochemical oxygen demand over 5 days (BOD_5),* chemical oxygen demand *(COD):* Indicators of whether excessive loading or fouling of adsorption filters might occur, and whether biological growth might occur within the treatment unit, such as in the packed column air stripper where conditions are favorable for rapid growth of microorganisms when degradable compounds are present.
- *Total suspended solids:* Matter that has the potential to plug carbon filters and ion exchange systems.
- *Odor:* The smell of the treatment system, which is a concern because odorous volatile compounds can cause a public nuisance when released from the treatment system.

The identities of the contaminants in the ground water are determined through laboratory analysis. The goal of the treatment designer is to use the properties of the compounds to advantage in the selection of a treatment technology and thus to make the treatment economical. The disposition of the contaminants to treatment depends upon their properties, specifically: volatility, solubility, adsorption potential, and degradation potential. Potential treatment technologies are: stripping, biological treatment, carbon adsorption, chemical precipitation, and ion exchange.

Volatility. The potential for a compound to evaporate—to pass from the liquid phase to the gas phase—is the compound's volatility. It is quantified by the equilibrium constant or the Henry's law constant. For cases of ground water contamination when the concentration of contaminants is low, Henry's law states that the ratio of the concentration of a compound in water to its concentration in air is, at equilibrium, a constant; this condition is maintained independent of the concentration of the solution. The Henry's law expression for equilibrium is:

$$y = \frac{H}{P} x$$

where y = concentration of compound in air
x = concentration of compound in water
P = total system pressure
H = Henry's law constant

The water and air must be in contact for an extended time in order for the equilibrium to manifest itself; in addition, the amounts of solute, water, and air must be constant. If a compound initially is not at equilibrium, it will diffuse from water into air—this is *stripping;* or it will diffuse from air into water—this is *scrubbing.* Migration of a compound between the two phases is *mass transfer.* The likelihood that mass transfer will occur depends upon the difference between the initial concentration and the equilibrium concentration of the compound. This difference is the *driving force* of mass transfer. When the Henry's law constant for a compound is greater than 1000 atmospheres (atm [1.03×10^7 kg/m^2]), the compound is considered highly volatile. When the constant is between 1 and 1000 atm (1.03×10^4 and 1.03×10^7 kg/m^2), the compound has moderate to high volatility. Compounds with a Henry's law constant greater than 1 atm (1.03×10^4 kg/m^2) are amenable to treatment through stripping.

Air Stripping. Moderately and highly volatile compounds usually are removed from ground water by means of air stripping. The compounds are transferred into the air and discharged into the atmosphere under conditions controlled by air pollution regulations. Air stripping, as a treatment method, can tolerate a wide range of concentrations in the influent. (Low concentrations may be more economically treated with carbon adsorption.) However, the efficiency of the treatment operation is inferior at temperatures near the freezing point of water. Therefore, air stripping units expected to operate year-round in middle latitudes require special considerations in both design and applicability; alternatively, air stripping may be combined with another process, such as carbon adsorption, and treatment modes may be alternated during the year.

An aeration basin can provide air stripping for ground water that contains highly volatile compounds, if high removal efficiency is not required. Such a basin is simple and relatively inexpensive. It consists of a continuous-flow tank outfitted with bubble diffusers placed at the bottom. Compressed air is diffused into the column of water, and the bubbles provide a surface area for the mass transfer of volatile compounds. The aeration basin provides roughly 90 percent removal of volatile compounds and, moreover, operates as a completely mixed reactor. Thus, the driving force for mass transfer is low because of low average concentrations.[19] Performance can be improved with baffles in the basin, as they promote plug flow and increase the driving force for mass transfer.

Cross-flow towers and draft towers increase the surface area of the ground water available for mass transfer. The water cascades down a tower of slats or structured packing material while a fan draws air through the tower, but the air does not truly flow countercurrently to the water. Smaller volumes of air are required for cross-flow and draft towers than for aeration basins. The treatment efficiency is moderate because of plug flow of the water and because a significant area for mass transfer is provided by cascading droplets. The energy required to move the air is low because the pressure drop through the tower is low. Mist drifting from the tower can create a nuisance in the environs of the treatment facility unless controlled. This treatment technique has limited application when temperatures are below the freezing point of water.

Packed columns commonly are used to remove volatile compounds from ground water.[20] Air and the contaminated ground water contact each other in countercurrent flow within a vertical column packed with specially shaped pieces, the packing, that improve the effective surface area for mass transfer. The packing can be composed of individual pieces, or it can be a structure stacked in the column. The air is blown upward, and the water flows down the column; both travel through the packing material. The size of the column is dictated by the necessary flow rates of contaminated ground water and of air. A perforated support plate holds the bed of packing in the column and allows the air to enter the packing while water drains from it. A demisting section is placed atop the packed bed to remove entrained water droplets from the air exiting the column. Packed columns can achieve removal efficiency greater than 99.9 percent. Because the flows are countercurrent, the air and water approach equilibrium, and the volume of air is minimized. To reach true equilibrium, however, an infinitely tall column would be needed. The height of the column is determined by a cost analysis, including the capital cost for column construction and the operating costs for the blower.

The surface area of contaminated ground water exposed to the air can be increased by spray formation, with nozzles or mechanical agitators in the tanks or impoundments. Such a treatment mode requires more energy and land area than basins or towers need.

Steam Stripping. Volatile compounds can be transferred to gases with steam used as the gaseous agent. Heat from the steam raises the water temperature and increases the mass transfer and the partitioning of the volatile compounds into the vapor phase. Steam stripping is limited to circumstances where the volatile compounds have low water solubility and are highly concentrated; compounds miscible with water are not removed with this treatment technique. The energy costs of steam stripping are higher than those of air stripping.

Steam is introduced at the base of packed or tray columns, and the contaminated ground water enters the top. Heat from the steam raises the temperature of the water, raises the diffusivity, and creates a more favorable equilibrium, thereby driving the volatile compounds into the vapor phase. Vapors from the top of the column condense in a heat exchanger and collect in a decanting receiver tank. Compounds immiscible with water are separated as a distinct liquid phase. The water phase, saturated with the compound, is recycled to the column.

Solubility. The solubility of a compound is assessed in ground water treatment investigations because solubility provides an indication of the potential mobility of the compound within the ground water. Water-soluble compounds dissolve in ground water until either the water becomes saturated with the solute, or the compound is completely dissolved. The concentration at which water becomes saturated defines the *water solubility* of a compound. Solubility values are a function of temperature; and when mixtures of solutes are present, the solubilities of individual compounds may be affected by interactions of the solutes or by interactions of the solutes and water. In such cases, data on pure compounds are useful only as a guide, and laboratory testing is necessary to determine actual parameters.

The portion of a compound that is dissolved in water will flow with the ground water, whereas liquid compounds with slight solubility will form a separate liquid phase. A compound whose density is less than that of water, such as gasoline or fuel oil, will migrate atop the ground water; but one whose density is greater than the density of water, such as the common metal degreasing solvents trichloroethylene and perchloroethylene, will sink through the ground water to the depth of the first confining stratum. If two or more distinct liquid phases are present, the separation techniques using gravity, discussed above in the section on "Collection Practices," are applicable.

Certain metallic elements—for example, lead, chromium, copper, iron, cadmium, arsenic, barium, and silver—become charged ionic species when dissolved in water. A change in the valence, or charge, of the ion can radically alter the solubility of the compound. For example, the conversion of hexavalent chromium (Cr^{+6}) to trivalent chromium (Cr^{+3}) by the addition of a reducing agent diminishes significantly the water solubility of the metal because the two ions have different properties. Most often, metal ions are precipitated in the form of a metal hydroxide by elevating the solution pH with sodium hydroxide or lime. Chromium can be removed from solution by adjusting the pH.

A general chemical equilibrium equation describes the reaction of metal ions with anions in solution to form a neutral solid:

$$n\,M^{+}_{(aqueous)} + m\,A^{-}_{(aqueous)} \rightleftarrows MA_{(solid)}$$

Precipitation occurs when the reaction described in the equilibrium equation proceeds to the formation of a solid, and can be initiated by the addition of either ionic species. The balance of the equilibrium between anions and cations is described by the solubility product expression:

$$K_{sp} = [M^{+}]^{n}\,[A^{-}]^{m}$$

where $[M^{+}]$ = cation concentration
$[A^{-}]$ = anion concentration
K_{sp} = solubility product constant

The solubility product constant is a property of each organic compound. The actual concentrations of dissolved species depend on the equilibrium of all the interacting anions and cations.[21]

A precipitation system is designed to remove specific compounds. Inorganic compounds can be removed from wastewater by the addition of polymers, alum, ferric hydroxide, or caustics, which cause metals to precipitate. In order for precipitation to be an effective treatment, the addition of the precipitating agent is adjusted to match the concentration of the influent; without sophisticated controls, an influent with a variable concentration can affect an operation adversely. Suitable precipitation depends upon dissolved species present, intensity of the mixing, and availability of suspended solids for flocculation and settling. In devising a precipitation treatment system, the designer considers control of the precipitating agent, intensity of mixing, storage and handling of the agent, and handling and disposal of sludge.

Ion Exchange Treatment of Inorganic Compounds. Treatment techniques utilizing ion exchange can remove dissolved inorganic compounds by capturing the ions on special resinous materials. Ion exchange treatment systems are classified by the resins used; there are cation exchangers, anion exchangers, and mixed-bed exchangers. The cation and anion exchangers are filled with select resins, whereas the mixed-bed exchanger contains both. Specialty resins are available that selectively remove ions, such as valuable heavy metals. A resin is obtained normally in the form of beads with a nominal dimension of 0.5 to 1 mm. Used resins are flushed with a solution of brine, acid, or caustic and thus regenerated. The waste regenerate solution tends to have an extreme pH level and to contain a high concentration of dissolved material, but it can be treated to recover precious materials by means of precipitation and neutralization, concentrating the waste into a sludge for disposal.

Ion exchange units generally cannot remove organic compounds; in fact, organics generally foul the surfaces of the resins and reduce the effectiveness of the treatment. The exchangers also are subject to clogging by suspended solids, but such solids usually are removed by means of filters in advance of the ion exchange treatment.

An ion exchange system usually provides treatment with two or more columns set in parallel; one unit operates while the other is regenerated. Exchanger vessels contain internal accessories: a water feed header to distribute influent over the resin, a porous support for the resin bed to retain the resin and allow treated water to drain from the unit, and a distributor for water to backwash the unit during regeneration. Accessory equipment includes chemical feed pumps for regeneration chemicals, storage for raw chemicals, and storage for waste regenerant solution. The treatment units can be operated manually or with automatic controls.

A precipitation treatment system is advantageous because it removes organic compounds, requires only moderate capital cost, and is a tested and known technology. The system can be mobile, it has a relatively short start-up time, the concentrations in the effluent are low, and it requires only a moderately skilled operator. However, precipitation has disadvantages also; it is of limited use with influent of a low concentration, cannot remove organic compounds, requires the storage and handling of chemicals, requires that the sludge be handled and dewatered, and generates a sludge that must be disposed of because there is limited opportunity for waste recovery.

Adsorption Potential. Compounds may have an affinity to adhere physically or chemically to the surfaces of certain solid substances, termed *adsorbents.* In the adsorption process, a solute can partition itself between the water and sorbent phases, and an equilibrium is established for the partitioning of the solute. The preference of a solute for the sorbent phase (surface of the solid) depends generally on its solubility, as compounds with low solubility tend to partition to that phase. This equilibrium is expressed as follows:[22]

$$\log_{10} X/M = \log_{10} K + (1/n) \log_{10} C_S$$

where X/M = mass solute adsorbed/mass adsorbent
C_S = equilibrium concentration of solute in water
$1/n$ = measured constant
K = partition coefficient

The partition coefficient K and the constant $1/n$ are evaluated from a double-log plot of test data for X/M versus C_S, utilizing a specific adsorbent and a specific solute or solute mixture. The U.S. EPA has tested numerous adsorbents.[23] The relative adsorbability of organic compounds conforms with several general trends. Compounds with molecular weight less than 40 have low adsorption. For compounds with fewer than four carbons in a chain, the relative adsorption is ordinal, as follows: undissociated organic acids > aldehydes > esters > ketones > alcohols > glycols. The propensity of compounds with higher molecular weights to adsorb to carbon is closely related to the solubility of the particular compound.[24]

The adsorption of individual compounds can be influenced by the other compounds present in the ground water, either by solute-solute interaction or by competition for surfaces to which the compounds may adsorb. The adsorbents themselves generally are nonselective for specific compounds. An important aspect of the adsorption reaction is that it is reversible, so that the adsorbent potentially may be regenerated and reused.[25] A wide range of both organic and inorganic compounds can be removed from ground water by means of carbon adsorption. The technique is capable of removing even very dilute concentrations.

Carbon particles are confined within a contacting vessel, and contaminated ground water is passed through the particles. Depending upon the arrangement for the contact, application rates range from 2 to 8 $gal/min/ft^2$ (120 to 470 $L/s/cm^2$). Efficiency is lost at the low end because of limitations of mass transfer; at the high end, efficiency is lost because of the excessive pressure drop and energy loss through the bed of carbon particles.[26] The carbon vessel is arranged to accommodate the dynamics of the adsorption process and to maximize the use of the carbon. Residence time is typically set at a minimum of 15 to 20 minutes, but certain compounds that have low water diffusivities require a longer contact time.[27] Carbon used in carbon adsorption units can be regenerated; however, ground water treatment facilities typically do not utilize carbon at an adequate rate to justify regeneration on site. For short-term projects, carbon adsorption vessels that also can serve as suitable disposal containers often are used to avoid the handling associated with loading and unloading carbon from contactors.

Potential arrangements for carbon contacting units include: downflow, gravity/pressure fixed-bed; upflow, expanded-bed; continuous countercurrent; series or parallel.[28] Downflow units operate with gravity or pressure. The contaminated ground water is distributed over the top of the carbon bed and flows through the carbon via screened drain ports. These units also filter suspended solids from the ground water. If suspended solids exceed roughly 50 mg/L, the carbon bed becomes plugged, and backwashing is necessary to remove the solids. Gravity units typically are limited to operation at 2 gal/min/ft^2 (120 L/s/cm^2) based on the pressure drop through the bed. Pressure units can operate at higher rates, typically 5 gal/min/ft^2 (290 L/s/cm^2).

Water feeds into the bottom of the carbon bed in upflow units, and the effluent is drawn from the top. Upflow units operate with loading rates of up to 8 gal/min/ft^2 (470 L/s/cm^2), and they are more tolerant of suspended solids than fixed-bed units. Fine carbon particles are carried with the effluent, and nominal filtering may be necessary following the adsorbers. Upflow units permit deeper beds and longer contact times than fixed-bed units.

The continuous countercurrent contactor is an alternative type of upflow operation. A cone-bottom vessel is used to draw evenly spent carbon from the bottom of the contactor while fresh carbon is added at the top. Only if the consumption of carbon is low and contact time must be of maximum duration is this type of treatment facility applicable.

Series and parallel configurations of fixed-bed units can be used to provide the required residence time or the required loading rates. Fixed-bed units in series are often used to optimize use of the carbon. Lead beds are spent first and taken off line while fresh units are added at the end. The operation is similar to a continuous countercurrent contactor.[29]

Degradation Potential. In this treatment process, the enzymes of organisms can break large contaminant molecules into fragments, or organisms can modify the chemical nature of these molecules as they obtain energy for cell growth. The resulting by-products generally have lower toxicity than the original substances. Many compounds are susceptible to degradation by biological action. The relative biodegradability of compounds present in ground water can be determined by evaluating the following ratio:

$$BOD_5 : COD : TOC$$

If the ratio is 1 : 2 to 3 : 1, the organic compounds are degradable. A lower BOD_5 value indicates that part of the organic material resists degradation; a lower TOC indicates that inorganic compounds making oxygen demands may be present.

One may review the literature to determine the predilection of a compound to degrade, although conflicting appraisals may appear.[30] The structural properties of a molecule that influence the ability of microorganisms to degrade or modify it are molecular weight, polymeric nature, aromaticity, halogen substitution, solubility, and toxicity. The size of a molecule can affect its transport through the cell wall and into an organism. In the case of polymeric compounds, organisms must excrete extracellular enzymes to metabolize the molecules. The presence of certain molecular subunits, or functional substitution groups, can influence the route an organism will use to metabolize a compound.

If it is not metabolized, the molecule will cause toxic effects in the organism—toxicity being the inhibition of cellular metabolism or growth by a compound when present in a concentration above a tolerable limit. A compound that is degradable by an organism also can be toxic if its concentration exceeds a threshold value. Some organisms have been cultivated to have an especially high tolerance to ordinarily harmful concentrations. Thus, they are capable of metabolizing toxic substances in conditions that would destroy ordinary organisms. Several environmental factors influence the metabolism of organisms and limit biological degradation:[31] the presence of compounds, in addition to the target compound, that also are toxic; the presence of nutrients or trace elements; pH; temperature; the concentration of microorganisms; acclimation or adaptation of the bacteria; and agitation of the cell suspension. (A specialized aspect of this subject is discussed further in Chapter 11.)

If a single reference in the literature indicates that degradation of a particular compound has been observed, further investigation and possibly bench-scale testing may be justified. The source and history of the seed culture used in the investigation cited are important information in evaluating the potential of an organism for a particular application. During bench-scale testing, the environment of the organisms is firmly controlled to foster the desired action. If, during the test, the compound disappears, the cause may be biological degradation; alternatively, it could be that the compound has volatilized, adsorbed to the organism, or adsorbed to the container. One weakness of testing the degradability of individual compounds is that one compound may act as the primary metabolite and another as the secondary metabolite. Secondary degradation of small concentrations of halogenated organic compounds—trichlorobenzene, for example—has been observed.[32]

Bacteria can metabolize certain organic wastes and thus provide treatment to contaminated ground water. In an aerobic biological reactor, wastewater is brought into contact with organisms, the biomass, and the organisms concentrate and destroy the organic compounds. Adequate contact time between the biomass and the organics is necessary to permit the organisms to react with the waste constituents. The rate at which the bacteria metabolize waste depends primarily on the concentration of the contaminants. In order not to overwhelm the biomass, the flow of the organics into the reactor is equalized. Following contact, the biomass is separated from the effluent in a clarifier unit. Sand filters or carbon filters polish the effluent after it leaves the clarifier if a higher final effluent quality is necessary. The reactor's size is determined by the expected hydraulic load; thus the practical size of the biological reactor may limit effluent quality.

There are two fundamental types of biological reactors: suspended-growth and fixed-growth. The two differ in the method used to maintain the biomass within the reactor. The suspended-growth system maintains the biomass as a mixed suspension. The organisms are supplied with oxygen by extended aeration, or activated sludge, systems in which compressed air is introduced by mixers or blowers, and oxygen dissolves in the water. Concentrated sludge from the clarifier is recycled to the reactor to maintain adequate loading of the organic material.

Fixed-growth systems retain the biomass by providing a large surface area to which the organisms adhere. Examples of fixed-growth reactors include trickling filters and rotating biological contactors. Because the organisms reside within the reactor for extended times, those organisms that reproduce and thrive may be better adapted both to metabolize the organic compounds introduced and to resist any loading shocks. The growth of the biomass is limited by the fixed-growth system's capacity to supply oxygen to the organisms, as the bacteria accumulate in thick layers. Dense patches of biomass separate from the tank surfaces when cells close to the wastewater surface die from lack of oxygen or food.

In a trickling filter system, organisms are attached to rocks 1 to 4 inches (2.5 to 10 cm) in diameter, and wastewater is distributed over them by means of a rotating arm. The water exits through a drain in the bottom of the bed, and air is supplied by natural convection. In a rotating biological contactor, the organisms adhere to discs with large surface areas. The discs are mounted on a shaft and rotate; they are always partially submerged in the wastewater. The rotation agitates the biomass film, enhances mass transfer, and maintains the desired thickness of film by shearing heavy films. Oxygen is available to the organisms when they are raised above the wastewater. Rotating biological contactors retain a greater volume of biomass than other biological reactors. Therefore, they are more stable in response to shock hydraulic and organic loadings; they are more flexible for operation in series to promote plug-flow treatment; and they have the capacity to produce effluents with low concentrations of organics compared with other biological reactors.

Clarifiers are used for both suspended-growth and fixed-growth systems to separate suspended biomass from the effluent. The clarifiers act through gravity sedimentation and are designed to permit the biomass particles to settle. The sludge may be 0.5 to 2 percent solids by weight. It is usually dewatered and may require further treatment before disposal.

Summary. At the planning and conceptual design steps of a remediation project, ground water treatment technologies are reviewed. The essential measure of a treatment technology is the character of the chemical waste.

Volatility, solubility, adsorption potential, and degradation potential are the characteristics of interest.

The object of ground water treatment is to separate the contaminant or contaminants from the water. The chosen treatment mode must be effective in reducing the concentration of the contaminant to regulatory standards, and it must be designed so that it is compatible with the environment in which it will operate. The design criteria of economical operation and convenient maintenance also apply. Common ground water treatment methods are: stripping, both air and steam; biological treatment; carbon adsorption; chemical precipitation; and ion exchange.

DESIGN OF TREATMENT FACILITIES

After remediation alternatives have been formulated, the design phase of the remediation project may proceed. Classical engineering design, regardless of the nature of the project, is undertaken in three logical and sequential steps:

- *Conceptual design:* Alternative remedial actions are devised, and the criteria for the design of the ground water treatment facility are instituted at the outset. The circumstances of the contamination are investigated sufficiently to assess, on a preliminary level, both the technical feasibility and the economy of the remedial alternatives posed. At the conclusion of the conceptual design phase, a single alternative or group of options is identified as worthy of additional analysis.
- *Preliminary design:* Next, the technical feasibility of the specific alternative or the set of options defined during conceptual design is investigated further, and additional information and details describing the implementation of the remedy are formulated. The results of the research during the preliminary design provide the basis for a final design.
- *Final design:* During the final phase of a project, the actual physical requirements of a remedial plan are defined. The engineer puts forth the prerequisites of the remediation program and the manner in which it will be implemented to the parties of interest, including the government regulatory agencies, in construction documents and specifications.

The design of hazardous waste remediation facilities, specifically ground water collection and treatment facilities, provides no exception to the classical design process. To explain this design process, each step and its correlative engineering considerations are presented below, and a case study is offered to illustrate the process.

Conceptual Design. During the conceptual design phase, the potential treatment alternatives are reviewed and evaluated in a broad and preliminary fashion. Although this phase of the project repeats part of the planning step in which remediation alternatives were evaluated, it is neither redundant nor superfluous. In the conceptual design phase, key decisions are made relative to the future direction of the project; in addition, the work of this phase amplifies the definition of the alternatives through further investigation. The goals of the conceptual design phase are to recommend one or a limited number of ground water treatment options and to describe in general terms the prerequisites of the necessary treatment facility. A schematic of the treatment process also is developed.

In the screening process and conceptual design, information is gathered about the contamination. Important issues reviewed are:

- The quality and quantity of the ground water to be treated.
- Location of the ground water collection system—pumping wells, for example.
- Site topographical survey.
- Subsurface conditions.
- Regulations and codes that apply to the situation at hand.

Treatment options also are reviewed to determine which have the capability to manage the contamination of the site. Each of the treatment options is reviewed according to the following criteria:

- Effectiveness of treatment.
- Reliability of the process.
- Capital and operations costs.
- Constraints of the site, including availability of nearby land.
- Aesthetic considerations, including visual qualities, noise, and odors.

The treatment options are screened by means of these criteria to include only the modes that are practicable. The number of treatment options that meet the criteria is further reduced by determining which treatment modes are applicable to the specific circumstances of the contamination at hand. For each treatment option, parameters are developed for the conceptual design. These parameters include treatment flow rates, number of process units, area required for construction, detention times and operating conditions, chemical requirements, and utility requirements. These parameters are developed in detail sufficient only to permit preliminary cost estimates to be formulated. Having developed them, the engineer and the parties of interest must make a decision; only those treatment modes that are both capable of addressing the specific contamination at hand and deemed economical are judged suitable and feasible for application, and they are passed on to the preliminary design phase.

Preliminary Design. Conceptual design provides the engineer with a set of feasible alternatives. The preliminary design process follows, and, through it, the alternatives are developed further. When the preliminary design phase begins, all the data describing the site and the specific project are reviewed again. Prerequisite data about the site are collected and analyzed. Design considerations also are evaluated further. Through the preliminary design, the engineer specifies the site of the treatment facility, size of the process equipment, layout of the facility, materials to be used in construction, and operational considerations. The engineer also estimates the cost. At the conclusion of the preliminary design, the selected treatment option is confirmed; the land required for the facility is identified; and both capital and operating costs are projected.

In order to determine a location for the ground water treatment facility, the site is examined in detail. It is surveyed to map its elevations, property lines, and physical features, notably proximity to water features and water supplies. Information about the types and characteristics of soils and the elevation of ground water is also essential, and meteorological conditions often are taken into account. Additional concerns are the health and safety of workers and the security of the facility during operation.

The current land use of a proposed site and its environs is an important consideration in selecting a location for a treatment facility. If the proposed site is proximate to water courses, water supplies, or population centers, unique design features may be dictated. For example, in one specific design instance, the prevailing wind pattern would have taken any volatile organic compounds released from an air stripping unit toward a nearby residential area. Therefore, the gases from the unit required treatment. Had this site been in a remote area where wind currents could have effectively dispersed the contaminants, the treatment of the gas could have been omitted.

When all the data about the site and about the design of the treatment facility are analyzed, a plan for the facility can be developed. The plan is drawn on the same base map that displays the topography and other pertinent site information. In addition to the facility, the location of collection wells and conveyance piping are represented on the map. Adjacent facilities, property lines, access roads, and other pertinent features also are shown.

Basis of Design of Treatment Process. The process schematic, developed when alternatives were evaluated, is used together with preliminary information about the sizing of process equipment as the basis of design for the facility. A first priority is to confirm the flow rate and the concentration of the contaminant in the ground water in order to determine treatment requirements. In many instances, a treatment process or a specific technology may appear feasible and economical as alternatives are evaluated, and so will warrant further investigation through bench-scale testing or testing in a pilot plant. These tests not only affirm treatability but also establish design parameters for loading rates, detention times, and chemical dosages. Based upon the flow rate of ground water, the treatment process schematic, and the results of any bench-scale or pilot plant testing, the processes of the treatment facility are sized, and a basis of design is developed. The data tabulated are listed in Table 10-5.

A schematic of the facility itself is drawn in accord with the proposed site plan and the basis of design. The constraints on the layout are the sizes of the equipment, concerns about operation, and access to the equipment. The layout shows the locations of items of equipment and of the processes. The design accounts for storage of chemicals and equipment to feed the chemicals, process equipment, instrumentation and control facilities, monitoring and sampling apparatus, and accommodations for the personnel who will operate the plant. The specific requirements for spill containment and leak detection are dictated by the degree of ground water contamination, the specific contaminants of concern, and the treatment technology employed.

TABLE 10-5. Elements of basis of design.

Waste characteristics	Functions of equipment
Type of treatment process	Materials of construction
Average and maximum flow rates	Pipeline sizes, function, and materials of construction
Number of process units	Management and disposal of residuals
Capacities and dimensions of equipment	Accessories or appurtenances
Loading rates	Provisions for spill containment and leak detection
Detention times	Security and fire protection measures
Chemicals and dosages	Utility requirements
Pumping capacities and horsepower	Electrical and instrumentation requirements
Operating weights of major items of equipment	Heating, ventilation, and air conditioning requirements
Horsepower ratings of major motors	

TABLE 10-6. Contents of preliminary design report.

Background of project	Layout of facility
Purpose of design and scope of work	Estimated capital and operating costs
Process schematic and description	Proposed design and construction schedule
Results of bench-scale or pilot plant testing	Discussion of particular design issues
Site plan	Recommendations for future action
Basis of design	

Pipe-in-pipe conveyance systems, specialized coatings or liners, and sophisticated leak detection systems are common components in ground water treatment facilities.

The information developed during the preliminary design phase is reported to the client for whom the treatment study is conducted. The regulatory agency supervising the ground water remediation program also may require such a report, to record the data and conclusions of the treatment study. The report's main purposes are to support the feasibility of the selected treatment process and to furnish the necessary details of the design to permit final design of the facility and its subsequent construction. The topics covered in the report are listed in Table 10-6.

Final Design. The first task of the final design process is to ascertain if sufficient data are available for preparation of the detailed construction documents. Because of the specificity required of the final design regarding the location of the treatment facility and the equipment to be used, precise and specific information is necessary. Data required but not in hand are developed at the outset of the final design phase.

The foundation of the final design is the preliminary design, as the basis of design is transformed into detailed construction documents. During final design, the complete design documents—the construction drawings and the design specifications—are prepared. In addition, a design specification document is composed to assist in the construction of the facility. Alternatively, depending upon the relative complexity of the project or the client's preference, the specification information is placed on the construction drawings. See Table 10-7 for a listing of the design specification document's contents.

The type of construction drawings prepared for ground water treatment facilities varies, depending upon the nature and complexity of the facility, the regulatory requirements, and the client's preferences. The drawings usually cover all the systems of the facility. Simple ground water treatment facilities may contain just a single collection well and an individual carbon treatment column and not require elaborate architectural, electrical, or

TABLE 10-7. Typical contents of design specification document.

General working conditions

General working requirements

Special conditions for safety:
- Site security
- Protection of the environment

Clarification of construction drawings

Technical specifications for:
- Construction materials
- Methods
- Equipment

Information about the contract:
- Bidding procedures
- Legal requirements
- Methods of payment
- Procedures for altering scope of work and schedule

TABLE 10-8. Specialties and disciplines participating in design of treatment facility.

GENERAL
Site work, piping, roads, utilities

ARCHITECTURAL
Buildings and appurtenances

STRUCTURAL
Foundations, structural supports, walls, roofing

MECHANICAL
System layouts, chemical feed and storage systems, inside piping, treatment process equipment

ELECTRICAL
Power supply and distribution, motor controls, instrumentation and control, electrical one-line diagrams

HEATING, VENTILATION, AND AIR CONDITIONING
Building services systems

PLUMBING
Roof drainage, internal water and sanitary lines

HVAC design documents. Complex, multiple-process treatment facilities require substantial detail, and the construction drawings are produced by numerous engineering specialties. See Table 10-8 for a listing of typical drawings.

The documents produced during the final design provide the basis for construction of the facility and are the guidelines by which the construction contractor proceeds. It is essential that these documents furnish the contractor with adequate and competent information to permit timely, economical, and technically sound completion of the treatment facility.

Applying the Classical Design Method. An example best illustrates how the classical engineering design sequence is applied in the field of hazardous waste remediation. An extensive hydrogeological investigation revealed that the site of a major manufacturing facility was grossly contaminated. Ground water monitoring and sampling disclosed that the underlying aquifer had become contaminated with high concentrations of volatile organic compounds. Moreover, a computer model of the ground water system projected a potential for contamination of a public water supply if the polluted ground water were not controlled. Therefore, a remedial plan was required expeditiously to prevent the migration of the ground water.

Having defined the nature and extent of the ground water contamination, engineers developed feasible alternatives for treatment. The alternatives were screened on the basis of land requirements; proven effectiveness of the treatment mode; capital, operating, and maintenance costs; ease of implementation; and potential for automated operation. The government regulatory agencies established effluent limits for total toxic organics (usually equivalent to all priority pollutants), total suspended solids, chemical oxygen demand, and pH. The main objective was to remove or reduce the concentration of organic compounds in the ground water; therefore, treatment methods reviewed during the conceptual design phase were air stripping, steam stripping, phase separation, carbon adsorption, and biological treatment.

The organic substances were sufficiently concentrated in the ground water to form two liquid phases: the organic substance itself and the organics in solution in the ground water. Because there was a clear demarcation between the two phases, the first treatment approach indicated was phase separation by gravity settling. Phase separation, then, would be followed by either air stripping, steam stripping, biological treatment, or carbon adsorption. The conclusion of the conceptual design phase recommended that bench-scale and pilot testing be conducted to determine the best treatment process.

The objectives of the preliminary design phase of the project were to select the best mode of treatment and to determine specific design values. Both the efficiency of each treatment method and the design parameters were identified through the bench-scale and pilot tests. The conclusion of the tests were:

1. Phase separation by gravity settling would be effective in concentrating the volatile organic compounds. Thus, the organic compounds could be withdrawn from the soil and disposed of.
2. Both air stripping and steam stripping could reduce markedly the concentration of volatile organic compounds in the aqueous phase; of the two, air stripping was projected to be the more economical approach.
3. Carbon adsorption could remove a major proportion of the contaminant; however, this treatment alternative would be expensive were it the pri-

mary means of ground water treatment because of the quantity of carbon required and the frequency of carbon regeneration. Were it deemed necessary, carbon adsorption could be employed to polish the effluent derived after another form of treatment.

4. If the ground water were pretreated with air stripping, activated sludge could effect a significant reduction in the nonvolatile organic content.

A treatment process was selected, based upon the results of the bench-scale tests, that proceeds as follows: After gravity separation, the ground water is pumped to a packed tower for air stripping. Vapors emitted are treated with activated carbon in a carbon adsorption unit. The effluent is then pumped to a biological treatment unit where it is neutralized with sodium hydroxide; it is then aerated with the addition of nitrogen and phosphorus as nutrients; next, a polymer is added; then, the sludge is withdrawn to a thickener. The liquid then flows through a sand filter, and that effluent is polished with a carbon adsorber; the sludge is backwashed, and particles are discharged to the sludge thickener. See Figure 10-6 for the process schematic.

A basis of design was developed, which is presented in Table 10-9. From it, the construction documents were developed, the cost of construction was

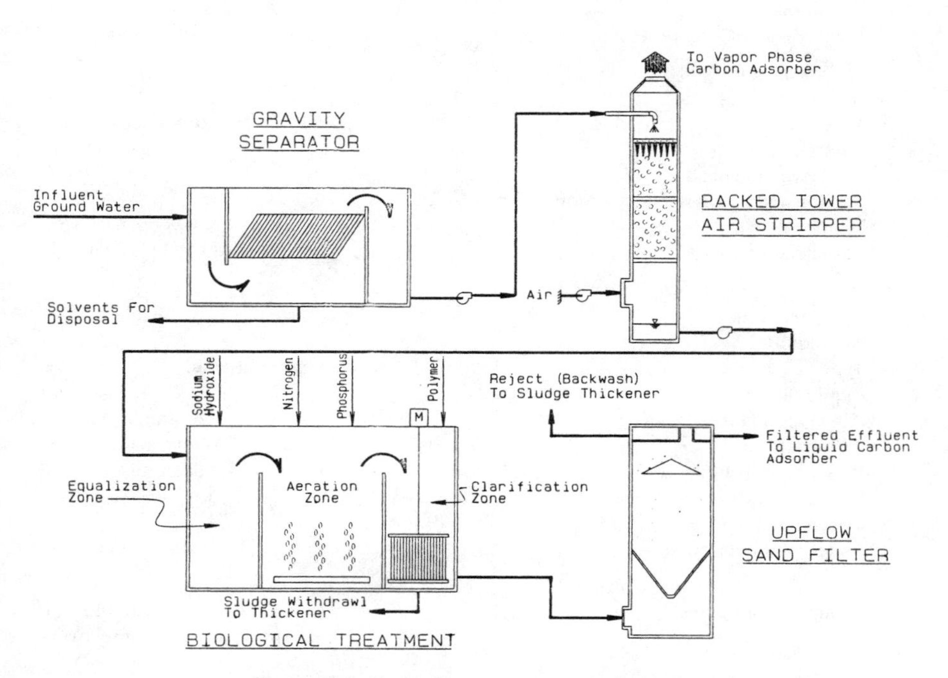

Figure 10-6. Ground water treatment schematic.

TABLE 10-9. Basis of design for ground water treatment. (Refer to Figure 10-6.)

1. Ground Water Characteristics	
Flow rate	150 gal/min
Waste characteristics	1000 mg/L VOC (max) 1500 mg/L COD 1000 mg/L BOD 50 mg/L TSS
2. Major Unit Processes	
Liquid phase separation	Chemical addition
Fluid transfer	Filtration
Air stripping	Liquid phase carbon adsorption
Vapor phase carbon adsorption	Sludge handling
Biological treatment (equalization, aeration, clarification)	
3. Liquid Phase Separation	
Treatment unit type	Gravity phase separation
Vessel type	Rectangular compartmentalized tank with inclined plates
Loading rate	0.5 gal/min/ft^2
Effective settling area	300 ft^2
Dimensions	5 ft $w \times$ 14 ft $L \times$ 10 ft h
Retention time	30 min
Material of construction	316 stainless steel
4. Fluid Transfer	
No. pumps	6
Type	Horizontal, centrifugal
Capacity	150 gal/min
Total dynamic head	100 ft
Suction and discharge connection	3 in.
Motor	7½ hp, 1750 rpm
Power requirement	480 V, 3-phase, 60 Hz
5. Air Stripping	
Treatment unit	Packed tower
No. required	1
Type	Vertical, cylindrical; countercurrent flow
Liquid flow rate	150 gal/min
Air-to-liquid ratio	20 standard ft^3/gal
Air flow rate	3000 standard ft^3/min
Pressure drop across tower	4 ft diameter $\times$ 40 ft h
Packing depth	30 ft
Packing size	1 in.
Material of construction	316 stainless steel
6. Vapor Phase Carbon Adsorption	
Treatment unit type	Carbon adsorption with in-place desorption and solvent recovery

TABLE 10-9. *Continued*

No. required	1
Type	Vertical, dual bed (one standby)
Air flow rate	3000 standard ft^3/min/bed
Carbon requirement	2000 lb/bed
Dimensions	7 ft $w \times$ 14 ft $L \times$ 10 ft h
Utility requirements	Water: 50 gal/min, 30 psi (max) Power: 120 V, 1 phase, 60 Hz
7. Biological Treatment	
Treatment unit type	Extended aeration, biological treatment
Vessel type	Circular, compartmentalized (4 zones)
Aeration hydraulic retention time	1.25 d
Sludge age	25 d at 20°C
Mixed liquor suspended solids	3000 mg/L at 20°C
Recycle rate	60%
Oxygen required	2400 lb/d at 20°C
Overflow rate	300 gal/d/ft^2
a. Equalization zone	Volume: 140,000 gal ea., retention time: 7.8 h
b. 2 aeration zones	Volume: 140,000 gal ea., retention time: 15.5 h ea.
c. Clarification zone	Volume: 80,000 gal
Clarifier mechanism:	Center drive, bridge supported
d. Sludge return system	Airlift type, capacity: 250 gal/min (max)
Vessel dimensions	70 ft dia × 15 ft side wall depth
Clarifier zone	Concentric, 30 ft dia
Material of construction	Carbon steel
Accessories	Aluminum-dome cover
8. Chemical Addition	
a. Phosphorus	(as phosphoric acid)
i. Daily volume required	10 gal/d (10% solution)
ii. Pump	Pos. displacement, diaphragm type, manual stroke adjustment
iii. Material of construction	Polyvinyl chloride
iv. Power	requirements 120 V, −phase, 60 Hz
b. Nitrogen	(as ammonia)
i. Feed system	Gas (vacuum)
ii. Capacity	100 lb/d (max)
iii. Control	Manually adjustable rate; automatic supply switch-over
iv. Power	requirements 120 V, −phase, 60 Hz
c. Polymer	
i. Feed system	Liquid polymer blender
ii. Capacity	5 gal/h concentrated polymer; 10 gal/min (max) blended solution
iii. Control	Variable pump speed control
iv. Power	requirements 120 V, −phase, 60 Hz
d. Sodium hydroxide	(neutralization)
i. Pump	Positive displacement, diaphragm type
ii. Material of construction	Polyvinyl chloride
iii. Power	requirements 120 V, −phase, 60 Hz

(continued)

TABLE 10-9. *Continued*

9. Filtration	
Treatment unit type	Continuous upflow sand filter
No. of units	2 in parallel
Filtration area	76 ft^2 (total)
Loading rate	2 gal/min/ft^2
Dimensions each	7 ft dia × 18 ft overall *h*
Reject rate (backwash)	10 gal/min (continuous)
Pressure drop	24 in.
Materials of construction	Carbon steel; stainless steel; FRP internals
Utility requirements	Compressed air; 5 standard ft^3/min at 25 psi
10. Liquid Phase Carbon Adsorption	
No. filters	2 in parallel
Type	Vertical, cylindrical
Total carbon required	10,000 lb each
Dimensions each	12 ft dia × 15 ft overall *h*
11. Sludge Handling	
a. Sludge thickener	(existing)
Type	Circular with rake mechanism
Dimensions	7 ft dia × 15 ft overall *h*
b. Sludge dewatering	(existing)
No. of units	4
Type	Plate and frame filter presses
Capacity each	100 ft^3

estimated, and the sequence of construction was projected. All aspects of the conceptual design and of the preliminary design were reported, both to the client and to the regulatory agencies reviewing the project.

Final Design of the Treatment Facility. Once the client and the regulatory agencies approved the preliminary design, the final design phase of the project commenced. The contamination was potentially serious because it was moving toward a potable water supply, and the construction schedule adopted was arranged to insure completion of the facility within the shortest practicable time. A critical path method of scheduling was developed. The schedule identified milestones in the project and indicated where bottlenecks might occur during construction. The schedule revealed that the time required to fabricate and deliver the packed tower air stripper might delay construction. Therefore, specifications were prepared for it before the final design phase was completed to permit the client to order it as early as possible.

The first component of the final design phase was design and construction of the ground water collection wells. They were installed to impede the migration of the contaminated ground water and prevent it from reaching the water supply. The collection wells channeled the contaminated ground water to the phase separation tank; there, the concentrated phase of the volatile organic compounds was removed.

During construction of the collection wells, the construction plans and specifications were prepared for the packed tower air stripper. The last phase of the design was engineering of the activated sludge and the carbon adsorption units. When the final design phase was completed, the contract for the treatment facility was bid for the activated sludge and carbon adsorption processes. The entire treatment facility was constructed in a timely and economical fashion.

Summary. Through the logical progression of conceptual, preliminary, and final design, a remediation project evolves from problem definition to implementation of a technically sound and economically feasible solution. In the conceptual design phase, available technologies are reviewed and screened. The product of this phase is the set of feasible alternatives that warrant further evaluation. The criteria and parameters that will determine the remedial design are investigated and tabulated during the preliminary design phase. Plant location, treatment effectiveness, capital and operating costs, and social and environmental concerns are explored in the depth necessary to estimate costs and to formulate a preliminary basis of design. This phase is the basis for the final design. During final design, all the preliminary data are utilized and, if necessary, augmented. From that information, the contract documents are developed. The project is implemented according to the contract documents.

The specific circumstances of each effort to remedy hazardous substances vary with each project. However, the logical, phased design approach generally leads to a well-conceived and sound engineering solution.

NOTES

1. American Water Works Association. *Standard for deep wells.* Denver, Colo.: 1967. AWWA-A100-66.

2. F. G. Driscoll. *Ground water and wells,* 2d Ed. St. Paul, Minn.: Johnson Division, 1981. For specific pump information, the well-design engineer contacts a pump supplier and furnishes the anticipated yield, the head conditions, and the required pump efficiency.

3. American Petroleum Institute. API specifications for casing tubing and drill pipe. API Spec. 5A. Dallas, Tex.: 1976.

4. American Society for Testing Materials. *Manual on industrial water.* Philadelphia, Pa.: 1977.

5. Driscoll.

6. U.S. Department of the Interior, Bureau of Reclamation. *Ground water manual.* 1981.

7. Johnson, 1985.

8. U.S.D.I., Bureau of Reclamation.

9. F. G. Driscoll and D. K. Todd. *Ground water hydrology.* New York: John Wiley & Sons, Inc., 1980.

10. P. J. Powers. *Construction dewatering: a guide to theory and practice.* New York: John Wiley & Sons, Inc., 1981. Table 6.1, pp. 100-101.

11. L. W. Canter and R. D. Knox. *Ground water pollution control.* Chelsea, Mich.: Lewis Publishers, Inc., 1985. See also Todd, 1980.

12. Todd, 1980.

13. C. Kufs et al. Procedures and techniques for controlling the migration of leachate plumes. Ninth annual research symposium on land disposal, incineration, and treatment of hazardous wastes. 1983.

14. See M. S. Hantush. Growth and decay of groundwater-mounds in response to uniform percolation. *Water Resources Research.* 3(1967):227-234. Cited by H. Bouwer. *Groundwater hydrology.* New York: McGraw-Hill Book Company, 1978. pp. 283-287.

15. Powers.

16. S. B. Blake and R. W. Lewis. Underground spill recovery. *Proceedings* of the second national symposium on aquifer restoration and ground water monitoring. National Water Well Association, 1982. pp. 69-76. Cited by L. W. Canter and R. C. Knox. *Ground water pollution control.* Chelsea, Mich.: Lewis Publishers, Inc., 1985. p. 85.

17. D. Molden, D. K. Sunada, and J. W. Warner. Microcomputer model of artificial recharge using Glover's Solution. *Ground Water.* 22 (Jan-Feb. 84):73-79.

18. Todd, 1980. See also Ch. 3, Analyzing hydrogeological conditions.

19. E. K. Nyer. *Ground water treatment technology.* New York: Van Nostrand Reinhold Co., 1985. p. 47.

20. Nyer. p. 49. W. D. Byers and C. M. Morton. Removing VOC from ground water; pilot, scale-up, and operating experience. *Environmental Progress.* 4(1985):112-118. R. L. Gross, and S. G. TerMaath. Packed tower aeration strips trichloroethylene from ground water. *Environmental Progress.* 4(1985):119-123. W. P. Ball, M. D. Jones, and M. C. Kavanugh. Mass transfer of volatile organics in packed tower aeration. *Journal Water Pollution Control Federation.* 56(1984):127-135.

21. R. A. Conway and R. D. Ross. *Handbook of industrial waste disposal.* New York: Van Nostrand Reinhold Company, 1980. p. 234.

22. The Freundlich isotherm equation was taken from Metcalf and Eddy. *Wastewater engineering treatment/disposal/reuse.* New York: McGraw-Hill Book Company, 1979. p. 280.

23. U.S. EPA, MERL. Carbon adsorption isotherms for toxic organics. Cincinnati.

24. Conway and Ross. p. 192.

25. Conway and Ross. p. 176.

26. Conway and Ross.

27. Nyer.

28. Conway and Ross.

29. Conway and Ross.

30. D. F. Kincannon, E. Stover, V. Nichols, and D. Medley. Removal mechanisms for toxic priority pollutants. *Journal Water Pollution Control Federation.* 55(1983):157-163. P. Pitter. Determination of biological degradability of organic substances. *Water Research.* 10(1976):231-235. E. L. Stover and D. F. Kincannon. Biological treatability of specific organic compounds found in chemical industry wastewaters. *Journal Water Pollution Control Federation.* 55(1983):97-109. Conway and Ross. p. 88.

31. G. S. Sayler, A. Breen, J. W. Blackburn, and O. Yagi. Predictive assessment of priority pollutant bio-oxidation kinetics in activated sludge. *Environmental Progress.* 3(1984):153-163.

32. E. J. Bouer. Secondary utilization of trace halogenated organic compounds in biofilms. *Environmental Progress.* 4(1985):43-45.

Chapter 11
In Situ Biological Treatment of Ground Water

Because nearly half the U.S. population depends on ground water for potable water, the contamination of soils and aquifers by harmful chemicals, especially organic chemicals, is a growing concern. Organic compounds may be halogenated—for example, solvents, paints, varnishes, refrigerants, herbicides, and pesticides; or they may be nonhalogenated—for example, petroleum hydrocarbons. Organic compounds may be difficult to degrade by conventional physical or chemical treatment processes.

Using current methods, contamination of an aquifer is treated either by removing the contamination or by isolating the contaminated water from use. In order to remove contaminants that are organic and relatively insoluble, the free contaminant is taken to the surface, treated, and disposed of. It is impossible to capture all of the contaminant, and a large volume of water generally must be managed through either reinjection or disposal. This approach may adversely affect the local ground water hydraulics. Containment, the alternative, sometimes only delays the effects of contamination because it does not remove the source of the contamination.

A technology that holds promise for the future is in situ biological treatment. Organic compounds can be metabolized by suitable microorganisms placed in the ground water system. In an ideal application, the microbial activity would mineralize the organic substances completely. (*Mineralization* is the conversion of an organic compound to an inorganic product; conversion to carbon dioxide and water is complete mineralization.) In situ biological treatment is currently under study and development, and the many issues that still require definition are presented in Table 11-1. Until the factors listed in the table are better defined, the technology cannot be considered for application.

The primary issues surrounding the utility of in situ biological treatment of ground water are divided into three levels of complexity:

1. *The biochemical scale:* The ability of microorganisms to mineralize organic compounds; that is, metabolism on the molecular level.

This chapter was developed by Stuart J. Spiegel of O'Brien & Gere Engineers, Inc.

TABLE 11-1. Issues requiring definition before utilization of in situ biological treatment.

Availability of organisms able to degrade anthropogenic compounds (xenobiotics).

Practicable availability of suitable organisms at sites of contamination.

Rates of biodegradation.

Effects of physical factors, such as temperature, on efficiency of the process.

Effects of chemical factors in the environment, particularly:
- Low concentrations of nutrients.
- Conditions conducive to production of methane.
- Sulfate reducing conditions.

Consequences of both aerobic and anaerobic conditions.

Biochemical pathways of the metabolism of organisms alien to the ground water system.

Relative advantages of pure and mixed cultures.

Practical limitations of the method due to technology, hydrogeology, cost, and scheduling.

Efficacy of adding nutrients to organisms in situ compared with seeding organisms in the ground.

Possibilities available due to recombinant genetic techniques.

2. *The microbial scale:* The microbial ecology of biodegradation activity.
3. *The ecological scale:* The application of theory and laboratory results in a field setting. Because the concentration of contaminants in ground water typically is low, the ability of organisms to recognize and utilize them as a substrate are important.

This chapter presents a perspective on each of these levels. Because in situ biological treatment is new, the information presented here is drawn from research literature. For reference, Figure 11-1 displays the structures of some of the compounds discussed in this chapter.

THE BIOCHEMICAL SCALE

There is no question that microbes can successfully mineralize many organic compounds that are toxic to humans.[1-4] Microbes have evolved a system to degrade a wide range of organics including aromatics and aliphatics, even their halogenated forms. All naturally occurring compounds can be utilized by a microorganism under suitable conditions.[5] The physical and chemical characteristics of a compound, including its solubility and volatility, affect the availability of that compound in aqueous solution. Compounds insoluble in water are less likely to be available for degradation. Environmental factors also act to favor or inhibit biodegradation; they include oxygen concentration,

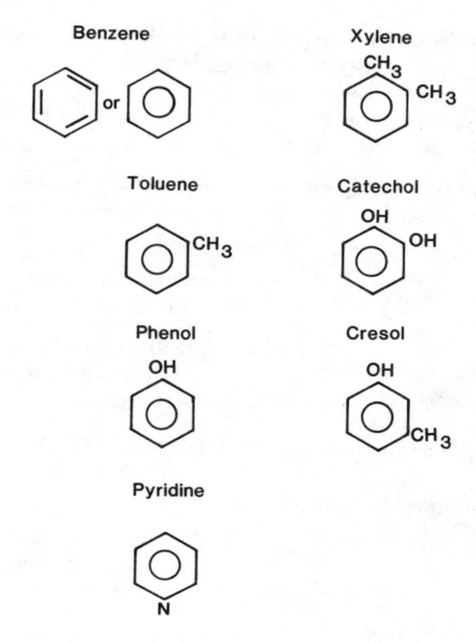

Figure 11-1. Common organic compounds.

oxidation-reduction potential, temperature, pH, availability of co-metabolites or nutrients, salinity, particulate matter, competing organisms, concentrations of viable organisms and substrate, and the presence of favored substrates or metabolic inhibitors.

Of especial interest is the competence of microbes to degrade anthropogenic compounds such as the halogenated organics. Many of these compounds have been in the environment for less than 50 years and probably were not produced at any time in the past, although there is some evidence of naturally occurring halogenated organic compounds. Anthropogenic organic compounds present a unique challenge on the molecular level. Organisms not previously exposed to them cannot be expected to have the ability to reduce them. However, some researchers believe that given the proper conditions, microbes are capable of degrading any compound. Many synthetic organic compounds that have no natural analogs can be transformed, although some resist biodegradation.[6] Relatively common bacteria have been observed to degrade, at least partially, anthropogenic toxics.[7]

For over 80 years, the metabolism of petroleum and other natural hydrocarbons has been studied, and the metabolic mechanisms of many different organisms on hydrocarbons have been elucidated. As pollutants, petroleum hydrocarbons can be categorized as occupying a middle ground of biodegradability between the readily degradable biogenic compounds and the highly recalcitrant anthropogenic organics. Crude petroleum consists largely of alkanes, cycloalkanes, and aromatics. The refining process yields many of the alkenes associated with gasoline, fuel oils, and other distillates.

The crucial first step in the microbial degradation of petroleum hydrocarbons is oxygenase, specifically monooxygenase, activity on alkanes, yielding alcohol. The presence of these enzymes requires aerobic conditions, although biodegradation can continue through nitrate or sulfate reduction processes under anoxic conditions.

Co-metabolism also may occur. A second substrate, the co-metabolite, does not serve as a carbon or energy source but is incidentally metabolized by organisms that utilize similar compounds. A good example is the degradation of DDT. Reportedly, DDT is metabolized only by a single fungus, but other organisms co-metabolize it and yield intermediates that are further degraded by still other organisms. The rate of degradation of a co-metabolite may be a function of competition with the primary substrate, a limiting factor in the rate of transformation. For these reasons, mixed cultures are probably preferable to a pure culture for the degradation of recalcitrant compounds.

A problem in the success of the transformation of some alkanes and cycloalkanes as well as other compounds is the toxicity of the substrate compound to the organisms. Toxic solvents include the monocyclic alkanes such as cyclopentane, cyclohexane, and cycloheptane. Both aerobic[8-10] and anaerobic[11-15] transformations of aromatic compounds have been reported, but the aerobic processes have been more often and more completely studied than the anaerobic. Phenolics, in particular, and polycyclic aromatic hydrocarbons have been observed to be susceptible to biotransformation.[16] Aromatic hydrocarbons such as alkyl benzenes were believed to be resistant to microbial degradation in the absence of molecular oxygen,[17] but there is evidence that such activity does indeed occur. The details of species identification and mixed cultures are limited. Both catechol and phenol reportedly undergo total degradation by a mixed culture of methanogens under anaerobic conditions, and the increase in methane production closely corresponds to the disappearance of the substrate spike.[18] All three isomers of xylene were observed to degrade under anoxic denitrifying conditions in a laboratory column system simulating saturated flow conditions typical for a river water-ground water infiltration system. The xylenes were used by the bacterial culture as the sole carbon and energy source.[19]

Persistence of compounds is often linked to the presence of chemical substitutes, including amines, methoxy groups, sulfonates, nitro groups, substituted benzene rings, ether linkages, and branched carbon chains. Of particular interest to environmental technologists is the presence of halogenated substitutes in alkyl and aryl compounds, particularly the chlorinated compounds present in solvents, pesticides, plasticizers, plastics, and the trihalomethanes. Many, if not all, halogenated aromatics can be acted upon by microorganisms. However, concern over dual toxicity arises in the degradation of haloorganics: substrate haloaromatic toxicity and metabolite toxicity. In the latter, metabolites such as chlorocatechols or chloro-substituted ring fission products may severely inhibit growth and may also suppress the development of the marginal members of the microbial community that utilize haloaromatics.[20] When a mixture of chlorophenol isomers is degraded, hydroxylation of 2- and 3-chlorophenol generates 3-chlorocatechol. This metabolite is critical because it is an inhibitor for catechol 2,3-dioxygenases. Therefore, as chlorophenols increase in the medium, this substrate is metabolized to the inhibitory intermediate, and further degradation rapidly decreases. In order to bypass this problem, diverse organisms must be present that will effect complete biodegradation.

The biodegradation of haloorganics can result in the formation of unwanted intermediates. The sequence of transformation of tetrachloroethylene results in the sequential formation of the following:[21] trichloroethylene, *cis*-1,2-dichloroethylene, *trans*-1,2-dichloroethylene, 1,1-dichloroethylene, and vinyl chloride; see Figure 11-2. Three conclusions have been made regarding the ease of dehalogenation of haloaliphatics: polyhalogenated compounds are more easily reduced than monohalogenated compounds; short chains are more easily reduced than long chains; and the strength of the reducing environment (methanogenesis, sulfate reduction) favors degradation.

THE MICROBIAL SCALE

Traditionally, biological activity below the root zone was thought to be minimal. However, recent studies have demonstrated the presence of subsurface microorganisms,[22-25] which are reported to be similar to surface organisms in their ability to degrade anthropogenic wastes.[26-31] In a study of the effect of depth on microbial numbers in a pristine environment, samples from a depth of 5 m were found to contain 4.8 million cells/g dry material.[32] The organisms from several depths were able to degrade toluene rapidly and chloroform slowly in microcosms prepared from materials obtained at the depths investigated. The important finding of this study was that indigenous organisms at a carefully selected pristine site could readily adapt to degrade anthropogenic wastes to which they never previously had been exposed.

Heterotrophic bacteria from samples of ground water, well depths ranging from 3 to 12 m, were evaluated for their ability to degrade anthropogenic substances including nitrogenous short chain acids (glutamic acid, sodium nitrilotriacetic acid), an aromatic acid (benzoic acid), and a halogenated

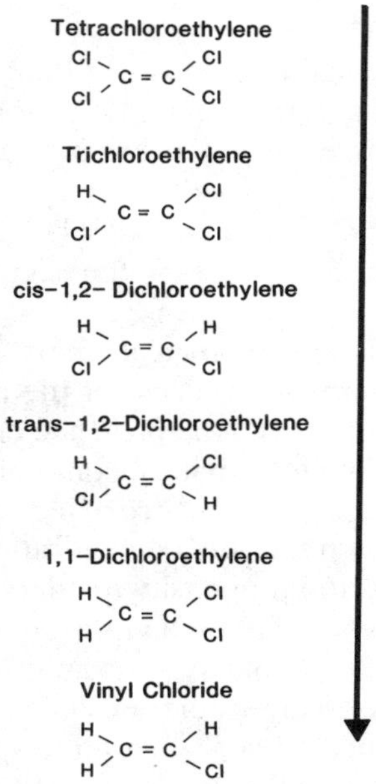

Tetrachloroethylene

Cl, Cl > C = C < Cl, Cl

Trichloroethylene

H, Cl > C = C < Cl, Cl

cis-1,2- Dichloroethylene

H, Cl > C = C < H, Cl

trans-1,2-Dichloroethylene

H, Cl > C = C < Cl, H

1,1-Dichloroethylene

H, H > C = C < Cl, Cl

Vinyl Chloride

H, H > C = C < H, Cl

Figure 11-2. Sequence of decay.

aromatic acid (2,4-D).[33] The biodegradation kinetics were in the range reported for oligotrophic surface water. This study supports the contention that ground water bacteria are specialized for the nutrient-poor habitats in which they are found. The apparent implication for in situ treatment methods is that a lack of nutrients will influence the level of the microbial population and reduce the rate at which a contaminant can be degraded.

The microflora at sites of contamination have been studied for the presence of ongoing microbial metabolism. In one situation where the ground water had been contaminated during operation of a coal tar distillation and wood treatment facility, a study found phenolic compounds in ground water being converted to methane and carbon dioxide by anaerobic bacteria.[34] Another study found that the microorganisms present in an aquifer material producing methane were able to metabolize chlorophenols after replacing the halogen substitutes with a hydrogen molecule.[35] In another situation, *meta*-, *para*-, and *ortho*-xylene were undergoing reduction in a plume of methane-producing leachate from a landfill. The aromatic hydrocarbons originally present suggested that the source of contamination was petroleum leaching water-soluble compounds. Most of these compounds were persistent and moved with the plume, except for the preferential removal of the xylenes.[36]

THE ECOLOGICAL SCALE

Ecological as well as physical and chemical limitations may be present in biodegradation studies of natural systems. Antagonistic interactions among microorganisms may occur, especially between bacteria and fungi where the bacteria are known to be antagonistic to fungal survival. Bacteria play a much greater role in degradation than fungi. Unfortunately, little is known about fungal adaptation and processes in the degradation of anthropogenic substances.

In the application of in situ biological treatment of ground water, the specific environment governs the success of the process. Basic considerations include the hydrogeology and the presence of chemicals in the subsurface that may inhibit metabolic activities, contaminant concentrations that may be toxic to microbial life, the presence of indigenous organisms that can degrade the target contaminant, low concentrations of contaminants, the presence of nutrients, and the propensity to adapt to novel substrates.

For concentrated plumes of contaminants in ground water, aerobic metabolism is limited by the solubility of oxygen,[37] and anaerobic processes are an attractive alternative. Pilot-scale studies designed to determine the growth rate of indigenous microbial flora at a gasoline spill site reported that the natural microbial population could be increased a thousandfold by aeration and the addition of inorganic nitrogen and phosphate salts to ground water.[38] The hydrocarbons were mineralized in proportion to the growth rate of the microbial population.

The hydrogeological conditions at a site are critical factors in the success of any applications, and are considerations independent of, though related to, biological factors. The hydrogeology must be such that the nutrients, microbes, and oxygen can contact the contaminant. In addition to accommodating soil characteristics, water flow must be essentially horizontal to optimize in situ biological treatment. A confining layer must be present to minimize vertical seepage and the transport of contaminants. There are two basic types of horizontal ground water flow: one in which there is significant regional flow, and a plume of contamination is formed; and a second in which the regional flow is minor, and the movement of contaminants in the ground water is radial. In either case, a slurry wall can be used under certain conditions to channel contaminated ground water to a suitable subsurface location where in situ biological treatment would be feasible.[39]

Local conditions can be such that indigenous flora capable of degrading a target contaminant are not present in the subsurface. In one study of the efficacy of inoculating systems with exogenous bacteria, the biodegradation of parathion in soil was greatly accelerated by adding a mixed culture of parathion-utilizing bacteria to the contaminated soil.[40] In another study, soil was inoculated with a pentachlorophenol-degrading *Flavobacterium*.[41] The study showed that soil conditions such as temperature, water content, pentachlorophenol concentration, and bacterial density must be maintained within optimum ranges in order to sustain satisfactory rates of biodegradation.

The effect of the addition of a second exogenous microorganism has also been studied.[42] Such an addition increased primary and complete biodegradation of polychlorinated biphenyls (PCBs) tenfold in soil. An inoculation of this system with *Acinetobacter P6* increased the degradation rate further. The study concluded that the indigenous microflora were unable to degrade the PCBs to any significant extent, but alternative sources were able to provide organisms to reduce them.

Very low substrate concentrations pose several problems for microbial transformations. Their slow metabolic kinetics may furnish insufficient energy to sustain the microbial population. In addition, trace concentrations of a contaminant may be insufficient to induce metabolism. Inadequate data exist to permit extrapolation from the laboratory to the field.

A key phenomenon in the microbial degradation of an anthropogenic substance is adaptation, which can be defined as the lag time before degradation starts in organisms that have never before encountered the particular compound. Lag times vary significantly with different microbial populations, target compounds, and environmental conditions. Adaptation may be linked to several mechanisms:

- Selection, which cannot be explained by studies of kinetic systems involving the increased growth of favored organisms.
- Emergence of new metabolic pathways.
- Evolution of new organisms.

There are two ways rapid evolution can be achieved. First, if the anthropogenic substance is similar to a natural substrate, existing enzyme action can be used. Second, new enzymes can be derived from preexisting genes that code for related enzymes or proteins.[43] In sum, the mechanism of adaptation is unknown, yet its qualitative aspects have been well documented. More information is needed regarding specific mechanisms and the threshold of adaptation to anthropogenic substances.

A study on adaptation reported that a lag phase usually occurs before the rapid utilization of a chlorinated substrate. It offered the following guidelines for improving the probability of obtaining acclimated cultures and minimizing the lag phase:[44]

- A high concentration of active cells is desirable to minimize the apparent lag.
- The initial substrate concentration should be in the low parts per million level to reduce inhibitory effects of the substrate. However, if the concentration of the target substrate is too low, acclimation may not occur.

- Microorganisms with diverse metabolic capabilities are desirable in order to increase the probability of successful degradation of the contaminent.
- When nutrient broth was added as an alternate substrate, concurrent substrate utilization occurred; the overall rate of chlorinated organic substrate removal was enhanced except at low concentrations of the substrate. Enhancement under these conditions was ascribed to the production of higher concentrations of active cell mass.

A detailed study of adaptation in a contaminant plume at a creosote waste site concluded that once acclimation occurs, biotransformation is so rapid that it can be considered instantaneous within the context of slow ground water flow.[45] Therefore, quantitative prediction of the rate of the biological activity or the fate of the pollutant shifts from the kinetics of the microorganism's utilization of the substrate to the utilization allowed by geochemical constraints on metabolism and the rate of supply of that substrate.

PROGRAM SUGGESTIONS

There has been little experience with in situ biological treatment of contaminated ground water, and there have been virtually no field applications. However, from the limited record of research and bench-scale tests performed, it is apparent that implementation of in situ biological treatment requires a comprehensive study of the site where application is proposed to provide knowledge of the hydrogeological parameters and the subsurface chemistry. During application, the technique may require an extended acclimation period and an equally long degradation phase. Therefore, in situ treatment may be practicable only if the long-term environmental or human health effects of the contaminants are extraordinary and preclude consideration of the costs, or if alternative long-term remedies are more expensive than the in situ treatment.

If implemented, a program of in situ biological treatment would proceed according to the following steps, incorporating interaction between field characterization and laboratory studies:

1. *Characterization of site subsurface hydrogeology:* Factors such as permeability, rate of ground water flow, plume formation, oxic and anoxic conditions, and contaminant concentration affect the applicability of an in situ solution to a particular site. A site with multidirectional horizontal ground water flow potential may give rise to unpredictable flow regimes; inhomogeneous subsurface conditions could confound the prediction and control of nutrient availability, microbial activity, and contaminant concentration. Remediation by in situ techniques then would be impracticable.

2. *Characterization of site subsurface chemistry:* Nutrient deficiencies may limit the potential for microbial growth. Given unidirectional or radial ground water flow, the addition of nutrients by injection may be possible. However, the chemical constitution of the soil may be selective for unfavorable organisms.
3. *Identification and presence of key organisms or groups of organisms:* Laboratory studies are required to determine the presence of indigenous organisms that can transform the target contaminant. The effect of secondary contaminants on the microorganisms could be positive, providing a favorable secondary substrate, or they could be negative and toxic. Metabolic factors must be considered, particularly whether the organisms require aerobic or anaerobic conditions to effect optimum biotransformations.
4. *Selection of microbial groups:* Where local microbial populations are unsuccessful at transforming the target contaminant, organisms from other sources are necessary. Such organisms may be found in industrial or municipal microbial wastewater treatment systems or in soil systems containing the same contaminants. The selected organisms must be tested for compatibility with the site's soil characteristics; that is, the presence of sulfate militates against organisms that produce methane.
5. *Subsurface inoculation:* When conditions require, microorganisms, nutrients, or oxygen can be injected into the contaminated zone to augment natural conditions. Injection could be accomplished with wells upgradient of the site.
6. *Monitoring:* The ground water must be monitored for:
   - *Acclimation:* Based on laboratory studies, it may be possible to predict the extent of the acclimation period prior to commencement of active metabolism. However, investigators have indicated that there does not appear to be a direct correlation between lag times in the laboratory and acclimation periods in situ.
   - *Biotransformation of contaminant:* A decrease in contaminant concentration coupled with an increase in methane production or the formation of metabolic intermediate and end products would be observed. When this is noted, the rate of transformation indicates the likely success of and time frame for remediation of the site.
   - *Chemical changes:* It is important to examine whether transformation occurs as a result of the biotic affect on the site or abiotic conditions due to biotic alterations in the contaminated zone.
   - *Long-term deviations in the character of the contaminant zone:* Potential alternations include modification from oxic to anoxic conditions due to a decrease in the depth of the water table, the movement of other chemicals into the contaminant zone, or an increase in ground

water flow rate due to removal at another location. These situations affect contaminant migration, microbial survival or activity, contaminant concentrations, and microbial residence time.

SUMMARY

In situ treatment is a novel technology, and parametric generalizations cannot yet be made. The technique will remain developmental and expensive for some time. Although it is evident that much is known about the factors involved at the biochemical and microbial levels, major questions still must be answered before it can be considered for routine application. First, more must be known about adaptation and its mechanisms, specifically an understanding of the processes on the cellular and molecular levels that regulate adaptation. That information will make it possible to perform genetic or environmental manipulations to remove rate-limiting steps and enhance the potential for success. Laboratory studies must determine the qualitative conditions necessary for applications in the field, although laboratory data are not directly applicable to field results. There should be an interchange between the investigators attempting to answer basic questions in the laboratory and the engineers attempting to apply the technique in the field. Without such feedback, there can be little efficiency in the review of field problems in the laboratory and little comprehension of the microbial or biochemical basis for success in situ.

NOTES

1. E. M. Davis, H. E. Murray, J. G. Liehr, and E. L. Powers. Basic microbial degradation rates and chemical byproducts of selected organic compounds. *Water Res.* 15(1981):1125-1127.
2. R. S. Horvath. Microbial co-metabolism and the degradation of organic compounds in nature. *Bacteriol. Rev.* 36(1972):146-155.
3. H. Kobayashi and B. E. Rittmann. Microbial removal of hazardous organic compounds. *Environ. Sci. Technol.* 16(1982):170A-183A.
4. H. H. Tabak, S. A. Quave, C. I. Mashni, and E. F. Barth. Biodegradability studies with organic priority pollutant compounds. *J. Water Pollut. Control Fed.* 53(1981):1503-1518.
5. J. F. McNabb, and W. J. Dunlap. Subsurface biological activity in relation to ground-water pollution. *Ground Water.* 13(1975):33-44.
6. McNabb and Dunlap.
7. For example, the partial conversion of the pesticide methoxychlor from a substituted trichloroethane to two dichloroethanes by *Klebsiella pneumoniae*, a common environmental enteric pathogen, has been observed. W. H. Baarschers, A. I. Bharath, J. Elvish, and M. Davies. Biodegradation of methoxychlor by *Klebsiella pneumoniae. Can. J. Microbiol.* 28(1982):176-179.

8. W. C. Evans. Microbial degradation of aromatic compounds. *J. General Microbiol.* 32(1963):177-184.

9. S. Dagley. The microbial metabolism of phenolics, pp. 287-317 in: A. D. McLaren and G. H. Peterson (Eds.). *Soil biochemistry.* London: Edward Arnold, 1967.

10. S. E. Herbes and L. R. Schwall. Microbial transformation of polycyclic aromatic hydrocarbons in pristine and petroleum contaminated sediments. *Appl. Environ. Microbiol.* 35(1978):306-322.

11. B. F. Taylor, W. L. Campbell, and I. Chinoy. Anaerobic degradation of the benzene nucleus by a facultative anaerobic microorganism. *J. Bacteriol.* 102(1970):430-437.

12. J. B. Healy, Jr. and L. Y. Young. Catechol and phenol degradation by a methanogenic population of bacteria. *Appl. Environ. Microbiol.* 35(1978):216-218.

13. D. Grcic-Galic and L. Y. Young. Methane fermentation of ferrulate and benzoate: Anaerobic degradation pathways. *Appl. Environ. Microbiol.* 50(1985):292-297.

14. E. P. Kuhn, P. J. Colberg, J. L. Schnoor, O. Wanner, A. J. B. Zehnder, and R. P. Schwarzenbach. Microbial transformations of substituted benzenes during infiltration of river water to ground water: Laboratory column studies. *Environ. Sci. Technol.* 19(1985):961-968.

15. B. H. Wilson and J. F. Rees. Biotransformation of gasoline hydrocarbons in methanogenic material, pp. 128-139 in: National Water Well Assoc. (Ed.), *Proceedings* of the NWWA/API Conference on Petroleum Hydrocarbons and Organic Chemicals in Ground Water—Prevention, Detection and Restoration. Dublin, Ohio: Water Well Pub. Co., 1986.

16. Herbes and Schwall, 1978.

17. L. Y. Young. Anaerobic degradation of aromatic compounds, pp. 487-523 in: D. T. Gibson (Ed.), *Microbial degradation of organic compounds.* New York: Marcel Dekker, 1984.

18. Healy and Young, 1978.

19. Kuhn et al., 1985.

20. E. Schmidt, M. Hellwig, and H.-J. Knackmuss. Degradation of chlorophenols by a defined mixed microbial community. *Appl. Environ. Microbiol.* 46(1983):1038-1044.

21. J. T. Wilson, J. F. McNabb, D. L. Balkwill, and W. C. Ghiorse. Enumeration and characterization of bacteria indigenous to a shallow water-table aquifer. *Ground Water.* 21(1983):134-141.

22. Wilson, McNabb, Balkwill, and Ghiorse, 1983.

23. W. C. Ghiorse and D. L. Balkwill. Microbiological characterization of subsurface environments. First International Conference on Ground Water Quality Research, Houston, Tex., 1981.

24. W. C. Ghiorse and D. L. Balkwill. Enumeration and morphological characterization of bacteria indigenous to subsurface environments. *Dev. Indust. Microbiol.* 24(1983):213-224.

25. R. M. Ventullo, and R. J. Larson. Metabolic diversity and activity of heterotrophic bacteria in ground water. *Environ. Toxicol. Chem.* 4(1985):759-771.

26. Wilson, McNabb, Balkwill, and Ghiorse, 1983.

27. Ventullo and Larson, 1985.

28. G. G. Ehrlich, D. F. Goerlitz, E. M. Godsey, and M. F. Hult. Degradation of phenolic contaminants in ground water by anaerobic bacteria: St. Louis Park, Minnesota. *Ground Water.* 20(1982):703-710.

29. E. M. Godsey, D. F. Goerlitz, and G. G. Ehrlich. Methanogenesis of phenolic compounds by a bacterial consortium from a contaminated aquifer in St. Louis Park, Minnesota. *Bull. Environ. Contam. Toxicol.* 30(1983):261-268.

30. J. M. Suflita and G. D. Miller. Microbial metabolism of chlorophenolic compounds in ground water aquifers. *Environ. Toxicol. Chem.* 4(1985):751-758.

31. M. Reinhard, N. L. Goodman, and J. F. Barker. Occurrence and distribution of organic chemicals in two landfill leachate plumes. *Environ. Sci. Technol.* 18(1984):953-961.

32. Wilson and Rees., 1986.

33. Ventullo and Larson, 1985.

34. Ehrlich et al., 1982.

35. Suflita and Miller, 1985.

36. Reinhard et al., 1984.

37. J. T. Wilson, J. F. McNabb, J. W. Cochran, T. H. Wang, M. B. Tomson, and P. B. Bedient. Influence of microbial adaptation on the fate of organic pollutants in ground water. *Environ. Toxicol. Chem.* 4(1985):721-726.

V. W. Jamison, R. L. Raymond, and J. O. Hudson, Jr. Biodegradation of high-octane gasoline in ground water. *Dev. Indust. Microbiol.* 16(1975):305-312.

38. Jamison, Raymond, and Hudson, 1975.

39. D. C. McMurtry and R. O. Elton III. New approach to in situ treatment of contaminated ground waters. *Environ. Prog.* 4(1985):168-170.

40. R. W. Barles, C. G. Daughton, and D. P. H. Hsieh. Accelerated parathion degradation in soil inoculated with acclimated bacteria under field conditions. *Arch. Environ. Contam. Toxicol.* 8(1979):647-660.

41. R. L. Crawford and W. W. Mohn. Microbiological removal of pentachlorophenol from soil using a *Flavobacterium. Enzyme Microb. Technol.* 7(1985):617-620.

42. W. Brunner, F. H. Sutherland, and D. D. Focht. Enhanced biodegradation of polychlorinated biphenyls in soil by analog enrichment and bacterial inoculation. *J. Environ. Qual.* 14(1985):324-328.

43. S. Ghosal, I.-S You, D. K. Chatterjee, and A. M. Chakrabarty. Microbial degradation of halogenated compounds. *Science.* 228(1985):135-142.

44. C. J. Kim and W. J. Maier. Acclimation and biodegradation of chlorinated organic compounds in the presence of alternate substrates. *Journal Water Pollution Control Federation.* 58(1986):157-164.

45. Wilson, McNabb, Cochran, Wang, Tomson, and Bedient, 1985.

Chapter 12
Correcting Leaking Underground Storage Systems

Petroleum products, industrial process chemicals, solvents, and their wastes have been stored routinely in buried storage systems since the early years of industrialization. If these liquids are stored improperly, or if the condition of their storage is neglected, there is a high potential for leakage. As a result, hazardous contaminants can enter the environment. An underground storage system includes not only the storage tank itself, sometimes termed a *storage vessel,* but also the piping, pumps, sumps, drains, fittings, and any other components that serve to contain the substance stored. Any of the components, not just the storage vessel itself, may leak. Underground storage, although not addressed by CERCLA, is pertinent to a discussion of the remediation of hazardous wastes, as hazardous waste sites sometimes include such facilities.

In the past, storage of liquids underground served a number of purposes: protection from possible fires, conservation of land, and improved aesthetics. Until the 1980s, the effects of underground storage facilities were often ignored. Few government regulations required owners of underground tanks to install the tanks correctly or to monitor the tanks' integrity; moreover, the potential adverse environmental effects of leakage were virtually unknown. Tanks, with or without the knowledge of the owner, might leak for decades, contaminating ground water and soil without remedial actions being taken. The U.S. EPA estimates that nearly 11 million gal (41,600 m^3) of gasoline alone may be leaking from underground storage tanks each year.[1] Other substances that are leaking include hazardous materials and wastes and nearly every type of petroleum product.

Most of the leaking storage tanks were installed during the industrialization of the 1940s, 1950s, and 1960s, and they usually were constructed of bare steel without protection against corrosion. The American Petroleum Institute estimates that such unprotected tanks have only a 15-year service life

---

This chapter was developed by Scott J. Adamowski, P.E. of O'Brien & Gere Engineers, Inc.

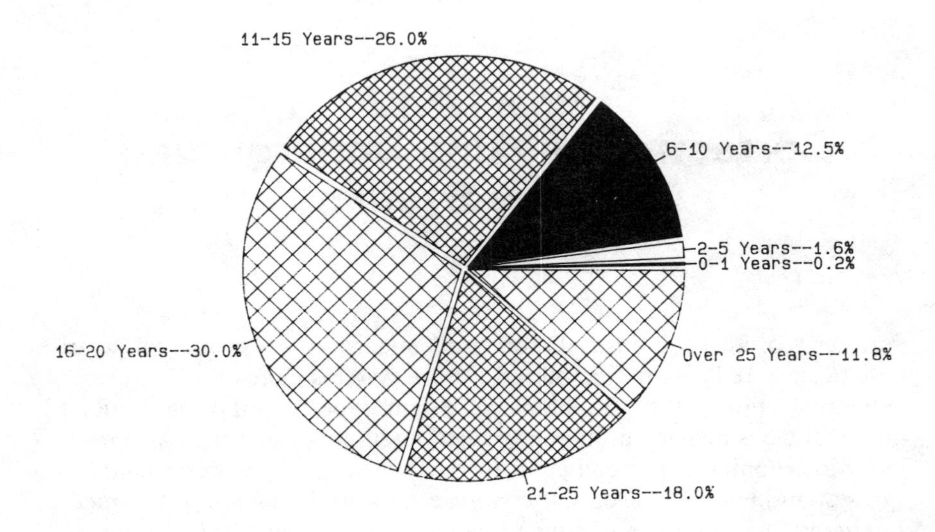

Figure 12-1. Incidence of tank leakage by age.

before leaking. Figure 12-1 displays frequency of leakage based on tank age for bare steel tanks.[2]

Since 1983, government regulators have focused attention on underground storage systems. These systems now are viewed as an obvious source of environmental contamination. Federal, state, and local governments have begun to implement laws and regulations governing underground storage systems. At a minimum, these programs require the registration of new and existing systems. The objectives are to:

- Define the volume of storage.
- Identify the substance stored.
- Describe the characteristics of the storage system: the age, methods of monitoring and testing, and problems experienced with the system.

Many states require that existing tank systems be tested for integrity, whether there is evidence of leakage or not.

This chapter first presents an overview of existing and proposed federal and state legislation governing underground storage. It then suggests an approach for predicting tank failure and for correcting leaking underground storage systems. The practices summarized here present a practical program for correcting the problem of leaking underground storage systems.

GOVERNMENT REGULATIONS

In the late 1970s, the federal government and many states adopted laws to regulate facilities that stored petroleum products. In general, these regulations stipulated that owners of facilities with more than 42,000 gal (160 m^3) of underground petroleum storage develop plans to prevent and control oil spills. During the early 1980s, the federal government and several state regulators began to appreciate the scope of the environmental problems associated with hazardous and toxic materials entering the environment from underground storage systems. Federal legislation subsequently was drafted governing facilities having a capacity over 1100 gal (4 m^3) and storing hazardous substances, as defined by CERCLA, and petroleum products. The rules govern both new and existing systems. The Hazardous and Solid Waste Amendments of 1984 amended RCRA[3] and contained provisions for the regulation of underground storage systems. In brief, these amendments require:

- Registration of new and existing tanks with designated state and local regulatory agencies.
- U.S. EPA regulation of all underground storage tanks with regard to the following factors of operation:
  - Leak detection, inventory control, tank testing.
  - Record keeping and reporting.
  - Corrective action.
  - Financial responsibility for corrective action and liability.
  - Closure.
- Interim standards for the installation of new underground tanks so that the tanks:
  - Are capable of preventing releases due to corrosion or structural failure for their lifetime.
  - Have cathodic protection against corrosion, are constructed with noncorrosive material, are steel-clad with a noncorrosive material, or are designed to prevent the release of any stored substance.
  - Are constructed or lined with a material compatible with the substance stored; however, a storage tank may be installed without corrosion protection if the soil resistivity is 12,000 Ω-cm or more.
- Permanent performance standards for the installation of new under-

ground systems to address design, construction, installation, release detection, and standards for compatibility.

- Allowance for states to administer the program in place of the U.S. EPA using the federal requirements or more stringent state requirements.
- Study of petroleum and other regulated substances contained in underground tanks.
- Enforcement of the established laws and regulations.

The U.S. EPA has also changed Subtitle J of RCRA in order to regulate underground storage of hazardous waste and provide a second means of containing the substance should the tank fail.[4] This secondary containment system would:

- Prevent waste or liquid from escaping to the soil or water for the life of the tank.
- Collect waste or leakage until material is removed.
- Be constructed or lined with material compatible with the waste and with sufficient strength to prevent failure from pressure, climate, traffic, and daily use.
- Have an adequate base or foundation capable of resisting settlement, compression, and uplift.
- Have a system capable of detecting leaks within 24 hours of occurrence.
- Be sloped or drained to permit removal of leaks, spills, and precipitation and contain provisions for such accumulation to be removed.
- Contain 110 percent of the design capacity of the largest tank within the containment boundary.
- Prevent run-on or infiltration of precipitation unless the collection system has excess capacity (beyond the 110 percent) to hold precipitation from the 25-year, 24-hour rainstorm.

Tanks already in operation do not need the secondary containment system if they are furnished with a system to monitor ground water and if the tanks are tested according to a schedule.

In addition to the federal initiatives, several states and municipal governments have begun to enact and administer regulations to police underground storage systems. In most instances, these requirements are more stringent than the federal regulations. For example, the federal tank registration program exempts all fuels used for consumptive heating on the premises, but New York State exempts only No. 5 and No. 6 fuel oil. Table 12-1 identifies several state requirements.[5]

TABLE 12-1. Summary of state regulations for underground storage of petroleum, hazardous substances, and hazardous waste, February 1987.

	STATUS OF UNDERGROUND BULK STORAGE[a]			
STATE	OIL	HAZ SUB	HAZ WASTE	NOTIFICATION FORMS
Ala.	N	N	EPA	E
Alaska	N	N	EPA	E
Ariz.	N	N	EPA	E*
Ark.	F	N	F	E
Calif.[b]	F	F	F	S
Colo.	P	N	EPA	E
Conn.	F	N	EPA	S
Del.	F	F	F	S
D.C.	N	N	EPA	E
Fla.[c]	F	F	F	S
Ga.	P	P	EPA	E
Hawaii				E*
Idaho	N	N	EPA	E
Ill.	FP	N	EPA	E
Ind.	N	N	EPA	E*
Iowa	N	N	EPA	S
Kans.	P	P	F	E
Ky.	N	N	EPA	S
La.	P	P	F	S
Maine	F	F	EPA	S
Md.	F	N	EPA	
Mass.	F	F	F	E
Mich.	P	D	F	E
Minn.	P	N	F	S
Miss.	N	N	EPA	E
Mo.	N	N	EPA	E
Mont.	N	N	EPA	E
Nebr.	N	N	EPA	E
Nev.	D	D	EPA	E
N.H.	F	F	F	E
N.J.	D	D	EPA	S
N.Mex.			F	E
N.Y.	F	F	F	E
N.C.	D	D	D	E
N.Dak.	N	N	EPA	S
Ohio	P	N	EPA	S
Okla.	N	N	EPA	E
Oreg.	P	N	EPA	
Pa.	N	N	D	E
R.I.	F	F	F	E

TABLE 12-1. *Continued*

STATE	STATUS OF UNDERGROUND BULK STORAGE[a] OIL	HAZ SUB	HAZ WASTE	NOTIFICATION FORMS
S.C.	F	N	EPA	S
S.Dak.	N	N	EPA	E
Tenn.	N	N	EPA	E
Tex.	N	N	F	E
Utah	N	N	EPA	E
Vt.	P	D	EPA	S
Va.	F	F	F	E
Wash.	P	N	F	S
W.Va.	N	N	EPA	E
Wis.	P	D	D	S
Wyo.	N	N	EPA	E

* = Supplementary form required
D = Draft
EPA = Effective date for 40 CFR 260-265, 270, and 271 was 1-12-87.
F = Final
N = None
E = U.S. EPA form
P = Partial regulations.
S = State form
[Blank]= No information.

[a] State fire marshal or municipal agencies may have additional codes.
[b] Individual city and county agencies enforce UST regulations.
[c] Dave, Browery, and Alachua counties have UST jurisdiction.

CAUSES OF LEAKAGE

Leaks of hazardous liquids from underground storage facilities may arise from any of several sources. Data indicate that corrosion and poor installation are by far the most common causes of storage system leaks.[6] Other sources include: tank or pipeline ruptures; slow deterioration, such as that caused by corrosion; spills when the product is transferred or when overfilling occurs; poor construction, installation, or design; use of storage system beyond design life; and vandalism.[7] The American Petroleum Institute (API) surveyed 1717 leaking underground tanks and piping systems to discover common sources of leakage. The results are presented in Table 12-2.[8]

According to this study, the majority of leakage occurs in steel storage vessels that are not protected against corrosion. Among tanks that are protected— for example, those using cathodic protection—the corrosion rate is significantly reduced or eliminated. The information of Table 12-2 illustrates why federal policy no longer permits the use of unprotected steel tanks underground.

Based on these API data, corrosion is responsible for approximately 90 percent of leaks in underground storage systems. However, corrosion does not manifest itself only in metallic tanks. Plastic tanks can also corrode—

TABLE 12-2. Sources of leakage in underground storage systems.

SOURCE	NO.	%
Unprotected steel tanks	913	62.0
Steel piping	454	30.8
Fiberglass piping	50	3.4
Fiberglass tanks	28	1.9
Steel tanks with impressed current	13	0.9
Interior coated steel tanks	7	0.5
Steel piping with impressed current	7	0.5
Steel tanks with sacrificial anodes	0	0.0
Subtotal	1472	100.0
Unspecified tanks	216	
Unspecified piping	29	
Total	1717	

TABLE 12-3. Factors affecting the likelihood and rate of corrosion.

Electrolyte acidity	Soil resistivity
Oxidizing agents	Moisture level
Temperature	Soil variations
Surface films	Chemicals present
Bacterial action	Adjacent underground metal structures
Physical stresses	Stray currents

swell, crack, or soften—when incompatible chemicals are stored in them. In all cases, the container and the materials stored must be chemically compatible. Table 12-3 presents the most important variables that drive corrosion.[9]

Metallic corrosion is due to an electrochemical process that results in a loss of electrons from the metal tank and thus corrosion. This electrolytic corrosion occurs when a direct current enters and leaves the metallic structure via an electrolyte. The surrounding conducting environment, soil, for example, can act as the electrolyte. The current can originate with machinery powered by direct current, such as a railway or a power generator. The current enters the structure at the cathode and exits at the anode, with corrosion occurring in the anodic areas where metal ions are carried away with current exiting the structure.

Galvanic corrosion occurs when a metallic structure is placed in an electrolyte, and a difference in the dielectric potential develops. Such corrosion occurs in three ways:

- A current is generated from one metal to the other when dissimilar metals are dielectrically connected in an electrolyte. Galvanic corrosion occurs in the metal from which the current leaves—that is, the anode. The other metal, the cathode, serves as the receptor of the current. The galvanic series of metals and alloys is displayed in Table 12-4.[10] Coupling two metals far apart in the series results in rapid deterioration of the more active metal. (Table 12-4 should be used only as a general guide because exceptions may be encountered.)

- Imperfections within a single metal can cause the metal to develop differences in potential. Portions of that metal are able then to become anodic relative to the remainder of the metal surface, and corrosion occurs in the anodic areas.
- Variability in the electrolyte—the soil—can produce corrosion too. Such variations along a metallic surface can promote galvanic activity, with corrosion appearing at the anodic areas.

Poor construction of the underground storage system causes approximately 10 percent of the leakage that occurs, according to the API survey.[11] Poor construction includes loose fittings, leakage during construction, and bro-

TABLE 12-4. Metals prone to galvanic corrosion.

Corroded end (anodic)	magnesium
	zinc
	galvanized steel/galvanized wrought iron
	aluminum
	cadmium
	mild steel
	wrought iron
	cast iron
	13% chromium stainless
	18-8 stainless type 304
	lead
	tin
	naval brass
	nickel (active)
	inconel (active)
	yellow brass
	aluminum bronze
	red brass
	copper
	silicon bronze
	nickel (passive)
	18-8-3 stainless type 316
	silver
	graphite
	gold
Protected end (cathodic)	platinum

ken lines. Therefore, tank owners should take precautions to insure that construction or installation does not degrade the performance of underground storage systems. For example, if the installation specifications for a tank require backfill of a specific compaction, the construction inspector should review carefully the performance of the installer. If the backfill is inadequately compacted and permits frost heave or differential settling, the fittings of the system can become loosened. Additional factors that require meticulous attention during installation are listed in Table 12-5.[12]

In addition to corrosion and construction practices, there are other causes of leakage, including wear on the tank at the dipstick point, tank overfilling, faulty instrumentation, mechanical or pump failure, and vibrations caused by surrounding construction or other activities. The quality and compatibility of the components used in the installation of an underground storage system are also important elements in the system's long-term integrity. Tank designers generally specify only components and materials that meet, at a minimum, the standards of the manufacturer of the tank, Underwriters Laboratories, Inc., the Steel Tank Institute, the American Petroleum Institute, the National Fire Protection Association, and the American Society for Testing and Materials. In addition, the designer follows applicable federal, state, and local regulations.

LEAK DETERMINATION

Leakage may be discovered in various ways: visually observed losses, inventory monitoring, detection of product in a secondary containment area. However, leakage is determined more accurately by means of various integrity tests. Before such a test may be conducted, the tank generally must be

TABLE 12-5. Construction factors influencing corrosion.

Loading/unloading of fiberglass tanks	Use of clean, high-resistivity backfill
Bedding between hold-down slab and tank bottom to prevent transmission of load	Placement of backfill around fiberglass tanks to avoid structural stresses
Collisions between tank and construction equipment	Installation of proper number of sacrificial anodes on tanks protected with cathodic systems
Concrete hold-down pads, straps	Tank cover material for protection from traffic loads
Dielectric bushing to segregate dissimilar metals of tanks and pipes	Replacement of tank-top "shipping" plugs with permanent leak-tight fittings
Damage to tank coating	
Welding practices	
Seals at pipe joints	

full. Because the tests cost between $300 and $1000—depending upon the type of test, size of the tank, and materials stored—the tank owner should weigh the costs for the testing against the probability of leakage. If it is highly probable that the tank is leaking, the economical course of action may prove to be removal of the tank without an integrity test.

The manner in which leakage of underground storage systems is investigated depends upon the number of systems. Leakage often can be inferred from product inventory records, as continual unaccounted losses disclose a leak. If only one or a few tanks are on the premises, the owner can assess the potential for leakage. If that potential is high, the tanks might be removed without integrity tests.

When dozens of tanks are to be reviewed for potential leakage, it is appropriate for engineers to conduct an organized investigation, taking the following steps:

1. Collect tank system data, including type of material stored; tank size; tank capacity; material of construction; types of leak prevention or leak detection facilities; past problems; frequency of testing or inspection; environs of the tank, including depth of ground water, soil type, and distance to nearest water supply.
2. Develop a mathematical model to evaluate the tanks using the collected data. The model could be a simple numerical ranking or a sophisticated statistical model.
3. Using the model, arrange the tanks in order of their probability of leaking.
4. Determine whether to test or abandon the tank, basing the decision upon the model's results and also upon an evaluation of the necessity of the tank to the owner's operations.
5. Test the integrity of the tanks, or abandon the tanks.

These steps are amplified below.

Data specific to the site of the underground storage systems are collected in order to assess the potential of each system to corrode and, hence, to leak. The survey may be conducted by means of a questionnaire directed to the operator of the storage facility. The following categories of information are solicited:

- Information required by 40 CFR 280, Notification Requirements for Owners of Underground Storage Tanks, and information required by pertinent state regulations.
- Tank physical data.

- Environmental data.
- Leak hazard data.

Details of the information requested are described in Table 12-6.

The data from the questionnaires are analyzed, and based upon that analysis and a site-specific mathematical model, the storage systems are classified according to their probability of leaking. The mathematical model assigns a value to each specific variable based on whether it increases or decreases a tank's potential for leaking. The various parameters are weighted according to their importance in determining a tank's potential to leak. The weighting factor is developed by dividing the tanks into two categories: steel and fiberglass reinforced plastic. For steel tanks, the tank's age and the methods of corrosion protection are considered. For fiberglass reinforced

TABLE 12-6. Data utilized in assessing leak potential of underground storage systems.

TANK DATA	
Tank age	API estimates half of bare steel tanks leak after 15 years; age is essential datum
Construction of tank	For example, carbon steel, fiberglass reinforced plastic, stainless steel
Internal protection	Coating and lining (glass, zinc, plastic, etc.)
Cathodic protection	Sacrificial anode or impressed current
Secondary containment	For example, double-wall tanks, concrete vaults, lined excavations
ENVIRONMENTAL DATA	
Depth to ground water	Average depth of ground water
Ground water use	Distance to nearest potable supply well
Surface water	Lakes or streams potentially affected by leak
Soil	Porosity and resistivity as they affect movement of leakage and affect corrosion of tank
LEAK HAZARD DATA	
Tank status	Active or inactive; full or empty
Repair frequency	History of tank repairs and methods used
Previous leak testing	Test methods used and results
Previous leaks detected	Frequency of past leaks
Material stored	Hazard rating of materials, including toxicity, flammability, carcinogenicity

plastic tanks, the following are considered: soil moisture; method of installation; proximity of traffic, ground water, and frost heave; and compatibility with material stored. After the weighting factors are applied, the scores are summed for each storage system, and thus the systems are ranked.

The storage systems are divided into a number of action categories depending upon their ranking; the following are possible categories:

1. Remove tank.
2. High hazard rating.
3. Medium hazard rating.
4. Low hazard rating.

The range of scores for each hazard ranking is derived by comparing the history of leakage of the underground storage systems on the site under study with known systems previously ranked by the model. Therefore, the ranking is both based upon and refined by empirical data. The tank ranking is important. The systems with the highest hazard ranking are marked and removed, and tanks below that class are tested for integrity in the order of their ranking.

A number of methods may be employed to test the integrity of an underground storage system. They are generally low-pressure hydrostatic tests (<5 lb/in^2 [<3500 kg/m^2]) using the fluid that normally was stored in the tank. Six methods claiming to meet the stringent requirements of the National Fire Protection Association (NFPA) are listed in Table 12-7.[13]

NFPA requires that the integrity test compensate for variables that affect the performance of the test itself, including temperature and tank-end deflection under applied pressure. Further, the test must have the ability to detect a leak of 0.05 gal/hr (0.05 ml/s).[14] Figure 12-2 shows some of the tools used in the tests.

TABLE 12-7. Tank integrity tests developed to meet standards of NFPA.

Test
Petro-Tite, formerly Kent-Moore Trademark of Health Consultants, Inc.
ARCO-HTC Trademark of ARCO Petroleum Products Company
Tank Auditor Trademark of Leak Detection Systems, Inc.
Horner EZY-CHEK Trademark of Horner Creative Metals, Inc.
Ainlay Tank Tegrity Tester Trademark of STI Services Corporation

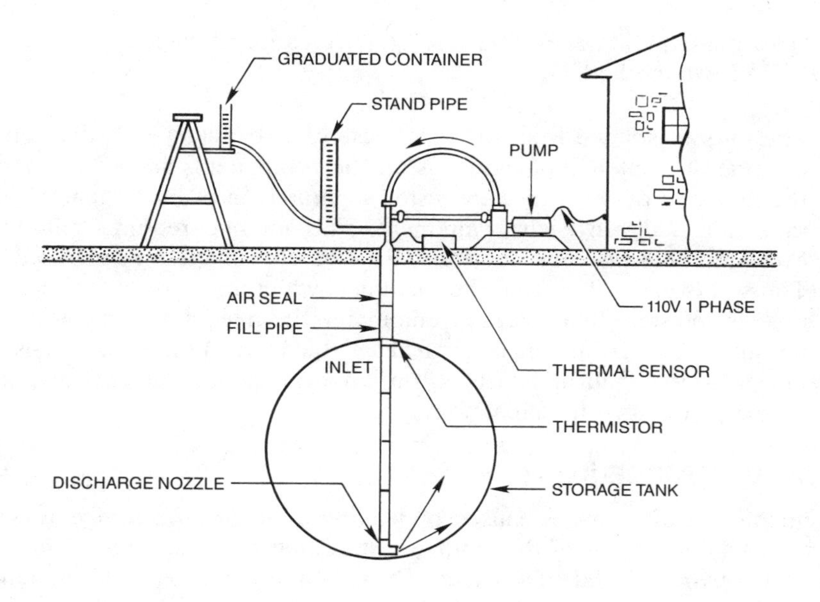

Figure 12-2. Examples of equipment used to test integrity of underground storage systems.

Once a system is known to be leaking, the location of the leak is identified by one of two methods:

- The tank is isolated from the pipelines either by installing valves or by capping the tank and pipelines. After the components are isolated, both the tank and the pipelines are tested separately in order to quantify the rate of leakage in each component. This method requires minimum excavation because only the ends of the pipeline need to be exposed.
- The soil is excavated from the pipelines, which are observed while the tank is retested. This method eliminates the need for a line test and permits visual verification of any leakage. Its disadvantages include use of low pressure (<5 lb/in^2 [<3500 kg/m^2]) to evaluate line integrity and high excavation costs if the pipeline is lengthy.

REMEDIAL PROCEDURES

Remediation of damages caused by leaking underground storage systems embraces both repair of the storage system that failed and redressing the adverse environmental effects caused by the failure. Because environmental remediation is treated in other chapters in this book, only remedies dealing specifically with the storage system are discussed here.

In general, acceptable procedures for the repair or replacement of underground storage systems are addressed by API, NFPA, Underwriters Laboratories, and the Steel Tank Institute, among others. Although repair of the system may range from simply replacing a valve or fitting to relining the entire tank, such repairs are always temporary, and eventually the tank itself will be replaced. (Design of a replacement underground storage system is discussed below.) Any new system, however, should specify use of materials compatible with the product to be stored; should intrinsically or by design prevent corrosion; and should include safety features such as secondary containment, leak-monitoring equipment, and devices to prevent overfill. Figure 12-3 depicts the design components of a modern storage tank system.[15]

Any tank found to be leaking, any tank with a high hazard ranking, or any tank no longer required for service should be corrected by abandonment and removal from the ground. In the past, tank abandonment meant "no longer using the tank," "emptied and left in place," "filled with water," or "filled with sand." According to current regulations, tank abandonment means either abandonment in place or complete removal of the storage system from the ground.[16]

Abandonment in Place. Although it is generally more desirable to remove the tank completely, in certain circumstances a tank is abandoned in place, as when it is indoors, under a building, or beneath a foundation, or if other

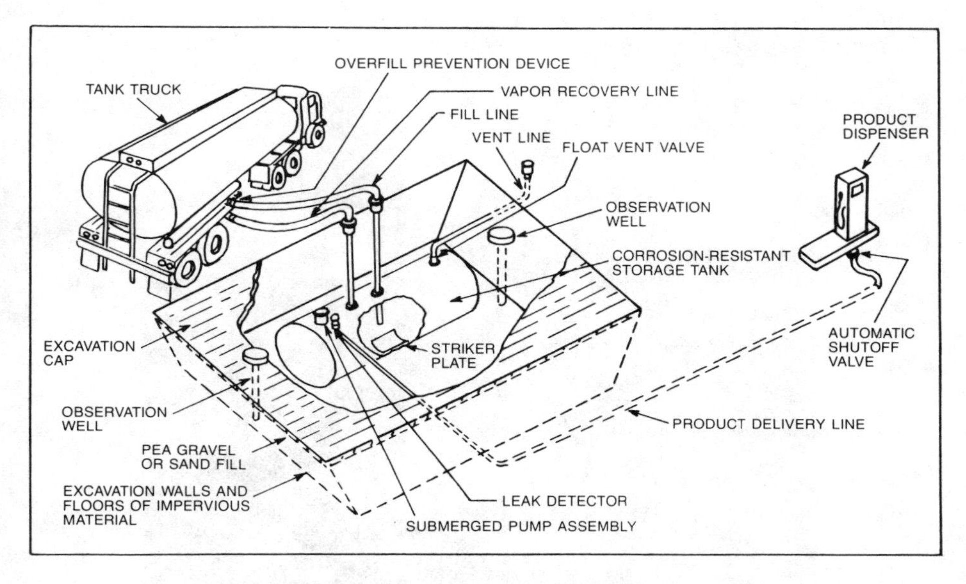

Figure 12-3. Elements of underground storage tank installation.

structural barriers are present. Before a tank may be abandoned in place, these steps are followed:

1. Assess the tank's integrity. A tank may be abandoned in place only if it has never leaked.
2. Remove all liquids.
3. Remove and dispose of sludges and residues.
4. Clean the tank and dispose of the cleaning residue.
5. Fill the tank with inert material such as sand, gravel, or concrete.
6. Disconnect piping and plug it with concrete or nonshrinking grout, or remove all piping.

If the tank is not cleaned before inert material is deposited, the potential for contact between the formerly stored substance and the environment remains. Figure 12-4 illustrates the complete removal of a storage vessel.

Future regulations may require that the integrity of tanks be investigated before they are abandoned. Such a test could be executed by installing soil borings or monitoring wells adjacent to the tank. Soil and ground water samples could then be collected and analyzed to determine if any of the material formerly stored in the tank had reached the environment. If contamination of ground water were discovered, regulatory agencies would be notified, and a broader remediation effort might be required.

Figure 12-4. Underground storage tank removal.

This procedure does not relieve the tank owner's liability for environmental damage, however, because the borings and wells can cover only a limited area. In addition, the soil or ground water data could easily be confounded by other industrial operations that had occurred on the premises. For example, if a spill had occurred on the surface of the ground near the tank, the ground water might show contamination when analyzed, yet that contamination might not have been caused by the underground storage system.

Alternatively, the integrity of the tank could be tested before it is abandoned. If leakage were thereby established, the tank would likely have to be removed. If no leakage were determined, abandonment in place could proceed.

Tank Removal. The abandonment alternative that effectively limits liability and environmental damage is complete removal of the tank. Ground water analyses generally are not required when a tank is removed,[17] and only obviously stained soils need be removed and disposed of. Tanks should be removed only by contractors familiar with pertinent government regulations and knowledgeable about the safeguards necessary to prevent environmental harm and to limit potential liability to the owner of the storage system. Removal is accomplished in the following steps:

1. Analyze tank contents according to the U.S. EPA hazardous waste characterization process[18] to determine the proper disposal procedure for the contents.
2. Empty the tank.
3. Clean the tank interior with high pressure water, steam, or solvent.
4. Purge vapors from the tank using air, carbon dioxide, or nitrogen.
5. Remove the tank from the ground.
6. Render the tank unfit for further use: puncture it repeatedly, crush it, or cut it into pieces.
7. Examine soil around the excavation for contamination; inspections are conducted by sight, by odor, or with a monitoring device such as a photoionization detector (HNU), which determines the concentration of organic vapor.
8. Remove any contaminated soil.
9. Obtain soil samples in the cleaned area for analysis. The results of these analyses will document the effectiveness of the clean-up effort.
10. Dispose of the tank and contaminated soils.
11. Backfill the excavation.
12. Document the tank removal and disposal of the tank and soils. If any spills occurred during the work, file a report with the controlling government agencies and with the tank's owner.

It is less costly to abandon a tank in place than to remove it. However, if during testing the tank to be abandoned in place is found to have leaked, it is removed, and the action is even more expensive than forthright removal would have been.

Replacement Tank Systems. Because of the history of liability and problems associated with underground storage of hazardous liquids, many tank owners now are choosing above-ground facilities for any new storage whenever it is feasible; but the advantages of underground storage and the frequent prohibition of above-ground tanks by fire regulations and local building codes favor the continued installation of underground tank systems. Government regulations and the potential liability of underground storage ordain that close supervision be exercised during the installation, however; and owners of leaking underground systems who are experiencing the cost of correcting a leak will, as a matter of course, install modern underground tank systems that are guaranteed by the manufacturer for at least 30 years. (The regulatory framework governing storage systems is dynamic, and federal and state programs are specifying the types of tanks that may be installed underground. Therefore, before introducing any tank system, it is important for the owner to check applicable regulations.)

The design of modern underground storage systems has two distinct features lacking in earlier tanks: leaks are minimized, and if a leak occurs, it is detected and contained. To reduce the occurrence of leaks, new tank systems use corrosion protection—often cathodic protection and protective coatings for steel tanks. Alternatively, nonmetallic tanks, such as fiberglass reinforced plastic, are used. The nonmetallic tanks are constructed with thicker walls than those used in the past, to enhance structural rigidity. However, these methods do not preclude the possibility of leakage; so new storage systems usually include measures for leak detection or leak containment.

Double-wall construction commonly is used now for both steel and fiberglass tanks. The construction is, essentially, a tank within a tank, and leakage is monitored between the two walls. If the space between the two tanks is a vacuum, the monitoring is effectuated by detection of a change in pressure. Alternatively, tanks may utilize a water- or product-sensitive probe between the walls. For other double-wall tanks, a port is furnished to permit observation of any leakage.

To insure protection of the environment in the event of a leak, the excavation into which a tank is installed may be lined with a natural impermeable soil or clay or with a synthetic liner. Any leakage from the tank or appurtenances is then contained within the liner. A sump is installed as part of the liner and is monitored by means of a well extending into it. Whatever the choice of material for the liner, it must be compatible with the substance stored.

A tank could be installed inside a concrete or other type of vault, one specially designed to collect leakage. The vault material must be compatible with the substance to be stored, and the vault must be constructed with no penetrable joints. Such a vault may be either open to permit visual inspection or backfilled with aggregate; and if backfilled, it is furnished with a sump that is monitored for leakage. Leaks are detected either by visual inspection or sampling of observation wells or by sophisticated sensors that detect the presence of various types of products. The sensors can be made part of alarm systems, thus permitting monitoring for leaks to be conducted remote from the tanks. In addition to leak detection devices and alarms, the environment can be protected with overflow prevention instruments, such as floats, ultrasound detectors, or pressure-sensitive devices. If the product overflowed, the devices would trip an alarm. In addition, fail-safe systems are employed that close the valve supplying the product at a predetermined level.

The typical design of a modern underground storage system is presented in Figure 12-5. Whenever a storage system for hazardous substances is installed, it is beneficial to have the work managed by a professional who is both knowledgeable and experienced in the design and construction of such systems. The appropriate standards and government regulations must be followed carefully, and the contract documents and the specifications for the installation should identify the components and the methods of assembly. After design and during construction, a representative of the owner should inspect the storage system and warrant the builder's conformance with the

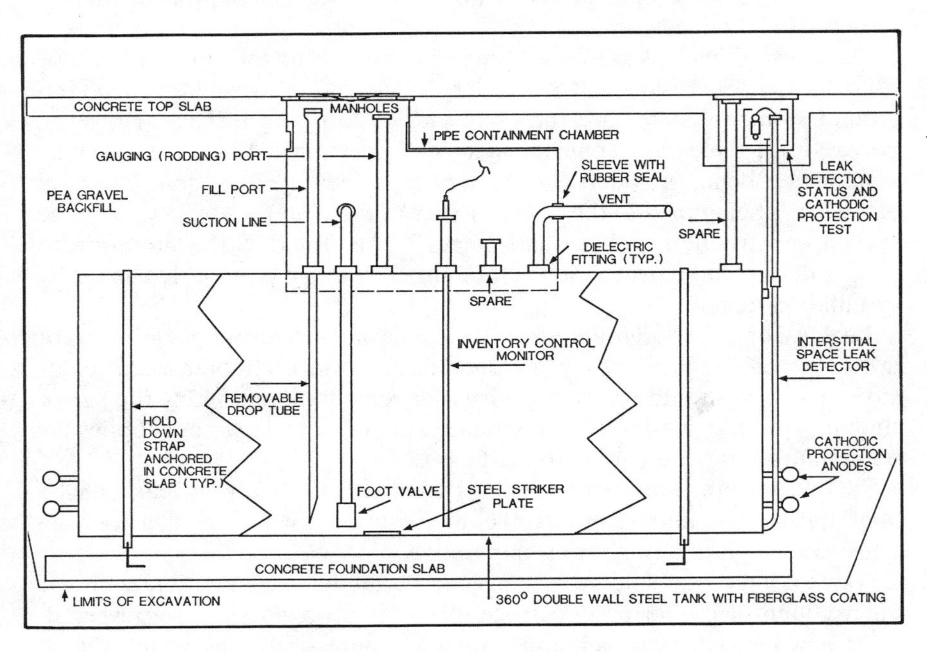

Figure 12-5. Modern design of underground storage system.

design specification. Such precautions will insure construction of an economical storage system that is benign to the environment.

SUMMARY

When addressing storage tank leakage and remediation, the engineer evaluates several factors:

- Requirements of regulations.
- Engineering detail, including tank removal and integrity testing.
- Environmental contamination and cleanup.
- Economy.

Profound environmental problems have been caused by leaks in underground liquid storage systems. In response to this predicament, governments enacted laws and promulgated regulations in the late 1970s ordering registration of certain underground storage facilities. Government policy tightened in the early 1980s. Government regulators demanded tank inventories and inspections, and, where leakage was detected, correction of the storage system—usually its removal. They also stipulated remediation of the environmental damage.

Data describing leakage were gathered to provide an information base for remediation. Corrosion and improper design and construction of underground storage systems are the chief factors fostering leakage. However, corrosion of tanks and appurtenances can be inhibited, and new storage systems are being provided with cathodic protection; in addition, careful attention is being paid to the compatibility of the materials used for storage system construction and the substances to be stored. Strict attention to proper design and construction is minimizing leakage previously caused by installation techniques.

Leakage of an underground storage system can be determined by means of a number of integrity tests. Many factors determine whether a particular storage system should be given priority for remediation, notably the age of the tank and the possibility of contamination of potable water supplies for communities in the environs of the facility.

Tanks that are found to be leaking or are quite likely to leak can be abandoned. If ground water or soil contamination is widespread, a comprehensive remediation program is imperative.

New underground storage systems can be rendered compatible with the environment if proper practices, standards, and regulations are observed. With new construction techniques such as double-wall tank systems, leak detection instruments, and overfill protection, hazardous substances can be stored safely underground.

NOTES

1. U.S. EPA, Office of Technology Assessment. *Draft report on leaking underground storage tanks.* Oct. 1984.

2. New York State Department of Environmental Conservation. *Technology for the storage of hazardous liquids: A state of the art review.* Albany, N.Y.: 1983.

3. Amendments of 1984 to RCRA, Subtitle I, Section 9002.

4. RCRA; Modifications to Subtitle J. July 1986.

5. O'Brien & Gere Engineers, Inc., Syracuse, N.Y. 13221.

6. U.S. Department of Transportation. *Hazardous materials emergency response guidebook.* Washington, D.C.: U.S. Department of Transportation, 1980. P 5800.2.

7. New York State Department of Environmental Conservation.

8. Table modified and taken from New York State Department of Environmental Conservation. p. 10.

9. Adapted from R. H. Perry and C. H. Chilton. *Chemical engineers' handbook,* 5th Ed. New York: McGraw-Hill Book Company, 1973. Cited by New York State Department of Environmental Conservation.

10. Adapted from Perry and Chilton.

11. New York State Department of Environmental Conservation.

12. Adapted from Perry and Chilton.

13. National Fire Protection Association. Underground leakage of flammable and combustible liquids. Pamphlet 329. Quincy, Mass.: NFPA, 1983.

14. NFPA. Underground leakage of flammable and combustible liquids.

15. New York State Department of Environmental Conservation.

16. The NFPA has fostered this approach to tank remediation, and many states have adopted regulations consonant with NFPA's guidelines.

17. For an example of state provisions, see Title 6 of the New York State Codes Rules and Regulations (NYCRR) Part 613.9 (a, b, and c).

18. 40 CFR 261, Identification of Hazardous Waste.

Chapter 13
Incineration as a Disposal Alternative

Incineration, one of the most effective measures for the disposal of hazardous wastes, is an engineered process in which thermal oxidation at a high temperature modifies materials. The consequences of incineration are a reduction in the volume of wastes and virtually the complete destruction of organic compounds. Carbon dioxide, water vapor, and ash are the principal products of incineration. Under the provisions of CERCLA, the responsibility for a waste remains with its generator for the lifetime of the waste or until it is destroyed or detoxified. Long-term liability for the adverse effects of disposing of hazardous wastes is virtually eliminated when incineration is the remedial measure of choice,[1] whereas such liability is uncertain when landfilling or deep well injection is the chosen remedial measure. Although a long-term liability does remain with regard to the disposal of the ash residue after incineration, the volume of ash is relatively small, typically one percent for highly incinerable liquids for example, compared with the original volume of waste. Any metals present in the original waste are concentrated in the residue, which often is treated further or deposited in a secure landfill as a hazardous waste.

Prior to 1985, incineration ordinarily was economical only for wastes with a high energy value such as solvents or wastes that could be burned in existing municipal or industrial incinerators; otherwise, it was an expensive alternative when compared with landfilling, and its higher cost was perceived as prohibitive to waste generators, even accounting for the advantage of reducing future liabilities. Because landfills were a readily available alternative disposal method, there was little incentive for design engineers to be innovative and to improve the existing incineration technology in order to make incineration more efficient and more economical. Now incineration is becoming more attractive economically because the ground rules are changing. The Hazardous and Solid Waste Amendments to RCRA, passed in November 1984, inlcuded a provision known as the "landfill ban."[2] In order to optimize the use of scarce landfill resources, specific liquid organic wastes amenable to treatment or incineration were to be gradually barred from landfills. This

---

This chapter was developed by Cheryl L. Cundall and Cindy M. Klevens with the assistance of John Rinko of O'Brien & Gere Engineers, Inc.

process began in May 1985 and continued through July 1987. These liquids are now banned from landfills in the U.S. Consequently, interest in incineration as an alternate disposal measure was renewed.[3]

Incineration technology is not new; it is widely used in Europe where limited landfill space has long been a reality. However, in the United States in 1986, there were just 200 incinerators possessing appropriate government permits; about 180, or 9 in 10 of them, were owned by waste generators in the chemical process industries, and the remaining 20 were commercial incinerators that burned wastes from a variety of customers.[4] About 25 percent of all U.S. incinerators recover some energy. The commercial facilities operating in 1986 had submitted permit applications to or had received interim permits from the U.S. EPA.[5] (The U.S. EPA had issued only interim permits to applicants, as it was just developing the procedures to grant final permits.) Six of the commercial facilities accept wastes containing PCBs.

Because chemical disposal areas or abandoned hazardous waste sites usually contain a variety of chemical constituents in various matrices, complete characterization of the specific wastes is mandatory before incineration can be examined for its use as a remedial measure. Whether or not incineration is practicable depends largely on the nature of the wastes, the volume, and the matrices. Although it may be suitable for the waste products, it may be inappropriate for the contaminated soils or ground water at a particular site. The designer should project the length of time the incinerator will be used. Many hazardous waste sites contain a fixed volume of waste, and it often is uneconomical to construct an on-site incinerator. In such cases, a mobile incinerator may be well suited to the task of remediation. If incineration is the chosen remedial measure, it is important that the designer select both the appropriate incineration technology and the applicable air pollution control equipment. The process also must be acceptable to communities in the environs.

GOVERNMENT REGULATION

Before a new incinerator may be built and operated, federal, state, and sometimes municipal regulatory agencies first must approve the project at the conceptual, or design formulation, stage. The design and operation of incinerators and the permits required to operate them are established in regulations amplifying RCRA that were promulgated in 1976.[6] These regulations do not specify design criteria but furnish performance standards, and the U.S. EPA has intentionally kept their wording general to encourage innovation in incinerator design. Figure 13-1 diagrams the steps involved in selecting the incineration process, together with the prerequisites for securing construction and operation permits.

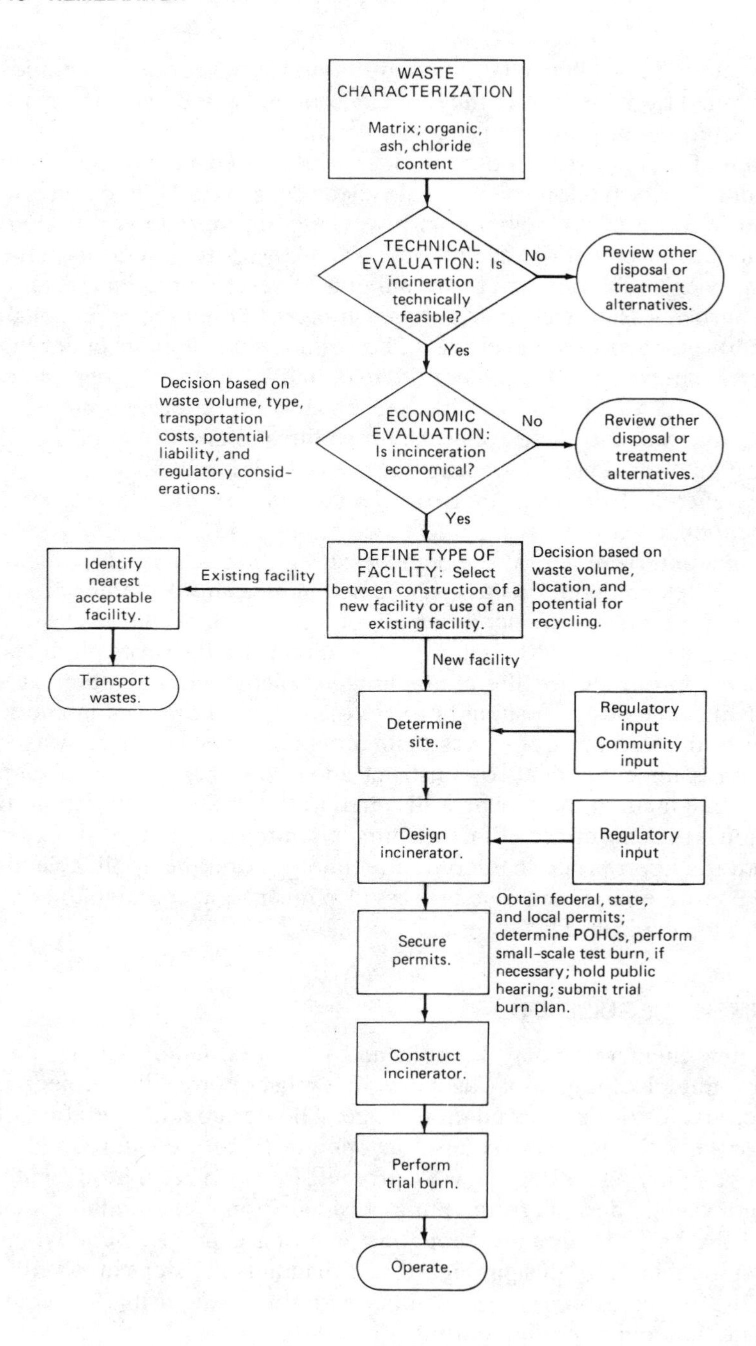

Figure 13-1. Flow chart of incineration review process.

Before considering incineration, the waste generator must characterize the waste fully. If it is to be transported off-site, a manifest is completed; for on-site disposal, an annual report is issued. The U.S. EPA selects the principal organic hazardous constituent (POHC) based upon waste analysis[7] and utilizes it as the basis for establishing the performance standard for the incineration process. The POHC, based on its concentration in the waste and its heat of combustion, dictates the conditions for incineration. The U.S. EPA lists 272 chemicals considered potential POHCs and has ranked them by their heats of combustion.[8]

RCRA regulations pertaining to the incineration of hazardous waste include the following provisions:

1. The Destruction and Removal Efficiency (DRE) for the POHC must be at least 99.99 percent (the "four nines"):

$$\text{DRE} = \frac{C_1 - C_2}{C_1} \times 100\%$$

where C_1 = concentration into incinerator.
C_2 = concentration exiting incinerator.

The DRE for PCBs, dioxins, and dibenzofurans must be at least 99.9999 percent (the "six nines").

2. If the incinerator is used to destroy chlorinated wastes and produces emissions of more than 4 lb/hr (0.4 g/s) of hydrogen chloride, pollution control equipment must be provided to remove 99 percent of any hydrogen chloride emissions. The removal of nitric oxides and sulfuric oxides from emissions is determined on a case-by-case basis. There are currently no regulations addressing the emission of products of incomplete combustion (PICs), but most permits are written to require continuous monitoring for carbon monoxide, an indicator of incomplete combustion.
3. Particulate emissions are limited to a maximum of 0.08 grain/dry standard ft^3 (dscf), corrected to 12 percent carbon dioxide. (The SI equivalent, and the measure specified in the regulations, is 180 mg/dscf, corrected to 7 percent oxygen.)

These criteria assume that destruction of the POHC assures destruction of all other organic constituents. During the start-up tests of all new facilities, this assumption is scrutinized, and a number of waste constituents are monitored. If a 99.99 percent DRE is achieved for all monitored organic constituents, or 99.9999 percent is achieved for PCBs, monitoring during the actual operation of the facility is limited to the POHC.

The single most important factor in siting an incinerator to process hazardous waste is land use in the vicinity of the proposed location. Siting an incinerator is difficult. Despite the technical merits of design, public opposition to incineration facilities often has been sufficiently persuasive to regulatory agencies to scuttle proposed projects. A frequent major objection is that once site remediation is finished, the incinerator will be converted to commercial use. Public concern appears not to be with the incineration technology itself but with the creation of permanent hazardous waste disposal facilities and the attendant transport of hazardous wastes through communities.[9]

A disadvantage of providing incineration at the hazardous waste disposal site is the time and effort required to obtain the necessary permits for construction and operation. Then, after the delay incurred in securing permits, construction may take three to five years. Such an extended time frame—four to eight years for acquisition of permits and construction—usually is impracticable for remediation of a hazardous waste site because of the limited use of the incinerator once constructed. If such an incinerator could be converted to commercial use, the economics might be more favorable. At least one of the commercial incinerators operated by a major electrical manufacturer in the Northeast was originally built for remediation of a hazardous waste site.

Alternatively, if the incinerator is located remote from the site of the wastes, transportation of wastes to it presents risks and additional liability. The decision on whether to build an on-site incinerator or to transport the wastes to an existing incinerator depends upon land use in the environs, the volume of wastes to be burned, and the length of time the incinerator will be needed. If circumstances permit, utilization of a commercial incinerator off-site is advantageous because the design and operation of the unit would already have been approved by government agencies. However, there are only a few commercial incinerators available, and not all are proximate to identified hazardous waste sites.

An obvious alternative to the on-site versus off-site dilemma is to use a mobile incinerator. A number of mobile rotary kiln incinerators have been built, but few have been employed as a remedial alternative at hazardous waste sites. Operating difficulties and problems with securing permits have plagued many of the mobile incinerators produced to date. The U.S. EPA also has tested a mobile research unit, which has experienced a common difficulty with maintaining a tight seal in the kiln. Many states and municipalities demand that operators of mobile incinerators obtain the same permits required of permanent facilities, and the process includes the attendant public hearings and test burns.[10] The U.S. EPA is reviewing the impasse and may implement a permit valid in all states for the incineration of similar wastes by mobile incinerators.

After a suitable construction site for an incinerator is approved, the project sponsors must secure permits to build and operate the facility. Site data are collected to support the permit application, including information on wind velocities, background air quality, location of potentially affected receptors, traffic patterns, incinerator design, and operating conditions. These data are submitted to the regulatory agencies, which often hold public hearings to permit citizens in the project's environs the opportunity to voice their concerns. Without approval from government agencies, construction of the incinerator cannot begin.

Before a newly built incinerator may operate, a test burn must be conducted to determine if it meets relevant performance standards.[11] A protocol describing the test burn is prepared by the project sponsor and submitted to the regulatory agencies for their approval before the test burn may proceed. During the test burn, ranges of temperatures and residence times are reviewed, and either actual waste or a synthesized mixture spiked with additional hazardous constituents is tested. It is often preferable to use a synthetic waste because if the incinerator demonstrates the capacity to destroy constituents less incinerable than the anticipated POHC, the owners can apply for a permit allowing it to accept those additional wastes. If an incinerator replicates an existing incineration facility that holds a permit to operate, the test burn requirement may be waived. The time required for preparation and review of the permit application and the test burn protocol, as well as the time necessary for the test burn itself, can easily extend to two years.[12]

TECHNOLOGIES

It is necessary to characterize wastes fully in order to achieve the optimal design and size for the incinerator and to determine the relevant type of pollution control equipment. The economic feasibility of incineration for a specific waste depends upon the chemistry of both the waste and the matrix of the waste; the matrix could be liquid (water, oil, or a solvent), solid, or sludge. It is most important to characterize the organic content, ash content, and chloride content.[13] Organic constituents, which are oxidized during incineration, determine the combustibility of the waste. The ash and chloride content determine the volume and character of the solid residuals and the air emissions, and thus the requisites of the pollution control facilities. Emissions are a particular concern with chlorinated hydrocarbons, materials laden with heavy metals, and wastes with high sulfur content.[14] For example, a chlorinated feed—defined as containing at least 0.5 percent chlorine—produces corrosive hydrochloric effluent gas ("acid gas"); an ash content above 10 percent manifests itself as either particulate emissions or an ash residue after combustion.

If combustion is incomplete, organic constituents are only partially oxidized, and products of incomplete combustion (PICs) are yielded. PICs are compounds with different chemical structures from the organics that produced them, and they may exhibit very different chemical properties from those of the source substances; in some cases, PICs are more toxic than the original compounds. Dioxins and dibenzofurans, for example, are the PICs of polychlorinated biphenyls; 1,3,5-trinitrobenzene and 2,4-dinitrotoluene are the PICs of the explosive TNT (2,4,6-trinitrotoluene). Government regulatory agencies first became alarmed about PICs in the late 1970s when dioxins and furans were discovered in the emissions of incinerators burning municipal refuse and hazardous wastes. The U.S. EPA Science Advisory Board outlined concerns about PICs and counseled that future research should both identify PICs and establish criteria to insure the complete destruction of all constituents during incineration.[15] Because that information is not now available, government regulatory agencies have been able to address PICs only indirectly by mandating monitoring for carbon monoxide, itself a product of incomplete combustion. The generation of by-products can be minimized with appropriate control of temperature, residence time, turbulence, and oxygen during incineration.

Inorganic components, typically metals, are not destroyed by incineration but remain in the ash residue. Ash containing metals may be a hazardous waste and require disposal in a secure landfill. Stack emissions with metallic particles yield a scrubber blowdown that may require pretreatment prior to discharge into a public sewer. Wastes with high levels of nitrogen or sulfur produce emissions of nitric oxides and sulfuric oxides, respectively; these emissions can be controlled with a caustic scrubber.

Wastes with diverse characteristics sometimes are segregated; otherwise, they may compromise the incineration process. For instance, heat recovery is incompatible with chlorinated wastes, as the corrosive gaseous emissions damage the heat recovery equipment; heat recovery is also incompatible with wastes containing metals, as the metallic residues are deposited on the boiler tubes and reduce the efficiency of the heat transfer. An economic evaluation is performed for each situation because it may be less expensive to incinerate problem wastes in a specifically equipped incinerator elsewhere than to install specialized pollution control equipment to accommodate them.

During the test burn, the project sponsors often evaluate the incinerator's operating conditions—temperature, residence time, feed rate, fuel rate, and so forth—to determine optimal operating ranges. In addition, the DREs for a number of constituents are monitored during the test burn, and the stack emissions are monitored for hydrochloric acid, carbon monoxide, nitric oxides, sulfuric oxides, and particles. Any fugitive emissions—discharges from leaking pipe fittings, valves, or pumps—also are corrected.

Certain flammable wastes can be co-fired with conventional fuels in standard industrial boilers. Such wastes are currently exempt from RCRA regulation, though the U.S. EPA is developing performance standards for such co-firing.[16] Recent U.S. EPA studies indicate that under most operating conditions, co-fired wastes are as effectively destroyed as wastes subjected to other forms of incineration.[17]

The waste's matrix also influences the technical complications and economics of incineration. Matrices such as soil or ground water with low heating values require a greater heat input than matrices with higher heating values, such as petroleum. Soils can be burned either as solid feed or as a slurry. The material composing the matrix is evaluated on the basis of combustibility, water content, and viscosity. If containers or other materials are to be incinerated with the wastes, they also are characterized. A nonchlorinated waste, for instance, packaged in plastic containers may or may not be amenable to incineration with the container; certain plastics, polyvinyl chloride, for example, contain chlorine, and specialized pollution control equipment may be required for the acid gas that would be generated. Once characterized, the combustibility of the waste and containers is evaluated with a test burn on a pilot-scale unit; such a unit is also known as a *research combustor.* It is difficult to correlate test data with the operation of a full-scale incinerator; however, tests with research combustors provide preliminary information about waste combustibility, including the type of residues and emissions produced.

The DRE is, as a rule, a function of temperature, residence time, and turbulence. Although RCRA regulations do not specify actual operating conditions to achieve the DRE, incinerators are typically operated under the following conditions:

- Temperature of approximately 2250°F, slightly higher for PCBs.
- Residence time of 2 seconds, although several incinerators have been allowed to operate with a one-second residence time after test burns indicated that all performance standards were met.
- Excess air—25 to 35 percent for liquids and about 100 percent for sludges—compared to the stoichiometric volume needed for combustion, and air mixing systems such as baffle arrays or direction changes. These systems promote turbulence and ensure adequate contact among the fuel, air, and waste.

The DRE is monitored throughout the operating life of an incinerator to insure that fouling, corrosion, and other maintenance factors do not compromise the efficiency of the unit.

TABLE 13-1. Commercial incineration systems.

INCINERATION SYSTEM	OPERATING TEMPERATURE °F	°C	FEED
Rotary kiln	1500-2900	820-1600	Liquids/slurries/solids
Liquid injection			
Land-based	1200-2900	650-1600	Liquids/slurries/gases
Ocean-based	300-2900	150-1600	Liquids/slurries/gases
Fluidized bed	840-1800	650-1600	Liquids/gases/solids
Multiple hearth	Drying zone: 600-1000	320-540	Sludges/liquids/gases mixed with sludge
	Incineration: 1400-1800	790-980	
	Cooling zone: 400-600	200-315	

The essential elements in conventional incineration systems include a waste feed system; an air- or oxygen-fed burner system, though some systems operate with little or no air; a combustion chamber, with monitoring controls; ash removal equipment; an air pollution control device, though ocean-based incinerators operate without air pollution controls. The incineration system best suited for a particular disposal application depends on the character of the waste and its matrix. Basic design principles suggest that certain systems are efficient for the combustion of solids and sludges, whereas others are better for the destruction of liquid or gaseous wastes. Table 13-1 lists the four types of incinerators that are in use or have significant potential for use in the destruction of hazardous wastes.[18]

Rotary Kiln. The rotary kiln is the most versatile of incinerators, capable of burning a broad range of hazardous and nonhazardous liquids, solids, and slurries. It burns liquid and solid wastes efficiently because, as the cylindrical combustion chamber rotates, wastes become well mixed with oxygen. However, gaseous wastes are not usually processed in kilns because they must be fed at a higher rate, and the residence time available is generally insufficient for complete oxidation.

The typical rotary kiln consists of a cylindrical, refractory-lined, horizontal shell mounted at an incline; see Figure 13-2. Generally, the length-to-diameter ratio ranges between 2 and 10, and the speed of rotation varies between 1 and 5 revolutions/min; both factors are decided by the character

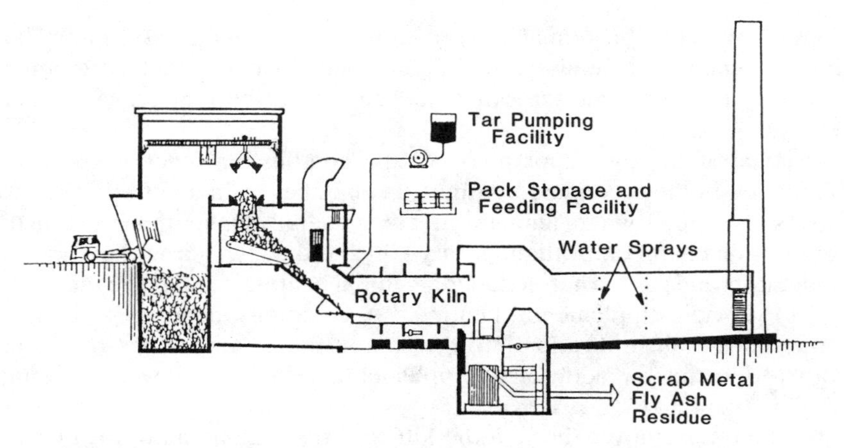

Figure 13-2. Rotary kiln design. Source: U.S. EPA Hazardous Material Incinerators Design Criteria. October 1979. EPA-600/2-79-198, NTIS PB80-131964

of the waste burned, and, in general, both factors determine the residence time of the waste.

Batch feeding through the elevated end is common for rotary kiln systems, although other feed mechanisms are employed. Most industrial-scale kilns are designed to accept both solid and liquid feed. Typical feed rates for solid wastes are 1300 to 1400 lb/hr (160 to 175 g/s). A bulk-feed mechanism, such as an overhead crane or a conveyor belt, is commonly used to feed bulk containers and drums of solid waste into the incinerator. All liquid wastes are first transferred from drums to a receiving tank. The liquid is strained to remove large solids while being pumped from the receiving tank to a burning tank. In the tank, it is blended with other liquids in order to create a mixture with a uniform heating value. The liquid waste mixtures are atomized with steam or air and burned in suspension in the main combustion chamber. The rate of feed for the atomized waste stream typically ranges between 22 and 79 ft^3/hr (170 and 620 cm^3/s). Liquid and solid wastes are burned simultaneously in the kiln: as the solid wastes are lowered by the bulk-feed mechanism into the charging hopper of the kiln, liquid wastes are injected horizontally into the kiln. Solid and liquid wastes are vaporized through a series of partial combustion reactions in the main combustion chamber and then enter the secondary combustion chamber for further oxidation. Often, the first and second combustion chambers are operated at different temperatures to avoid problems with specific waste. For instance, salt-laden wastes may be fired first at temperatures under 1400°F (760°C) to destroy bulk solids without causing molten salts to deposit. Next, the waste is fired in a second chamber with combustion at a temperature greater than 1800°F (980°C) to fire the remaining gaseous wastes. The second chamber usually is mounted vertically to permit molten salt to collect at the bottom for convenient removal.

In many units, the vapor phase of the combustion reaction usually is completed in the secondary combustion chamber or in an afterburner, for those so equipped, where temperatures are generally higher than those in the combustion chamber. Both the rotary kiln and the afterburner have natural gas ignitors and gas burners to facilitate initial heating, to stabilize the flame, and to provide supplemental heating.[19] A common operating problem of rotary kilns is air infiltration into the combustion chamber; the results are both a demand for additional supplementary fuel and lowered thermal efficiency.

As the refuse moves through the kiln and the organic matter is burned, inorganic matter—slag, drums, and other metallic material—collects as ash at the bottom of the unit. Ash is discharged at the lower end of the kiln and quenched in a water trough. If the remaining ash contains hazardous constituents, it ultimately is deposited in a secure landfill. Stack particulate emissions can be controlled with air pollution abatement systems such as a venturi scrubber.

Capital and operating costs for the rotary kiln are usually higher than those of other types of incinerators. The installation cost is generally about 200 percent of the equipment cost. Operating costs depend primarily on the volume of auxiliary fuel required, a factor controlled by the characteristics of the waste.

Liquid Injection. Currently, liquid injection is the most common method of incinerating liquid hazardous wastes, especially among manufacturing industries.[20] With this technology, liquid, gaseous, or slurry waste is injected into the combustion chamber through a nozzle or burner. The nozzle mixes the liquid with air, atomizing it into a suspension of droplets that is quickly vaporized. The combustion chamber of a liquid injection unit normally consists of a cylindrical horizontal or vertical vessel lined with refractory brick. An auxiliary fuel burner system fires either at one end or tangentially into the cylinder. To foster added turbulence, some liquid injection units are supplied with forced drafts. See Figure 13-3.

The liquid waste can be atomized in either of two ways:

- *By mechanically using rotary cups:* Open cups are mounted on a hollow shaft. The cups rotate rapidly, and liquid is admitted through their shafts.

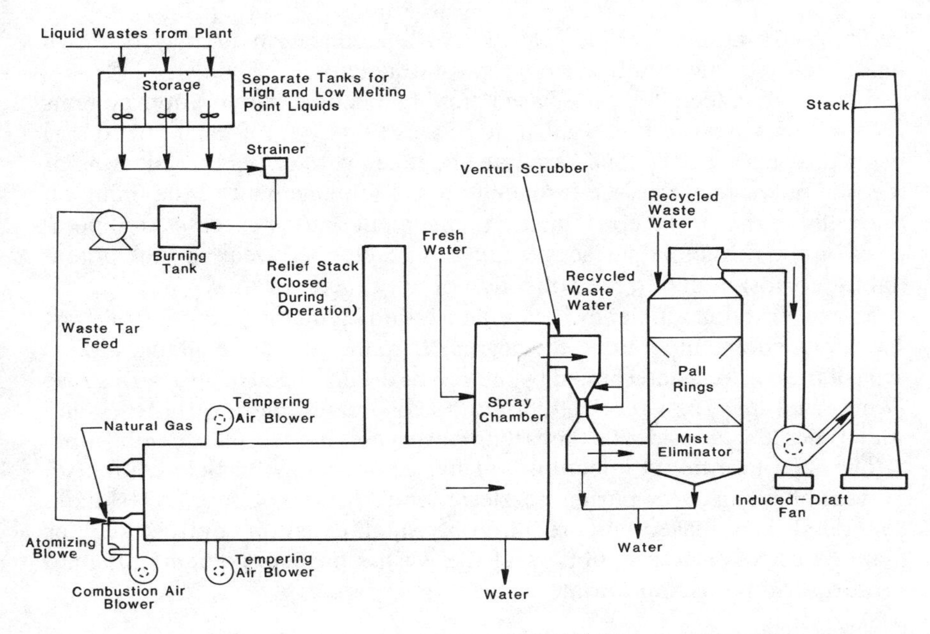

Figure 13-3. Design of liquid injection unit. Source: U.S. EPA Hazardous Material Incinerators Design Criteria. October 1979. EPA-600/2-79-198, NTIS PB80-131964

Centrifugal force tears a thin film of the liquid from the lip of the cup, and surface tension re-forms the liquid into droplets.
- *By means of high pressure air, nitrogen, or steam:* The liquid is directed by internal tangential guide slots through the center of gas fluid nozzles. Suitable atomization can be achieved at pressures between 100 and 150 lb/in^2 (70,300 and 105,000 kg/m^2).

Before a waste is fed into a liquid injection incinerator, pretreatment in the form of filtration, degassing, neutralization, or mixing may be required, especially to reduce particles or to modify the viscosity of the liquid. The waste should have a viscosity generally below 10,000 Seconds Saybolt, Universal (SSU) units—equivalent to a kinematic viscosity of 2200 centistokes (22 cm^2/s)—to be pumped and handled in pipes. For atomization, the viscosity should not exceed 750 SSU—equivalent to approximately 160 centistokes (1.6 cm^2/s) at 100°F (38°C). High-viscosity wastes are often pretreated with in-line heaters or tank coils, or they are blended with a miscible liquid of lower viscosity. If insufficiently prepared waste products are burned, impurities may interfere with atomization or may result in deformation of the thin film from the rotating cups.

The rate of feed for land-based liquid injection incineration systems ranges between 50 and 1000 gal/hr (0.05 and 1.05 L/s), 150 gal/hr (0.16 L/s) being a typical rate. Special care must be taken with respect to injection of the fuel or waste within the furnace to avoid impingement of the flame on the walls of the furnace, as flame impingement causes excessive temperatures on the refractory surfaces leading to accelerated wear and the potential for corrosion due to the collection of unburned residues.

The destruction efficiency in liquid injection incineration is controlled by the combustion temperature, the degree of mixing, and the available surface area of the waste as determined by the size of the atomized droplets. Because liquids first must be vaporized, liquid wastes require more turbulence and sometimes longer residence times than waste gases for complete combustion.

The feed to a liquid injection unit must have a low particle content to prevent plugging of the burner nozzles; therefore, little or no solid residue is generated. Land-based units require only stack emission controls, such as plate or packed-tower scrubbers. If the waste's particle content is high, a venturi scrubber is appropriate.

Ocean-Based Liquid Injection Incineration. Since 1969, European countries have been applying liquid injection technology for burning wastes at remote ocean locations.[21] In 1974, the United States began trial burns in the Gulf of Mexico using vertically mounted liquid injection units on board the ship *Vulcanus.* The *Vulcanus* was constructed from a tank ship by Ocean Combustion Services, Inc. in 1972 and has since been purchased by Chemical Waste Management, Inc. During the time the trial burns were being conducted on board the *Vulcanus,* the U.S. EPA developed permits specific to ocean incineration of hazardous wastes. The U.S. EPA did not require complex air pollution abatement systems because the agency adjudged that the emissions of particles and acid gases were rapidly buffered by the ocean, and the adverse effect on the environment was minimal. Basing a liquid injection incinerator on the ocean rather than on land is advantageous because, in the absence of strict stack emission regulations, a higher rate of waste combustion can be utilized. The average feed rate capacity for such units is 1650 gal/hr (1.7 L/s), compared with a mean of less than one-tenth that for land-based units.

Chemical Waste Management and a second company, At Sea Incineration, have also launched the *Vulcanus II, Apollo One,* and *Apollo Two* ships, each with incineration capabilities. Notwithstanding the diligence of the private sector, the U.S. EPA has been slow in completing regulations or final permit requirements for operation of ocean incineration facilities. The agency has had to address vociferous public concerns and requires additional tests to resolve key technical issues about its environmental effects. At the time of this writing, ocean-based incineration was not an available option for the routine disposal of hazardous wastes.

Fluidized Bed. Incineration with fluidized bed technology has been applied in a limited fashion in the petroleum and paper industries and has been used to process nuclear wastes and sanitary sludge. It is a relatively new method of disposal; its first commercial use in the United States dates only to 1962.[22] The fluidized bed technology promotes turbulence and facilitates superlative mixing of the waste with hot air and hot media. The sustained agitation of the media allows larger waste particles to remain suspended until the combustion is complete, thus enhancing combustion efficiency.

A fluid bed incinerator has a vertical refractory-lined reactor vessel containing a shallow bed of an inert granular material. The media of choice for the bed are usually silica or alumina. Figure 13-4 shows the components of this incineration technology. Forced draft air is introduced at the bottom of the combustion chamber at, typically, 5 to 7 ft/sec (1.5 to 2 m/s) and 3 to 5 psig (2100 to 3500 kg/m^2). As the air rises through a distributor plate and upward through the bed, it promotes strong agitation and causes the bed to mimic the physical properties of a liquid. The relatively low velocity of the air results in a low pressure drop across the media and thus lowers the energy

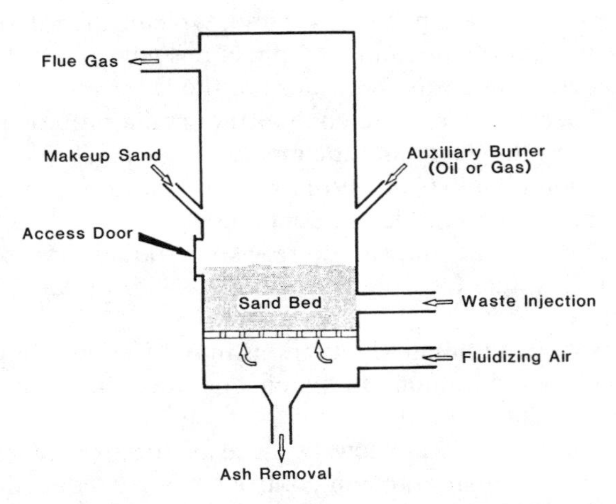

Figure 13-4. Schematic of fluidized bed incineration. Source: U.S. EPA Hazardous Material Incinerators Design Criteria. October 1979. EPA-600/2-79-198, NTIS PB80-131964

requirement for the blower. The bed is preheated to start-up temperature by a burner located above it. The waste is injected radially into the preheated media, and combustion takes place within the bed as heat is transferred from the media to the waste. Upon combustion, the heat released is returned to the media.

Auxiliary fuel, when required, is fed radially into the bed through nozzles located along the circumference of the reactor.[23] Dissimilar wastes, ordinary municipal refuse, for example, occasionally demand more auxiliary fuel than uniform wastes; waste preparation—shredding, sorting, drying—also is more complex when the waste is diverse. Homogeneous wastes can be injected directly and dispersed evenly into the bed; the overall operation is thereby optimized, and bed volume is minimized. Temperatures within the media range between 840 and 1500°F (450 and 815°C) and can be as high as 1800°F (980°C) above the media.[24] Auxiliary fuel is usually added to elevate the temperature above that of the media for completion of the gas phase combustion; thus, there is no need for an afterburner.

Residual fines and ashes become airborne and exit with the air stream through the top of the unit. Emission gases usually contain a considerable particulate load that can be reduced by treatment with a venturi scrubber. The spent scrubber water generally requires treatment before discharge.

The advantages of fluidized bed technology over other types of incineration are as follows:

- The high rate of heating per unit volume, typically about 16,000 Btu/ft^3 (114 kg-cal/m^3) affords the unit a compact design.
- Combustion efficiency is high because of the large surface area for heat transfer and because the turbulence of the media promotes mixing and contact between the waste and the media.
- The combustion chamber has no moving parts; thus, low maintenance costs and long equipment life are common.
- The process is versatile and can destroy solid, liquid, or gaseous wastes; however it is ill-suited for irregular or bulky solids or wastes with a high ash content.
- The incinerator can tolerate fluctuations in the feed rate or composition of waste such as continuous or batch operation because of the great quantity of heat stored in the media.
- Excess air requirements are low because of the excellent combustion efficiency; therefore, air emission control costs are reduced.

Fluidized bed incinerators also have certain drawbacks compared with other incineration technologies: the bed media require regular maintenance; inert residuals must be removed from the media to prevent operating problems; the cost to operate the incinerator is high; energy costs are particularly high; and wastes that would corrode or react with the bed media must be excluded from the incinerator feed.

Multiple Hearth. The multiple hearth incinerator has been in use since 1934 as a specialized incinerator for sewage sludge. It continues to be used for that purpose and is the most widely used sludge incinerator in the United States.[25] It is also applied to the incineration of industrial chemical sludges and tars, and it is employed to regenerate activated carbon. Pilot tests are under way to evaluate the applicability of the multiple hearth technology to the incineration of hazardous wastes. Currently, if liquid or gaseous wastes are burned, they are burned with sewage sludge and are used as an auxiliary heat source for combustion.[26]

As Figure 13-5 illustrates, this technology consists of horizontal refractory-lined hearths arranged in a vertical structure. The diameter of multiple hearth furnaces typically varies between 4.5 and 29 ft (1.4 and 8.8 m); there may be from 4 to 14 individual hearths. Multiple hearth units can be operated using any one of a variety of fuels including natural gas, propane, butane, oil, coal dust, and solvents.

Rabble arms and teeth rotate on a central shaft to agitate and convey the

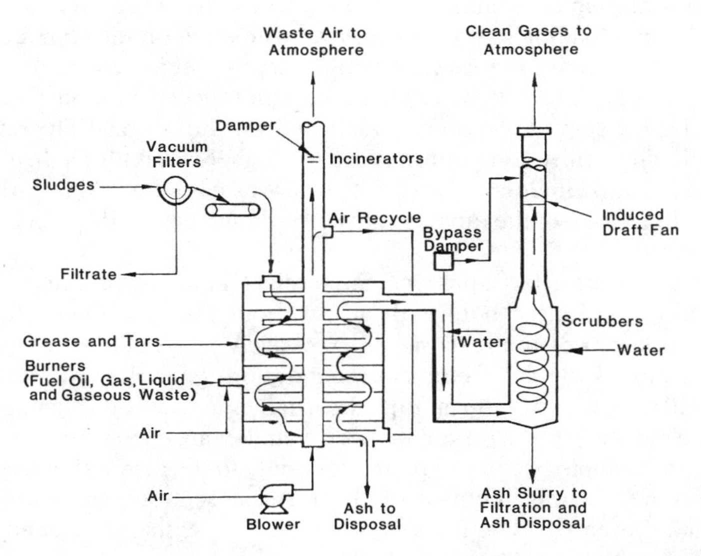

Figure 13-5. Multiple hearth incinerator design. Source: U.S. EPA Hazardous Material Incinerators Design Criteria. October 1979. EPA-600/2-79-198, NTIS PB80-131964

waste through the incinerator. Two or four arms are common in each hearth. Each arm has several teeth, or plows, that rake the sludge spirally across the hearth until it falls into the next chamber. The rabble arms and shaft normally rotate in the same direction, but they also can be reversed to induce more intense agitation. The liquid and gas wastes and auxiliary fuel are injected into the top of the main chamber through auxiliary burner nozzles. The waste material enters the top hearth, and it is heated to ignition temperature by means of several ignition burners. The amount of auxiliary fuel is typically high and is proportional to the amount of water in the waste.

The system has three operating zones: the top hearths are maintained at 600 to 1000°F (315 to 535°C) to expel excess moisture; the main combustion reactions are completed in the middle chambers, and they are maintained at 1400 to 1800°F (760 to 1000°C); the cooling zone is at the base, and hot ashes there radiate heat to the incoming combustion air. The waste is burned as it moves in a spiral path through drop holes within the hearths. The arrangement of these holes promotes mixing and surface contact with air to accelerate combustion. The rising exhaust gases also encourage turbulence and surface contact. Residual inorganic substances fall as ash to the bottom of the unit, where they are removed for disposal. Exhaust gases exit at 500 to 1100°F (260 to 590°C) from the top of the unit. The higher-temperature hearths encountered as the gases rise to the exhaust ducts function as an afterburner to complete the combustion reactions. The gaseous emissions are diverted through a scrubber to reduce particulate and gaseous pollutants.

The size and loading of the unit vary, depending upon the characteristics of the waste. Feed rates for sewage sludge vary between 7 and 12 $lb/ft^2/hr$ (9 and 16 $g/m^2/s$), 7.5 $lb/ft^2/hr$ (10 $g/m^2/s$) being most common. (The area specified is the sum of the areas of all hearths within a unit.) The retention time for low-volatility compounds is usually higher in a multiple hearth unit than that for other incinerator technologies. Waste retention time is dependent upon the design of the rabble tooth patterns and the rotation speed of the central shaft.[27]

The multiple hearth is a proven commercial technology, applicable to the incineration of sludges both with and without the addition of liquid or gaseous wastes. For the disposal of excessively wet sludges, the multiple hearth is one of the most efficient incineration technologies available. A singular disadvantage to the multiple hearth incinerator is its high operating costs; the moving parts in the combustion chamber are expensive to maintain. Before the technology can be applied routinely to the disposal of hazardous wastes, more accurate control of the furnace section temperatures are necessary. Presently, the unit is slow to respond to temperature changes and to adjustments of supplementary fuel rates.

Air Pollution Control. When wastes are incinerated, the common air pollutants are particulate matter, chlorides, sulfur, and nitrogen oxides. Other pollutants are normally not a concern because they are removed by the pollution control equipment designed to reduce the particles, chlorides, and oxides of sulfur and nitrogen. The waste burned in an incinerator dictates, in part, the type of air pollution abatement equipment. Other factors include the availability of water, sewerage facilities, and space limitations. Pollution control devices can use either wet or dry collection methods.

The efficiency of pollution control devices normally varies as a function of particle catch size. The removal efficiency is calculated from the mass of the pollutant removed divided by the total weight of the pollutant entering. Efficiencies are based on a specific range of particle sizes or on total particle loading. The size of particle to be collected also has a major bearing on the type of pollution control equipment required. Larger particles, those greater that 40 μm, can be removed using dry collection equipment such as bag houses or settling chambers. Particles as small as 15 μm are best removed by dry impingement or cyclone scrubbers. The smallest particulate matter, that less than 1 μm, usually requires wet collectors, and the particles are removed by contact with liquid droplets. Wet collectors include venturi, packed-bed, and plate scrubbers. High-efficiency air filters and electrostatic precipitators also have performed well in collecting particles smaller than 1 μm.

Wet scrubbers remove particles by producing several actions simultaneously: the solid and liquid particles intercept each other because of their relative motion toward one another; when a particle is passing an obstacle, gravity causes it to fall from its path and to settle on the surface of the obstacle; an obstacle placed in the path of the gas stream causes the gas stream to flow around it; larger particles are captured as they continue in a straight path due to inertial forces; and contraction and expansion promote condensation within the stream.

The effectiveness of a scrubbing system is usually directly related to the pressure drop across the scrubber. The higher the pressure drop, the greater the turbulence and mixing, so that the scrubbing action is enhanced.

Venturi scrubbers are widely used cleaning devices. They are efficient, but they consume prodigious quantities of energy, utilizing the kinetic energy of the high-velocity gas stream to atomize the scrubbing liquid into droplets.[28] For most applications to incinerators, venturi scrubbers are capable of more efficient particle removal than packed-bed or plate towers.

Packed-bed and plate scrubbers commonly are used when gaseous pollutants are to be removed,[29] as in liquid injection incineration facilities. Water is the solvent, and a chemical reagent is often added to it to increase the removal efficiency. The mechanisms in acid gas absorption involve solubility and chemical reaction equilibria; for example, the reaction between hydrochloric acid in the gas stream and sodium hydroxide in the liquid.

Electrostatic precipitators have proved effective for the removal of airborne particles. The gas stream passes through a series of negatively charged discharge electrodes that induce a negative charge to the particles, and the charged particles collect on a grounded collector electrode surrounding the discharge electrodes. Particles are removed from the collector plates for disposal.

Odoriferous agents smaller than 1 μm and residual hydrocarbons in the gas stream are treated by various methods. Direct flame or catalytic afterburners can be used, with the temperature elevated therein to destroy incompletely burned products. Gases are adsorbed on activated carbon, silica gel, or alumina. They also can be absorbed with a wet scrubber system, and they simply can be diluted with clean air.

CAPITAL AND OPERATING COSTS

The overhead cost of using incineration as a remediation measure at a hazardous waste site includes the design cost of the unit and its operation and maintenance costs. Expending the time and resources needed to obtain regulatory approval is risky because such approval may not be forthcoming. Should the permits be denied, the facility often must be located elsewhere and redesigned, or, perhaps, an alternate remedial measure must be selected. The technology and the size of an incineration facility are determined by the characteristics and combustibility of the waste and the volume of waste to be destroyed. The major equipment costs are for waste storage and loading facilities, for the combustion chamber, for air pollution control equipment (the cost of a scrubber is often the same order of magnitude as the cost of the combustion chambers), and for energy recovery equipment.[30] Other costs are incurred for the ancillary equipment: piping, storage tanks, pumps, fans, stacks, monitoring equipment, and instrumentation. In addition, the expense of acquiring permits, preparing the test burn protocol, and performing the test burns can be substantial.

The primary expenses in operating an incinerator are for labor and energy. The cost of energy depends on the heat input rate, the heat application method, the characteristics of the waste, and the heat recovery system. Additional costs are incurred to dispose of the ash residue, to operate the pollution control equipment and to pretreat scrubber blowdown, to monitor emissions, to analyze wastes, and to carry insurance.

The different technologies levy markedly different costs. For example, the capital cost of a hearth incinerator is approximately triple that of a liquid injection incinerator of the same input, but of the two the hearth destroys solid wastes more efficiently. The fluidized bed is usually somewhat less expensive than the multiple hearth.[31] The rotary kiln is more expensive but offers added flexibility in the variety of wastes that can be incinerated.

Capital and operating costs describing three major incinerators used to destroy hazardous waste are presented in Figure 13-6.[32] The costs presented are budget estimates and are accurate to ±20 percent of actual cost. Parts (a) through (c) of Figure 13-6 show equipment costs as a function of heat input for the rotary kiln, liquid injection, and multiple hearth incinerators. Freight on board (FOB) costs in the figure include waste storage and handling equipment, energy recovery equipment in the combustion chambers, and air pollution control systems. Installed costs are approximately 1.5 times FOB costs. Estimates of the heat input rate for a particular combustion temperature are presented in Figure 13-6(d). The cost of auxiliary fuel, based on 1981 prices, can be estimated from Figure 13-6(e); however, the world energy market is volatile, and these costs can vary significantly. Note that fuel consumption depends on the efficiency of the system, the types of waste, available heat recovery equipment, and other operating factors characteristic of a particular application.

Commercial incineration facilities typically charge for services on a tonnage basis. Surcharges are applied to halogenated wastes; wastes with high ash content; viscous liquids; and wastes containing nitrogen, sulfur, or metals. The cost to incinerate PCB wastes is typically 60 percent higher than that for other wastes. Transportation costs are an additional expense.

SUMMARY

Although incineration remains an expensive alternative in the choice of remedial measures to eliminate hazardous waste, it is the most effective method of disposal. Because it achieves intrinsically the complete destruction of the organic portion of the waste, it effectively limits future liabilities.

The decision on whether or not incineration is the appropriate remedial measure hinges on the nature of the waste, the waste's matrix, and the volume of waste. The technology is competent to incinerate any combustible waste; however, economic factors mandate that the waste have at least a moderate heat value. Major costs in incinerating wastes are the cost of energy and the cost of air pollution control equipment. All incineration processes require fuel for preheating; certain wastes can be incinerated with little or no auxiliary fuel, whereas others require substantial amounts.

Essential expenditures in utilizing incineration include the time and difficulty of securing the government permits for siting and operation of the facility. However, the regulatory climate is becoming more favorable toward incineration as other alternatives for disposal become more limited. When combus-

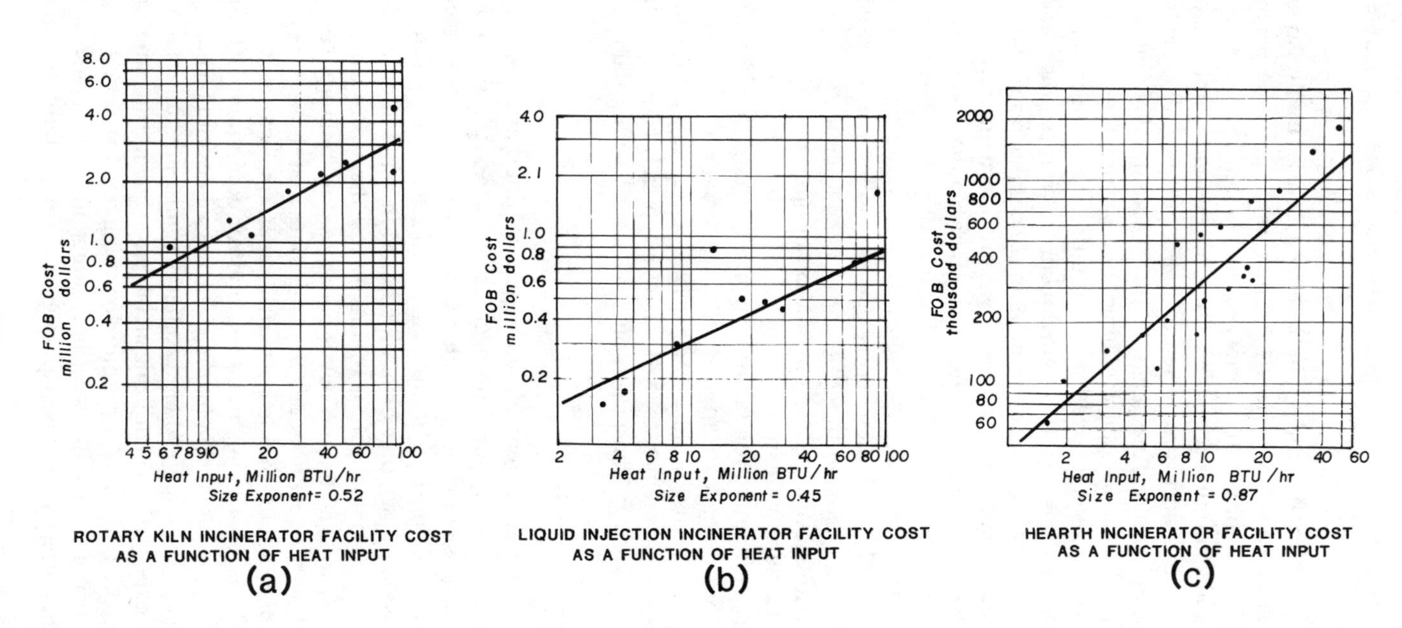

ROTARY KILN INCINERATOR FACILITY COST
AS A FUNCTION OF HEAT INPUT
(a)

LIQUID INJECTION INCINERATOR FACILITY COST
AS A FUNCTION OF HEAT INPUT
(b)

HEARTH INCINERATOR FACILITY COST
AS A FUNCTION OF HEAT INPUT
(c)

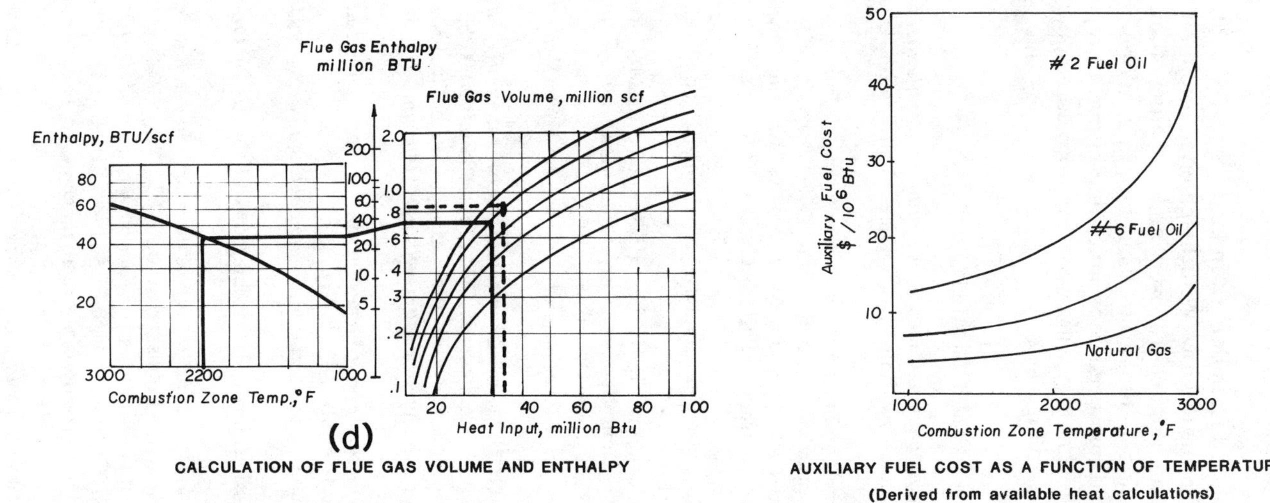

CALCULATION OF FLUE GAS VOLUME AND ENTHALPY

Source: U.S.EPA Incineration and Treatment of Hazardous Waste
Proceedings of the Eighth Annual Research Symposium. Apr. 1983.
EPA-600/9-83-003, NTIS PB83-210450

50
40
30
20
10
Auxiliary Fuel Cost $ / 10^6 Btu
2 Fuel Oil
6 Fuel Oil
Natural Gas
1000
2000
3000
Combustion Zone Temperature, °F

AUXILIARY FUEL COST AS A FUNCTION OF TEMPERATURE
(Derived from available heat calculations)

Basis: Stoichiometric combustion of fuels

	March 1981 Price	Heat Content
#2 Fuel Oil	$1.25 per gallon	137,000 Btu/gal
#6 Fuel	$ .86 per gallon	153,000 Btu/gal
Natural Gas	$2.69 per 1000 scf	1020 Btu/scf

(e)

Figure 13-6. Costs of incineration facilities.

tible liquid wastes were barred from deposit in landfills in 1984, interest in incineration technology began to grow. With heightened attention to and development of incineration, the heat recovery and operating efficiency of the technology are improving. In the future, especially in light of increasing liabilities to generators and improved economics due in part to regulatory restrictions, incineration technology will become a more common alternative for the disposal of hazardous waste.

BIBLIOGRAPHY

Ackerman, D. et al. *Destroying chemical wastes in commercial-scale incinerators.* Facility report No. 6., Rollins Environmental Services. Washington, D.C.: U.S. EPA, June 1977. PB 270 897.

Argonne National Laboratory. *Optimizing the energy efficiency of incinerators for the disposal of industrial waste.* Argonne, Ill.: June 1972.

Barnes, R. H., R. E. Barrett, A. Levy, and M. J. Saxton. *Chemical aspects of afterburner systems.* Research Triangle Park, N.C.: U.S. EPA, Apr. 1979. PB 298 465.

Basta, Nicholas et al. Hazardous waste incineration—a burning CPI concern. *Chemical Engineering.* 5(1986):21-26.

Bonner, T., B. Desai, J. Fullenkamp, T. Hughes, E. Kennedy, R. McCormick, J. Peters, and D. Zanders. (Monsanto Research Corp.) *Hazardous waste incineration engineering.* Park Ridge, N.J.: Noyes Data Corp., 1981.

Brunner, C. *Incineration systems: Selection and design.* New York: Van Nostrand Reinhold Company, 1984.

Castaldini, C., S. Unnasch, and H. B. Mason. *Engineering assessment report—hazardous waste co-firing in industrial boilers.* Cincinnati: U.S. EPA, 1985. EPA/600/S2-84/177.

Chemical Waste Management, Inc. Price schedule. Model City, N.Y.: 1985.

Dawson, R. Hazardous sludge criteria procedures. *Sludge Magazine.* 2(1)(Jan.-Feb. 1979):12-21.

Farb, D. and S. Ward. *Information about hazardous waste management facilities.* Cincinnati: U.S. EPA, July 1975. EPA-530/SW-145.

Hitchcock, D. Solid-waste disposal: incineration. *Chemical Engineering.* 86(11)(May 1979):185-194.

Hudson, W. and K. C. Lee. Design and operate incinerators. *Hydrocarbon Processing.* (Oct. 1984):47-49.

Kiaing, Yen-Hsiung and A. A. Metry. *Hazardous waste processing technology.* Ann Arbor, Mich.: Ann Arbor Science Publishers, Inc., 1982.

Lovell, R. J., R. A. Miller, and C. Pfrommer, Jr. *Trial burn testing of the EPA-ORD mobile incineration system.* Cincinnati: U.S. EPA, 1984. EPA-600/D-85-054.

Mackie, J. A. and K. Niesen. Hazardous waste management: The alternatives. *Chemical Engineering.* 16(1984):50-64.

Manson, L. and S. Unger. *Hazardous material incinerator design criteria.* Cincinnati: U.S. EPA, Oct. 1979. PB 265 541.

Ottinger, R. et al. *Recommended methods of reduction, neutralization, recovery, or disposal of hazardous waste,* Vol. III, Disposal process description, ultimate disposal, incineration, and pyrolysis processes. Cincinnati: U.S. EPA, 1973. PB 224 582.

Perry, R. H. and C.H. Chilton. *Chemical engineers' handbook,* 5th Ed. New York: McGraw-Hill Book Company, 1973.

Radimsky, J. and A. Shah. *Evaluation of emerging technologies for the destruction of hazardous waste.* Cincinnati: U.S. EPA, 1986. EPA/600/S2-85/069.

Resource Conservation and Recovery Act of 1976 (RCRA). PL 94-580. 21 Oct. 1976. 40 CFR 264 (Subpart O) and 265 (Subpart O).

Rhein, R. W., Jr. and J. Trewhitt. Time has run out for land disposal facilities. *Chemical Engineering.* 25(1985):30-32.

Scurlock, A. et al. Incineration in hazardous waste management. Washington, D.C.: U.S. EPA, 1975. PB 261 049.

Shultz, David W. (Ed.). Incineration and treatment of hazardous waste: *Proceedings* of Eighth Annual Research Symposium at Ft. Mitchell, Ky. Mar. 1982. EPA-600/9-83-003.

Sittig, M. *Incineration of industrial hazardous wastes and sludges.* Park Ridge, N.J.: Noyes Data Corp., 1979.

Star, A. M. Cost estimating for hazardous waste incineration. *Pollution Engineering.* 4(1985):21-24.

Time is slipping away for EPA deadlines. *Chemical Week.* (Apr. 17, 1985):24-27.

U.S. EPA. Office of Policy, Planning, and Evaluation. *Assessment of incineration as a treatment method for liquid organic hazardous wastes.* Washington, D.C.: Mar. 1985.

U.S. EPA. *Guidance manual for hazardous waste incinerator permits.* Washington, D.C.: July 1983. SW-966.

———. *Hazardous material incinerator design criteria.* Cincinnati: 1979. PB80-131964.

———. Destructing chemical wastes in commercial-scale incinerators, Technical summary, V. I (prelim. draft). Washington, D.C.: Mar. 1975. PB 257 709.

Vogel, G. A. et al. Hazardous waste incineration costs. Incineration and treatment of hazardous waste: *Proceedings* of Eighth Annual Research Symposium at Ft. Mitchell, Ky. Mar. 1982. pp. 14-21.

Wen, C. Y. and S. Uchida. Gas absorption by alkaline solutions in a venturi scrubber. *Industrial & Engineering Chemistry, Process Design & Development.* 12(4)(Apr. 1973):437-443.

Wolbach, C. D. and A. R. Garman. *Destruction of hazardous wastes co-fired in industrial boilers: Pilot-scale parametrics testing.* Cincinnati: U.S. EPA, 1985. EPA/600/S2-85/097.

Zurer, P. S. Incineration of hazardous wastes at sea: Going nowhere fast. *Chemical and Engineering News* 63(1985):24-42.

NOTES

1. Yen-Hsiung Kiaing and A. A. Metry. *Hazardous waste processing technology.* Ann Arbor, Mich.: Ann Arbor Science Publishers, Inc., 1982. p. 37. See, too, R. J. Barnes, R. E. Barrett, A. Levy, N. J. Saxton. *Chemical aspects of afterburner systems.* Research Triangle Park, N.C.: U.S. EPA, Apr. 1979. PB 298 465.

2. PL-98-616, Hazardous and Solid Waste Amendments of 1984, Section 201, 8 Nov. 1984.

3. Nicholas Basta et al. Hazardous waste incineration—a burning CPI concern. *Chemical Engineering.* 5(1986):21-26. See, too, Reginald W. Rhein, Jr. and Jeff Trewhitt. Time has run out for land disposal facilities. *Chemical Engineering.* 25(1985):30-32.

4. P. S. Zurer. Incineration of hazardous wastes at sea: Going nowhere fast. *Chemical and Engineering News.* 63(1985):24-42.

5. Basta.

6. 40 CFR 264.343, Performance Standards.

7. 40 CFR 261, Appendix VIII.

8. 40 CFR 260-265, 270.

9. R. J. Lovell, R. A. Miller, and C. Pfrommer, Jr. *Trial burn testing of the EPA-ORD mobile incineration system.* Cincinnati: U.S. EPA, 1984. EPA-600/D-85-054.

10. Basta.

11. 40 CFR 270.62, Hazardous Waste Incineration Permits.

12. A. M. Star. Cost estimating for hazardous waste incineration. *Pollution Engineering.* 4(1985):21-24.

13. The Hazardous and Solid Waste Amendments of 1984, PL 98-616, Nov. 1984.

14. M. Sittig. *Incineration of industrial hazardous wastes and sludges.* Park Ridge, N.J.: Noyes Data Corp., 1979.

15. Basta.

16. C. D. Wolbach and A. R. Garman. *Destruction of hazardous wastes co-fired in industrial boilers: Pilot-scale parametrics testing.* Cincinnati: U.S. EPA, 1985. EPA/600/S2-85/097.

17. Wolbach and Garman.

18. D. Hitchcock. Solid-waste disposal: incineration. *Chemical Engineering* 86(11)(May 1979):185-194.

19. D. Ackerman, J. Clausen, R. Johnson, and C. Zee. *Destroying chemical wastes in commercial-scale incinerators;* facility report No. 3. Systems Technology Inc. Washington, D.C.: U.S. Environmental Protection Agency, Nov. 1976.

20. T. Bonner, B. Desai, J. Fullenkamp, T. Hughes, E. Kennedy, R. McCormick, J. Peters, and D. Zanders (Monsanto Research Corp.). *Hazardous waste incineration engineering.* Park Ridge, N.J.: Noyes Data Corp., 1981.

21. Zurer.

22. Sittig.

23. Ackerman et al.

24. Ackerman et al.

25. U.S. EPA. *Hazardous material incinerator design criteria.* Cincinnati: 1979. PB80-131964.

26. R. Dawson. Hazardous sludge criteria procedures. *Sludge Magazine* 2(1)(Jan.-Feb. 1979):12-21.

27. Dawson.

28. C. Y. Wen and S. Uchida. Gas absorption by alkaline solutions in a venturi scrubber. *Industrial & Engineering Chemistry, Process Design & Development.* 12(4)(Apr. 1973):437-443.

29. Bonner et al.
30. G. A. Vogel, I. Frankel, N. Sanders, and E. Martin. Hazardous waste incineration costs. Incineration and treatment of hazardous waste: *Proceedings* of the Eighth Annual Research Symposium at Ft. Mitchell, Ky. Mar. 8-10, 1982. pp. 14-21.
31. Vogel et al.
32. Vogel et al. The costs were derived from a limited sampling of 25 incinerator component manufacturers, owners, and generators of hazardous waste.

Chapter 14
Closure Through Off-Site Remedies

In the course of addressing remedies for hazardous waste sites, there are instances when entombment of the waste on the premises of the waste generator is neither feasible nor desired. The site may be proximate to sensitive receptors such as a residential tract or even a busy commercial or industrial center; the value of the land might make on-site closure, and thus the removal of that land from the market, unacceptable; the volume of buried waste or the level of contamination may be small. In each of these cases, disposal of the waste off-site is the rational alternative to entombment on the site where the waste was deposited. Additionally, if on-site closure and consequent monitoring are more costly than off-site disposal, the off-site remedy is chosen.

The most common off-site remedial alternatives are:

- Relocation of the waste to a secure landfill.
- Removal of the waste to an off-site treatment facility for detoxification.
- Incineration, with correct disposition of the residue.
- Reclamation and reuse of the waste, as when solvents are recycled.

The selection of any particular corrective action for a hazardous waste emergency is based on factors specific to the site and on the physical and chemical characteristics of the waste; see Table 14-1.

When waste is treated off-site, landfilling is, by far, the most common remedy selected. Since the implementation of RCRA, there has been a prodigious increase in the volume of hazardous waste deposited in landfills. Consequently, the available capacity of the landfills has been utilized more quickly than had been projected when they were designed. Simultaneously, more stringent design requirements have been applied to the construction of new landfills by federal and state agencies. Landfill space is a premium commodity currently. Nonetheless, landfilling is becoming a less practicable option. The federal government has banned:[1]

This chapter was developed by Orest Hrycyk of OBG Technical Services, Inc. and John J. Lamanche of O'Brien & Gere Engineers, Inc.

TABLE 14-1. Typical factors evaluated to assess on-site versus off-site closure options.

SITE CHARACTERISTICS	WASTE CHARACTERISTICS
Depth to ground water	Volume of waste
Surface hydrology	Leaching potential
Soil types	Energy content
Climate	Toxicity
Receptor populations	Volatility
• Human	
• Animal	Migration mechanisms
• Plant	Recycling potential
Surrounding land uses and land values	

TABLE 14-2. Prohibitions against land deposition of certain hazardous waste, for selected states.

STATE	PROHIBITED WASTE OR ITEM
Arizona	Underground injection of hazardous waste
Maryland	Drums
New York	Solid waste with organic content >2% by volume
Rhode Island	Any substance that the state defines as an "extremely hazardous waste"
Wisconsin	Underground injection of hazardous waste

> . . . the placement of bulk or noncontainerized liquid hazardous waste or free liquids contained in hazardous waste (whether or not absorbents have been added) in any landfill.

Additionally, several states forbid the addition of a specific hazardous waste to landfills. For examples, see Table 14-2.

Incineration is nearly an ideal ultimate solution for organic hazardous waste disposal. No other technique commonly available provides more complete destruction of waste. However, incineration is probably also the most costly remedy. Only hazardous waste that is organic is destroyed by incineration; PCBs, nonchlorinated organic solvents, chlorinated solvents, organic sludges, and petroleum sludges are commonly treated by this process.

The cost of incineration is determined by the heat value of the waste, its ash content, and the presence of any halogens, nitrogen, or sulfur compounds that may generate gases or vapors requiring special treatment or handling prior to their discharge into the atmosphere. Any metals contained in the hazardous waste that is to be incinerated are not destroyed but remain with the ash residue after incineration. This residue is analyzed by the EP Toxicity test[2] to ascertain whether it contains certain heavy metals, pesticides, or herbicides. If the leachate exceeds the heavy metal or pesticide concentration limit,[3] the ash is adjudged hazardous, and it too requires special handling for disposal. (See also Chapter 13 for additional information on this topic.)

Hazardous waste also can be treated and thus rendered nonhazardous and easier to dispose of, but the various treatment processes are specific to individual waste types. There are diverse methods to treat waste: a particular substance may be coalesced chemically and then concentrated, as when heavy metals are coagulated and precipitated; the hazardous waste could be altered or destroyed by reduction-oxidation reactions; the substance could be removed by selective adsorption or absorption. Other technologies include encapsulation, a process in which the hazardous substance is chemically or physically bound to an impervious matrix such as cement or lime dust. Once bound, the hazardous constituent can no longer migrate into the environment. The resulting product is then hauled to a disposal site. (See also Chapter 9 for an explanation of several treatment approaches.)

One of the fundamental purposes of RCRA was to encourage recycling and reclamation of waste that otherwise would be discarded. In this regard, RCRA has been somewhat effective. Whenever off-site closure remedies are evaluated, serious consideration should be given to recycling and reclamation. The long-term benefit of recycling is conservation of natural resources; the short-term benefit is the salvage value of the recovered materials. Additionally, recycling and reclamation of hazardous waste avoid the long-term liability associated with landfilling. Waste cannot always be recycled or reclaimed successfully, but if a market exists for a specific waste, recycling usually is feasible. Other factors that bear upon the feasibility of reclamation are the availability and sophistication of technology to recover the product, the cost of recovery, and the concentration of the waste in the environment. Presently, used oils, precious metals, and chlorinated and nonchlorinated solvents are the likely candidates for recycling. The economic benefits of reclaiming and recycling oils and solvents are directly tied to the price of oil; as the price increases, reclamation and recycling become economically more attractive. Regulatory and economic incentives for recycling currently are not strong.

SPECIAL CONSIDERATIONS FOR OFF-SITE MEASURES

When an off-site closure remedy is employed, the engineer and the contractor apply management techniques different from those utilized when closure on-site is implemented. The waste first is isolated and extracted. Then it is contained, staged, and finally transported to the ultimate destination. Through the duration of the work, implementation of special safeguards insures that none of the activities contaminates the surrounding area.

Special efforts are made to insure the smooth coordination of the diverse activities. In most cases, each activity is executed by a different party, so that coordination is essential. Initially, the site managers coordinate their activities with government agencies and with the disposal facilities to which the waste will be sent. This early coordination insures that the waste will be disposed of in an acceptable manner. All personnel receive instructions to enable them to carry out their work safely and to insure that decontamination procedures are followed. The workers are grouped into crews in order to assign the responsibilities for specific tasks. All vehicles are scheduled in advance to insure that the hazardous waste will be removed expeditiously. Monitoring of the area surrounding the work site detects any stray contamination that might range beyond the work and decontamination areas. Because such contamination will be detected, it can be removed, and the total site can be remedied properly.

Precautions are taken to protect people, equipment, and the environment from contact with the hazardous waste during the time when the waste is removed, transported, and disposed of. Personnel are safeguarded from contamination by the proper use of safety equipment and decontamination facilities. The excavation, pumping, and other equipment used in the field is decontaminated with solvents or steam whenever it leaves the work zone. The area surrounding the work zone is shielded from contamination by insuring that no personnel or equipment leaves the work area other than through the decontamination area and that proper decontamination procedures are followed. Plastic sheeting, laid in specific areas of the uncontaminated zone where hazardous material might be spilled, serves as an additional safeguard.

Isolation of Waste

Before any hazardous waste material is removed from a site, a number of prerequisites are observed. The site first is surveyed, and a topographic map is prepared. A grid is then drawn on the map at 50- or 100-ft (15- or 30-m) intervals. Survey stakes are driven into the ground at points of intersection of the grid, and the stakes remain in the ground through the course of the work to assist crews in identifying the precise location of the hazardous waste. The site's topographic map also depicts present and future access roads, clean zones, hazardous waste staging areas, loading areas, and decontamination areas.

Wastes are characterized before they are removed from a hazardous waste site. Both the nature and the volume of the waste are determined before a remedial measure can be selected. A specific disposal method may be chosen when the characterization data are in hand. Sampling and analysis establish both limits of contamination at the site and a quantitative estimate of the amount of material to be excavated or recovered. Gas chromatography, atomic absorption, and the EP Toxicity test are methods used to aid in the analysis of hazardous waste. (For details of the analysis, see Chapter 5.)

A knowledge of the site's history, however fragmentary, is helpful in predicting what to expect at the hazardous waste site. Conversations with former owners or employees of the facility or with local observers can yield valuable information about the types of materials handled at the site. Investigators are careful when seeking information from sources external to the waste generator in order to protect the generator's confidentiality.

Once the disposal method is selected, several disposal sites capable of accepting the waste are located. Information about treatment, storage, and disposal facilities is available from state environmental agencies and from the regional offices of the U.S. EPA. In addition, private corporations print lists of these facilities.[4] Any facility that engages in the treatment or storage—for longer than 90 days—or disposal of hazardous waste requires a permit under the provisions of RCRA.[5] Such a facility is often known by the abbreviation TSDF, which stands for a treatment, storage, or disposal facility.

A TSDF operator requires an application for use of the facility and may charge a fee of $100 to $500 to process the application. The typical TSDF application form requests information about the waste generator and the waste to be disposed of. This information is essential for the TSDF in determining whether it can legally and, perhaps more important, safely dispose of the material. The TSDF operator may even ask that a sample of the waste be supplied together with the application form. Typical information requests found on the application are listed in Table 14-3.

It is important that the information requested by the TSDF be provided in as detailed a form as possible in order to avoid delays due to misjudgment of the waste. The TSDF's evaluation procedure typically will take six to eight weeks, but delays of six months can occur. Therefore, it is essential that the application process begin as early as possible in the remediation effort.

TSDFs vary in the waste they are authorized by their government permits and licenses to accept. For example, not all TSDFs can accept PCBs and no TSDF may accept dioxin waste for storage, disposal, or incineration. Once

TABLE 14-3. Information commonly required by TSDFs.

INFORMATION ABOUT THE WASTE
- Phase
- Viscosity
- Specific gravity
- General physical characteristics (to aid in determining possible disposal techniques)
- Flash point (to determine ignitability of an air mixture surrounding the waste)
- Energy content, Btu/lb (to determine the waste's heat value)
- pH (to determine the corrosive nature of the waste)
- Chlorine and sulfur by weight (to determine treatment of emissions if the waste is incinerated)
- Suspended solids, dissolved solids, layering (to determine if free liquid might separate from the waste)
- Organic and inorganic compounds
- Presence of any radioactive, explosive, pyrophoric, or shock-sensitive matter

LOGISTICAL MATTERS
- Quantity and frequency of delivery of the waste
- Mode of shipping
- The waste's U.S. Department of Transportation (DOT) shipping name and hazard class
- UN/NA Number

INFORMATION ABOUT THE GENERATOR
- Name
- Address
- U.S. EPA identification number
- State generator code
- Process that produced the waste

the TSDF determines that the material is compatible with its facility, it issues its approval, consenting to accept the waste. If the TSDF is able to handle the proposed waste, it will generally accept it for the proposed disposal method. If, on the other hand, the TSDF is unqualified to handle the waste, it will decline it.

Once the material is approved for disposal, the TSDF issues an authorization specifying exactly which waste material it will accept. It is commonplace for a TSDF to refuse to accept or return hazardous waste that does not meet the specifications of the contract. If the TSDF rejects a hazardous waste when it arrives at the facility, the generator is required to resubmit the application, and the new application process usually requires nearly as much time as the original application.

The contract of a TSDF stipulates the cost of the disposal. There are several components to the disposal cost:

- The cost charged by the TSDF, which covers its overhead and profit.
- Federal[6] and, if applicable, state Superfund taxes.
- State and local sales taxes.

The contract from the TSDF also should indicate that either a certificate of destruction or a certificate of disposal will be issued once the hazardous material has been properly disposed of or destroyed. These certificates are given to the generator of the waste by the TSDF; they attest that the hazardous waste was disposed of or destroyed by the TSDF in the manner agreed to by the generator and the TSDF.

Containment of Waste. In order to dispose of hazardous waste off-site, the material must be transported. In almost all cases, the volume of the waste is divided into quantities that can be conveniently and practicably moved. The waste may be transported in bulk, or it may be placed in containers and then transported. If waste is moved in bulk form, it may be placed in secured roll-off boxes, dump trailers, rail cars, or on barges. Containers for the waste include steel or fiber drums or custom-fabricated tote bins.

Movement of Hazardous Waste in Bulk. Roll-off boxes often are used to hold construction waste or any substantial quantities of waste. A roll-off box is a metal container that has been designed to fit on a specialized truck body, behind the cab. These boxes are convenient because they can be deposited at a hazardous waste site and recovered at a later time. The empty box is set on the ground by the carrier truck; after the box is loaded, the carrier truck retrieves it and moves it to the location where its contents are disposed of. Some roll-off boxes have low sides to make loading easier.

In order for a roll-off box to be used to transport hazardous waste, it is secured with double locks and made watertight. The rear doors are fitted with gaskets to insure that no contaminated water can escape during loading or transportation; alternatively, at a minimum, it is lined with a layer of polyethylene that is at least 6 mils (0.15 mm) thick. In addition, the roll-off box is inspected before it is moved to insure that there is no leakage. When the loaded roll-off box is transported, the top is covered to prevent any of the waste from blowing from the box and to prevent rain from mixing with the waste. Roll-off boxes can be overloaded in the field, and they are then difficult or impossible to load on the truck. Overloading is unacceptable because the transporter is required to comply with vehicle load limits for roads and bridges, limits that vary from state to state. When a roll-off box is overloaded in the field, the excess waste must be removed from the box until it is light enough to be loaded on the truck, or until it complies with state vehicle weight restrictions. The waste displaced from the box must be stored on a liner, have a berm placed around it, and be covered until it can be loaded into another roll-off box.

Dump trailers also can be used to move hazardous waste in bulk. Aluminum trailers are the most popular means of moving such waste because they are the lightest in weight and, given highway weight limitations, can carry the greatest payload. When used to move hazardous waste, a trailer is lined with at least one layer of 6-mil (0.15-mm) polyethylene. If the rear door of the dump trailer does not have gaskets for leak protection, the inside seams of the door are sealed with waterproof caulk or another waterproof material.

A vehicle transporting hazardous waste is required to adhere to payload limits and maximum axle weight restrictions for each state through which it passes en route to its destination. Care is taken in loading the vehicle to distribute the load properly and to keep the vehicle within weight limits. Portable scales can be employed at the loading site to give both the loader and the driver an estimate of the weights. Within those constraints, the vehicle is filled to capacity in order to maximize the economy of the transport. It is advisable to have the loaded vehicle's weight measured and documented at an independent weighing station to avoid penalties for exceeding state weight limits and to insure that the TSDF assesses the correct charge.

Rail cars and barges are used infrequently to move hazardous waste. Obviously, neither a rail car nor a barge could be considered for use unless a rail line or waterway, respectively, were convenient to the waste site. Moreover, given the economics, a barge could be considered only if at least 10,000 yd^3 (7600 m^3) of waste were to be removed. Unless the rail cars and barges are dedicated to transporting hazardous waste, they must be decontaminated before being returned to service, and that process can be costly.

Movement of Hazardous Waste in Containers. Placing hazardous waste in containers makes the waste more portable and enlarges the transport options. Decontamination is also less formidable when the waste is moved in containers rather than in bulk. When containers are used, they are placed on pallets and strapped together to reduce jostling. All containers are expected to maintain their integrity after they are deposited in a landfill. Liquid waste placed in a sealed container in cold weather must be safeguarded from freezing if the physical properties of the substance could cause the container to burst if the liquid solidified.

The choice of material for container construction depends upon the chemical properties of the waste. For example, a corrosive material will disintegrate a metal drum. Strong acids or strong bases require PVC or polyethylene-lined drums; extremely dangerous materials may require drums fabricated from heavier-gauge steel than that normally used to prevent rupture. The structure of the drum provides an indication of the material stored. If the entire lid of the drum is removable, the drum should store solid waste. A drum with a bung is typically used to store liquids. Before any reactive materials are placed in a container, the compatibility of the container or liner with the reactive material must be determined.

For many hazardous wastes, steel drums are the container of choice, as a steel drum has the strength to maintain its structural integrity when landfilled. Drums of standard sizes, such as 17C, 17E, and 17H, or over-pack drums may be used. Any steel drum must conform to the specifications of the DOT.[7] Over-pack drums are large containers in which leaking or damaged drums are placed for storage or shipment. All drums deployed must be clean, free of dents, and free of rust, and their lids must be secure.

Fiber drums may be used for specialized disposal purposes. For example, waste that is to be incinerated may be enclosed in fiber drums. In every case, the fiber drum is lined with at least one layer of polyethylene, 6 mils (0.15 mm) thick, before waste is placed in it. The seals of fiber drums must be secure. Fiber drums are not used in landfills because they cannot bear the weight of the fill without being crushed, thus losing their contents.

In some cases, the circumstances of the hazardous waste site remediation may be such that neither steel drums nor fiber containers are appropriate. It may be necessary to have customized containers fabricated for the single purpose of removing specific waste from a particular site. The purpose of such tailor-made tote bins is to insure absolutely the integrity of the hazardous waste during the entire time of its movement, to ultimate disposal. It may be prudent and advisable to fabricate tote bins to contain the waste if the substance is extraordinarily toxic—as in still bottoms or reactor bottoms containing mutagens or carcinogens, for example; or if the waste must be transported a long distance before disposal, perhaps with a change in the mode of transportation—rail to truck, for example—before final disposal. Each of these circumstances is extraordinary; therefore, each justifies the cost of fabricating tote bins.

Staging. Under ideal conditions, the excavated or recovered hazardous waste is immediately removed from the site and transported for ultimate disposal. However, the ideal case usually is not possible, and it is necessary to stage the material at the hazardous waste site for a time. As a result, a staging area is designated. This area is located as close as is practical to the area where the hazardous waste is being excavated, but there are constraints on the proximity of the staging area to the excavation site. Sufficient space is necessary between the staging and excavation areas to provide maneuvering room for the equipment. If the two areas cannot be arranged near each other, a haul road is either designated or constructed from the excavation area to the staging point. A turnaround is provided, if possible, to aid in the loading and removal process. Haul roads are sprayed periodically with water to keep dust down. If contaminated material is accidently spilled on a haul road, the affected area is excavated immediately and disposed of properly. All haul roads are monitored for contamination.

Within the staging area, special attention is given to preventing the release of any contaminant to the environment. To prepare for possible accidents, the area may require a liner, and a berm may be necessary around the perimeter. During the time when the hazardous waste is staged, it is covered with a tarp to prevent precipitation from percolating through it into the ground water.

Transport of Waste. Once the waste is appropriately packaged, whether in bulk or in containers, it is available for transport to the disposal site. While the waste is transported, the goal is to minimize its contact with human beings and the environment. In no event is the transport vehicle permitted into the contamination zone. If a transport vehicle becomes contaminated, it must be decontaminated before it may leave the site. During transport, the waste is in its least controlled condition; release of the waste material into the environment has the greatest chance of occurring during this time. Therefore, several precautions are taken to prevent accidents.

Government regulations of the U.S. EPA and the U.S. DOT police the movement of hazardous waste.[8] Each vehicle transporting hazardous waste is required to carry a manifest,[9] a document listing the contents of the haul vehicle's cargo. The manifest is designed to account for the waste throughout its passage from the point of origin to the point of disposal. Each state through which the transport vehicle passes may require a manifest. Information recorded on a typical manifest is displayed in Table 14-4.

TABLE 14-4. Data listed on hazardous waste manifest.

GENERATOR:
- Name, mailing address, and telephone number.
- U.S. EPA identification number.
- Identification number from state where waste originated.

TRANSPORTER:
- Name and U.S. EPA identification number.
- Identification number from each state through which the waste travels.
- Signature of transporter.

DESIGNATED DISPOSAL FACILITY:
- Name, site address, telephone number.
- U.S. EPA identification number.
- State identification number.
- Signature of the owner or operator.

WASTE DESCRIPTION:
- U.S. DOT name.
- DOT hazard class.
- UN/NA identification number.
- Number and type of containers used.
- Special handling instructions and other relevant information.
- U.S. EPA hazardous waste number.
- Discrepancies between the manifest and the waste products expected by the TSDF.
- Disposal technique.
- Total quantity of waste.

Most hazardous waste is moved by truck because trucks are the most flexible transportation mode, best able to reach both excavation and disposal sites. If rail transport is employed, the rail cars must be sealed. At each switch yard, the cars are inspected for leaks and then resealed. If a barge is to be used to transport the waste, a navigable waterway must be adjacent to both the excavation and the disposal site. It is sometimes necessary to change the mode of transportation, although it is never desirable to do so because costs are thereby increased, and the potential for accidental release of hazardous waste is heightened.

Before shipment, the vehicles are examined for safety. The tires, lights, tarp, and tailgate locks are inspected. The official state inspection and registration documents of the vehicle are checked to insure that they are current. In addition, the hauling companies are reviewed to establish that each has current permits from the appropriate government agencies. If the shipment will cross state lines, the truck is required to hold a permit from each state.

The schedule for transport and disposal is determined well in advance of the day when the waste is dispatched, as the disposal site must prepare the specific location to receive the waste, and it must be able to cover or otherwise secure it. TSDFs do not accept unscheduled deliveries of hazardous waste. Moreover, it is necessary to coordinate the arrival of the transportation vehicle with the operating hours of the disposal facility. At the excavation site, the vehicle should be loaded in time to arrive at the disposal facility during operating hours. Vehicle loading times should be estimated in order to stagger the arrival of the vehicles. Such a staggered schedule reduces the time that each vehicle consumes waiting. When the transport schedule is planned, time is allocated for loading—"poly-ing," sealing, and so forth. The cost of the loading crew is compared with the cost of the vehicles, and the project manager attempts to find the most economical mix of work time and unavoidable downtime for both cost elements.

Any event or route that may increase the risk of exposure of the transported hazardous waste to the general population is avoided. The shortest route available, the route that will affect the fewest people between the excavation site and the disposal site, is chosen for transportation of the waste. Interstate highways are used whenever possible, except for routes that traverse cities and other concentrations of people. The number of potentially affected people can be reduced by judicious scheduling of the shipments, too. For example, rush-hour traffic is avoided; in addition, if the route passes near an industrial complex, shipment is planned for those hours when the work force is not present. Travel during severely inclement weather is also avoided because of the increased risk of accident. While the hazardous waste is in transit, both the site manager and the driver of the haul truck have information on hand about the sources and availability of emergency cleanup services.

SUMMARY

In some instances, it is desirable and environmentally prudent to seal off or isolate hazardous wastes on the site where they are found. However, it is sometimes impracticable to do so, especially if hydrogeological factors, human development in the environs, or characteristics of the waste preclude on-site disposal. If a specialized waste treatment is chosen, such as incineration, it is common to move the waste to the site of the treatment facility rather than to develop the specialized treatment capacity at the location of the wastes. Because the federal government and various states increasingly are restricting the forms of disposal acceptable for hazardous wastes, specialized treatment is more frequently necessary now than in the past.

When a pernicious commodity such as hazardous waste is transported, extraordinary safety considerations are necessary in order to protect public health and the environment during each step in the preparation of the waste for disposal. The wastes first are isolated from the surroundings in which they are found. Then they are placed in containers with different types of containers used depending upon the physical and chemical characteristics of the wastes. The contained wastes are staged in preparation for shipment. All contaminated soil is disposed of, and the equipment used is decontaminated.

Transport of the wastes is firmly controlled by federal and state government regulations. It is important to move the wastes as quickly and safely as possible to reduce the risk of exposure to receptors along the transport route. Meticulous coordination with the disposal facility is essential to insure that the wastes will be compatible with the waste treatment process and that the facility will have the capacity to accept them when they arrive. Throughout the undertaking, it is necessary to maintain control over the schedule of both workers and equipment to insure that the remedial alternative is implemented economically.

NOTES

1. Amendment to RCRA, 14 July 1984; 40 CFR 264, 265, 18 July 1986.
2. U.S. EPA. *Test methods for evaluating solid waste—Physical/chemical methods.* WSW-84H. Apr. 1984. Method 1310.
3. 40 CFR 261.24.
4. See, for example, J. J. Keller and Associates, Inc. Hazardous waste services directory: Transporters, disposal sites, laboratories, consultants, specialized services. Neenah, Wis. Other references by J. J. Keller are: Identification, monitoring, treatment, disposal: Regulations, compliance, and management guidelines. Shipping, materials

handling, and transportation: Regulations, proposed changes, and reference data. State waste management programs: Generator, transporter, treatment, storage, and disposal standards.

5. 40 CFR 262.34. 1 July 1985.
6. CERCLA, Title II.
7. 49 CFR 178.115-178.120. 1 Nov. 1984.
8. 49 CFR, 40 CFR, generally. See also National Conference of State Legislators. Hazardous material transportation: A legislator's guide.
9. 40 CFR 123. 1 Nov. 1985.

Chapter 15
Implementing the Remedial Measures

There are many phases to the study of a hazardous waste site. Each study requires the talents of specialized technicians; each is costly; each is integral to the development of a defensible remedial alternative. Nonetheless, these studies are only preparatory to the actual remediation of the environmental dangers at a specific site. The purpose of this last chapter is to explain the administrative activities needed both to make the remediation of a hazardous waste site environmentally safe and to implement the remedial alternative in a thorough and successful manner. This chapter shows how the remedial alternative is executed.

A blueprint for the remediation is developed once the regulatory agency and responsible party (usually the owner of the site) have accepted a remedial alternative. To the casual observer, implementation of a remedial alternative appears straightforward, nearly identical to work that occurs on any construction site. However, it is qualitatively different. All aspects of the work—confidentiality, coordination, safety, environmental monitoring, and supervision of personnel—are planned deliberately and executed according to the letter of that plan by the responsible party's engineer or contractor. Most engineers and contractors familiar with construction sites and related work do not immediately appreciate the qualitative difference of the work of hazardous waste site remediation. Remediation of a hazardous waste site necessitates a discrimination in management not found on routine construction sites.

There are four major parties to the implementation of the remedial alternative:

- *The responsible party (or parties):* usually the owner or owners of the hazardous waste site; that is, corporations or individuals that bear the financial responsibility and legal liabilities of the site.
- *The regulatory agency:* in all cases of remedial actions under CERCLA, the U.S. Environmental Protection Agency. (Other agencies may be involved also, as discussed below.)

This chapter was developed by Terry L. Brown, P.E. of OBG Technical Services, Inc.

- *The engineer:* designer of the remedial alternative; not an individual but a proprietorship, partnership, or corporation.
- *The contractor:* again, a proprietorship, partnership, or corporation, a corporate structure being favored in most cases.

In addition, a state agency with responsibilities in environmental management or public health may be involved. Participation of states varies with each state's resources and interest; for example, the State of New York, through a memorandum of agreement with the U.S. EPA, participates thoroughly in the selection of the remedial alternative and its execution. Some states, on the other hand, are passive. In addition to the state, a municipal corporation (county, city, town, village, and so forth) may participate in the effort; municipalities usually act through the public health agency, the fire marshal, or an environmental management agency. Further, the well-being and opinion of neighboring communities are considered as the remedial alternative is implemented.

Work performed to remedy a hazardous waste site affects many sensitive public health and safety concerns. The contractor enters into an agreement with the responsible party limiting the disclosure of facts about the conditions of the remediation. A confidentiality agreement is a standard part of virtually every remediation contract, the reason for such an agreement being to protect the information of the responsible party. Confidential information is developed by the responsible party in the course of its study of the hazardous waste site through its legal counsel; this information is protected by attorney-client privilege, and the contractor, in order to perform its work, is privy to it. Other confidential information is information that could result in additional penalties or damages, and information related to trade secrets.

Certain responsible parties require blanket confidentiality agreements with their consultants. The following is an example of a comprehensive clause:

> Contractor shall not disclose the existence of this contract, any relationship with this Client [the responsible party], or any information provided to the Contractor in relation therewith.

Commonly, confidentiality agreements permit the contractor exceptions to the blanket prohibition against disclosure. Examples of exceptions are information that makes its way into the public domain through no fault of the contractor, information that the contractor must use to defend itself in a government proceeding, and information that the responsible party permits to be disclosed. (Although this last reservation appears obvious, it is important that the contract include it among the exceptions.)

All the employees of the contractor are bound by the confidentiality agreement. Therefore, any information about conditions at a hazardous waste site that is released to the public is firmly managed. A specific individual is assigned the task of dispensing information deemed material to the public. This precaution insures that no specious reports of events or conditions are broadcast. Extensive safeguards are exercised over the information disseminated because inaccurate information or negative and unsubstantiated claims could stop the project, with consequences to the project's schedule and budget and to the environment that probably would be detrimental.

PREPARATORY ACTIVITIES

In order to be implemented, the remedial alternative is developed into a remediation plan. Whereas the remedial alternative states goals and objectives, the remediation plan identifies the action to be taken on the site, action that will lead to achievement of the remedial alternative's goals. The ultimate responsibility for the development of the remediation plan lies with the responsible party. That same entity is answerable for the correct remediation of the site. If the remedial alternative is complex, the responsible party generally engages an engineer or a contractor to implement the remediation plan.

The provisions of the remedial alternative can be executed by several means:

- *Contract documents:* A promissory agreement is drawn up between the owner and a consultant that creates a legal relationship to effect the implementation of the remedial alternative. The remedial alternative, precise specifications for materials and standards, and administrative provisions—all written by the engineer—are generally contained in contract documents. These documents often are used for both competitive bidding and project control during the remediation effort.
- *Regulatory consent order:* An attorney of the regulatory agency may prepare the remediation plan and embody it in a formal document that is executed by both the responsible party and the regulatory agency. In most cases, the regulatory consent order describes only the major facets of the remediation plan and states performance standards. It is significantly less detailed than contract documents.
- *Cleanup plan:* The approach to remediation may be based upon an engineering and technical evaluation of a particular site. The document that incorporates this evaluation, perhaps written by the responsible party and approved by the regulatory agency, describes generally the activities

to be undertaken to complete the remediation. Such a cleanup plan would be cited in the agreement between the responsible party and the contractor.

In most circumstances, the owner holds a contract with a remediation contractor. The contractor may choose to perform all the work or may subcontract specific, specialized tasks to experts. There is no standard approach to the division of work between contractors and subcontractors; it is determined on a job-by-job basis. For an illustration of typical contract language, see Table 15-1.

TABLE 15-1. Representative contract specifications.

00814 WORK SCHEDULE
The Contractor shall prepare and submit to the Engineer for approval a network analysis of the construction progress schedule within ten (10) days after award of Contract. This method of analysis is generally referred to as the Critical Path Method (CPM).

This analysis shall include as a minimum a graphic representation of not less than 50 significant activities and events involved in the construction of the project, and a written statement explanatory thereof if necessary, for a complete understanding of the diagram.

The network graphic representation (Arrow Diagram) and statement must clearly depict and describe the sequence of activities planned by the Contractor, their interdependence and the times estimated to perform each activity.

In developing the project network, the Contractor shall use the following minimum information for each activity: the early activity start time, the latest activity start time, the earliest activity finish time, the latest activity finish time and the total float or slack. All winter-affected activities shall be noted.

All time shall be shown in "working days." The Contractor shall include with the initial schedule calendar dates for each activity.

The Contractor is required to obtain written approval of the Work Schedule from the Engineer prior to initiating work at the Contract Work Area and shall submit updates of the Work Schedule on a monthly basis to the Engineer.

00815 CONTRACT WORK AREA SECURITY
The Contractor shall be aware that existing chain link and snow fences are located within the Contract Work Area, as shown on the Contract Drawings. The Contractor shall remove the existing snow fences, shall relocate the existing chain link fences as shown on the Contract Drawings, and shall provide additional fence as required to enclose completely the Contract Work Area. The Contractor shall be responsible for all costs associated with relocating existing fences, providing additional fences, maintaining all fences during construction, and removing temporary fencing at the conclusion of construction.

00816 LINES, GRADES, AND ELEVATIONS
The Engineer shall set control lines and elevations to include a base line for the earthwork together with a suitable number of bench marks adjacent to the Contract Work Area. From the control lines and elevations, provided by the Engineer, the Contractor shall verify bench marks and develop and make all detail surveys needed for construction.

Because each waste site holds unique health hazards, both known and unknown, it is mandatory that the contractor implement remediation strictly in accordance with the specifications of the remedial alternative. Still, the contractor must retain the ability to react to unforeseen circumstances resourcefully. When the remediation is complete, it is the obligation of the responsible party to certify to the regulatory agency that the remediation has been completed satisfactorily. The certification of the responsible party usually is predicated upon the guarantee of the contractor's engineers, construction team, and scientists that the remedial effort is correct.

Codification of the Remedial Alternative. To restrict the jeopardy of the workers on the site and to guarantee that the known hazards at the site will be alleviated, the remediation plan is analyzed in detail. The first act of the contractor is to conduct a thorough and exhaustive evaluation of all the components of the remediation plan. Therefore, the contractor reviews or may develop a site operations and contingency plan, a sampling plan, a quality assurance plan, the site health and safety plan, and the engineering specifications. From that review, the contractor develops the operational components of the project: administrative procedures, the schedule for submittal of reports to the regulatory agency, the testing methodologies, the project organization, and the schedule of tasks.

The abstract remedial alternative and remediation plan thereby become concrete in the hands of the contractor. The contractor's process of synopsizing the remediation plan (the contract documents, consent order, or cleanup plan) and identifying all the specific activities required of the parties to the remediation is often termed *codifying.* During the contractor's codifying, all the requirements of the project are listed. The resulting compendium becomes the project schedule. The importance of a meticulous and deliberate codification of the remediation plan and of a comprehensive development of the project schedule cannot be overstated. If the schedule is deficient or contains oversights, the project can fall behind schedule or omit vital tasks. Any such errors will be costly and could turn a successful project into one with major problems.

Development of the Project Schedule. Because the contractor must anticipate the requirements of the project, it is able to incorporate all the necessary components into the work schedule from the beginning. Competent advance planning avoids unnecessary delays during execution of the project. As an aid to this planning, the contractor develops a summary sheet listing required actions. See Table 15-2 for an example of a contractor's planning sheet.

The summary sheet illustrated in Table 15-2 specifies not only the separate

TABLE 15-2. Planning review form, utilized during execution of remediation.

PRELIMINARY REMEDIAL MEASURES—ABC LANDFILL and ABC DUMP		*Legend:* A Action R Review			
REF.	ITEM	RESPONSIBILITY			
		Cntrt	Eng'r	EPA	Owner
BID Info Bid	B-1-1 Deliver executed contract within 15 days after award	A			
00202	Begin work 14 days after execution of contract	A			
00705	Notify intent to enter site 1 day in advance	A			
00702	Permits	A			
	Building Permits		A		
00718	Claims				
	File w/ engineer w/i 5 days of order or 5 days after begin work	A	R		
	Within 20 days detailed cost, and basis of claim	A	R		
00720	Safety and protection—appoint safety officer	A			
00725	Subcontractor statement 2 days prior to employment	A	R		
	Submit insurance evidence prior to entering site	A	R		
00735	Lump sum breakdown 10 days prior to employment	A	R		
00736	Dates of payment request		A		
	10 days prior to date, submit to engineer	A			
	10 days after receipt, advise contractor		A		
00738	Invoices	A	R		
	Evidence of payment within 60 days of receipt of payment	A			
00739	Payment within 30 days of payment request				A
00742	Substantial completion	A	A		A
00743	Final inspection	A	A		A

tasks but also the agent responsible for initiating each action and for reviewing and approving it. The contractor lists these allocated responsibilities in the schedule and thus apprises each party of its assignments. If shop drawings are to be reviewed by the engineer within ten working days, the contractor is obliged to furnish sufficient lead time within the schedule to permit the engineer to accommodate the task. It is incumbent, too, that the engineer then schedule the tasks so that a review can be made of the shop drawings within the ten days dictated by the contract.

The contractor codifies the remediation plan and develops the project schedule simultaneously. This schedule identifies the field and administra-

TABLE 15-2. *Continued*

PRELIMINARY REMEDIAL MEASURES—ABC LANDFILL and ABC DUMP		*Legend:* A Action R Review			
REF.	ITEM	RESPONSIBILITY			
		Cntrt	Eng'r	EPA	Owner
00744	Schedule of completion				
	Affidavit claims and liens	A			
	Written statement of surety	A			
00750	Authority to sign	A			
00752	Insurances: Proof of insurance prior to entering site	A			
	Copy of originals for which owner is named	A			
00754	Submission of performance bond same time as executed contract	A			
00756	Furnishing of contract documents 12 specs, 12 reduced, 12 full. 2 copies at site		A		
00757	Record drawings: 1 copy at site	A			
00810	Preconstruction meeting	A	A	A	A
00813	Right-of-way in advance of construction			A	A
00814	Work schedule prior to beginning work	A	R		
	Update every 14 days	A	R		
	Delays: notify engineer immediately in writing	A			
00815	Provide locks to work area: eight keys for each lock	A			
00818/ 01110	Grades		A		
00819	Cleaning				
	Location	A	R	R	
	Treatment of water location	A	R	R	
	Place solids at location within site	A	R		

tive activities in sequential order, and reflects the interdependencies of each. When developing the schedule, the contractor uses the planning review form (Table 15-2) and lists:

- Reference number: project-specific task number.
- Description of task or responsibility, for example "provide locks to work area."
- Responsible party.
- Start date.
- End date.

- Project items that precede each task or responsibility. For example, prior to entering the site, a project schedule must be submitted for review.
- Project items that follow each task or responsibility. For example, after the schedule is submitted, the contractor may enter the site and perform the remediation.
- Remarks about the possible critical nature of each task or potential slack time available.

All of this information is required for every task. When this list has been made, the interdependencies of each task become apparent, and the schedule can be formulated.

In developing the schedule, the contractor cannot permit concern with field activities to outweigh administrative tasks related to project submittals, organization, and timing. All field and administrative tasks are included on a single project schedule, and with good reason. Consider what would happen if, in the contract documents, the U.S. EPA had the duty to review and approve the liner system for a containment cell prior to installation—even though the particular liner system was specified in the contract documents—but, because of an oversight, that approval was not obtained. The U.S. EPA soon would become cognizant of the omission, and it would take corrective action. Not only would the contractor lose time, but the remediation project would become more costly and be placed in jeopardy.

A project scheduling information form is illustrated in Table 15-3. It displays a detailed listing of tasks with predecessors and successors, as well as the start dates and durations of the activities.

The information listed in Table 15-3 is used to develop a graphic schedule of the project. The critical path of this schedule is drawn to identify the tasks whose duration or other requirements set the minimum length of time required for the entire project. Figure 15-1 depicts a network diagram of a schedule. In this figure, observe that the critical path is described by tasks 1 through 9. It is these tasks that must be kept on schedule if the whole job is to be completed on time. Therefore, tasks 1 through 9 must be completed within 60 days. Tasks that proceed concurrently with the tasks on the critical path—but not synchronously—are:

- Erosion control, erection of a security fence, and removal of capacitors at a second site.
- Erection of a security fence at the primary site.

The timing for these concurrent tasks is:

- Duration of 60 days: tasks 1 through 9, the critical path.
- Duration of 21 days: tasks of erosion control, capacitor removal, and erection of the security fence at the second site.
- Duration of 35 days: erection of the security fence at the primary site.

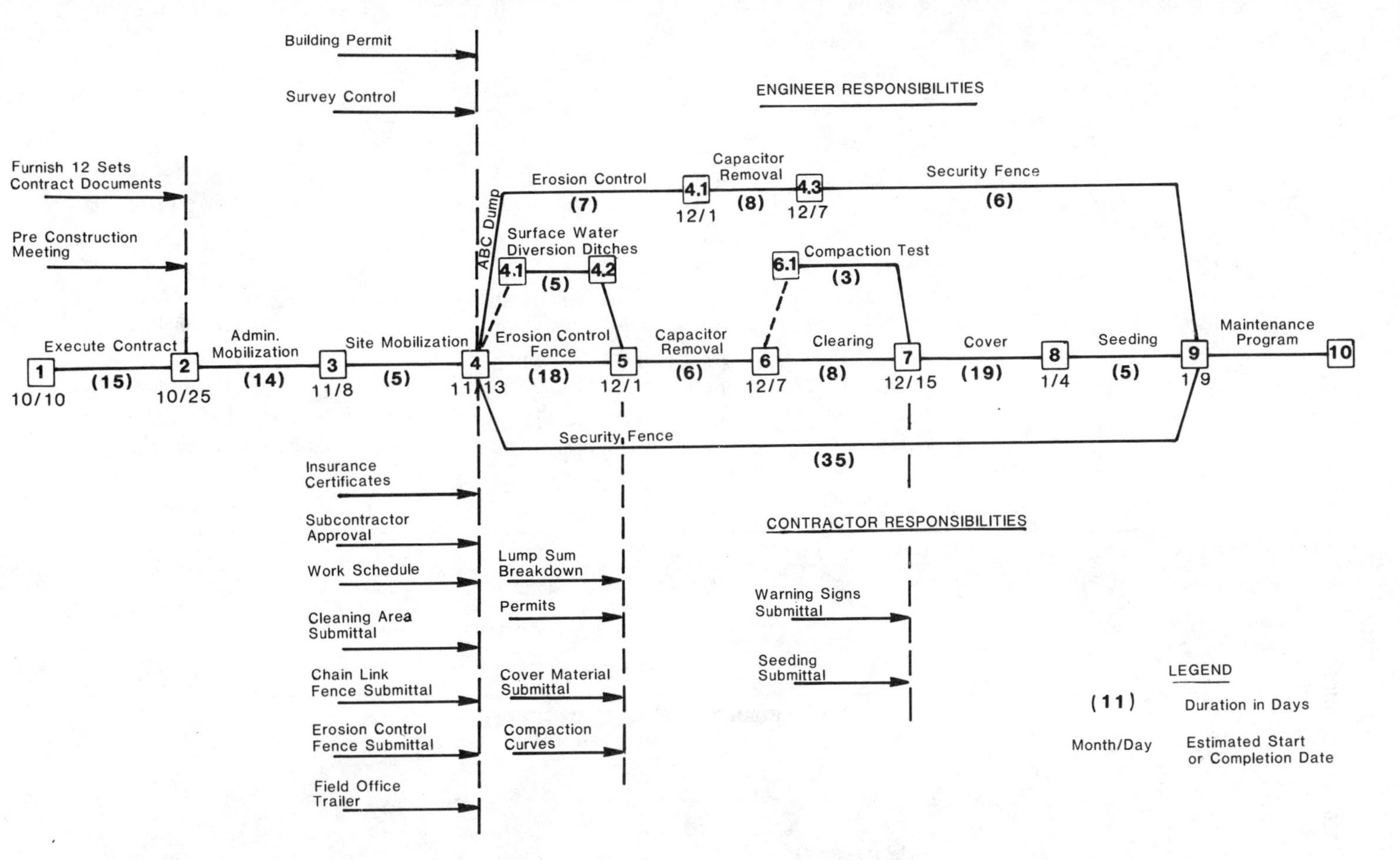

Figure 15-1. Critical path schedule for preliminary measures: ABC landfill and ABC dump.

TABLE 15-3. Project scheduling form.

SURFACE REMEDIATION—PROJECT SCHEDULE

PROJECT: ABC9A.ACT DATE: 09/09/88

#	ACTIVITY	DURATION (DAYS)	EARLY, LATE START	EARLY, LATE FINISH	SLACK AVAIL.	STATUS	DATE COMPLETED	ORIGINAL DATE ANTICIPATED
17	Compl Fence Reloc	5	06/17/1988 07/01/1988	06/21/1988 07/08/1988	10	Complete		6/21/88
18	Relocate Propane Pipe	3	06/11/1988 07/19/1988	06/13/1988 07/23/1988	27	Complete		6/13/88
19	Strip Top Soil	3	06/12/1988 07/15/1988	06/14/1988 07/17/1988	22	Complete		6/19/88
20	Monitoring Wells	15	10/01/1988 10/31/1988	10/21/1988 11/20/1988	22	25% Cmplt		8/01/88
21	SD Soils	1	06/06/1988 07/08/1988	06/06/1988 07/08/1988	21	Complete		6/11/88
22	Embankment	18	06/25/1988 07/25/1988	07/19/1988 08/19/1988	21	Complete	7/18/88	7/25/88

23	Clay Stockpile	12	06/24/1988 07/09/1988	07/10/1988 07/24/1988	10	Complete	7/29/88	7/10/88
24	Plce Bttm Clay	18	07/11/1988 07/25/1988	08/05/1988 08/19/1988	10	Complete	8/05/88	7/24/88
25	Lysimtrs Instal	4	06/17/1988 07/18/1988	06/20/1988 07/23/1988	22	Complete	7/11/88	6/25/88
26	Plce Bttm Liner	5	08/12/1988 08/20/1988	08/16/1988 08/26/1988	6	Complete	8/18/88	8/07/88
27	Const Temp Rds	3	06/17/1988 08/15/1988	06/19/1988 08/19/1988	42	Complete	7/26/88	6/14/88
28	Strt-up Trtmnt	3	06/20/1988 09/23/1988	06/24/1988 09/25/1988	65	Complete	7/22/88	6/24/88
29	Solidif Area	3	08/05/1988 08/20/1988	08/07/1988 08/22/1988	11	Complete	8/14/88	6/19/88
30	Dewater Ponds	40	08/05/1988 09/26/1988	09/30/1988 11/20/1988	37	New Unit On Site	Ongoing	7/16/88
31	Solidification	15	08/12/1988 08/23/1988	08/30/1988 09/13/1988	9	On Hold 09/10/88	Started 8/19/88	9/04/88
32	Silt Control	30	06/12/1988 10/10/1988	06/24/1988 11/20/1988	84	Ongoing		

There is no slack time for execution of the tasks on the critical path, by definition. For the concurrent tasks, there is slack time. Because the work at the alternate site should take only 21 days, there are 39 days of slack time at that site. For erection of the fence, there are 25 days of slack. Therefore, the contractor is free to allocate labor and equipment to these tasks as economically available. However, there is no leeway in the schedule of the tasks on the critical path—again, by definition.

The project schedule also can be displayed as a chart indicating the activities, their durations, the start and end dates, and the slack time available. This is sometimes known as a PERT chart, short for program evaluation review technique.

The project schedule is the cornerstone upon which the project's management rests. Considerable effort is spent in developing the schedule, and once it is in place, it is used continually. That is not to say that the schedule can be followed as first drafted. Given the unpredictable nature of hazardous waste sites, there are always exigencies to be met. The schedule is continually updated, revised, and improved; when an activity such as the submission of shop drawings, reporting deadlines, equipment purchases, or field programs must be delayed or advanced, the schedule is modified to reflect the changes. As the project proceeds, the current (revised) schedule is compared with the original one. If significant delays are apparent, action can be taken to accelerate the project to insure that the remediation will terminate on time. Because schedules often are revised, it is useful to have a computer maintain the information. A number of software packages with competent scheduling abilities and the capacity to draw the critical path are available, even for microcomputers. The schedule in Figure 15-2 is an example of a PERT chart produced by a microcomputer. A properly constructed, well-maintained schedule will save project costs, enhance efficiency, and expedite project administration.

Collection of Baseline Data. Before any work begins on the site, it is important that the contractor collect and record data describing the site and the physical condition of the people who will work there. This is done for the contractor's protection. These data are the baseline against which future data will be compared. Typical categories of data collected about hazards were given in Table 6-1. In addition, the characteristics of the ambient air and water are essential information because they are the background against which future samples will be compared. The central environmental factors gauged are:

- *Soil:* The area where the ground's surface has been contaminated is delimited before workers are allowed on the site. During remediation, dust

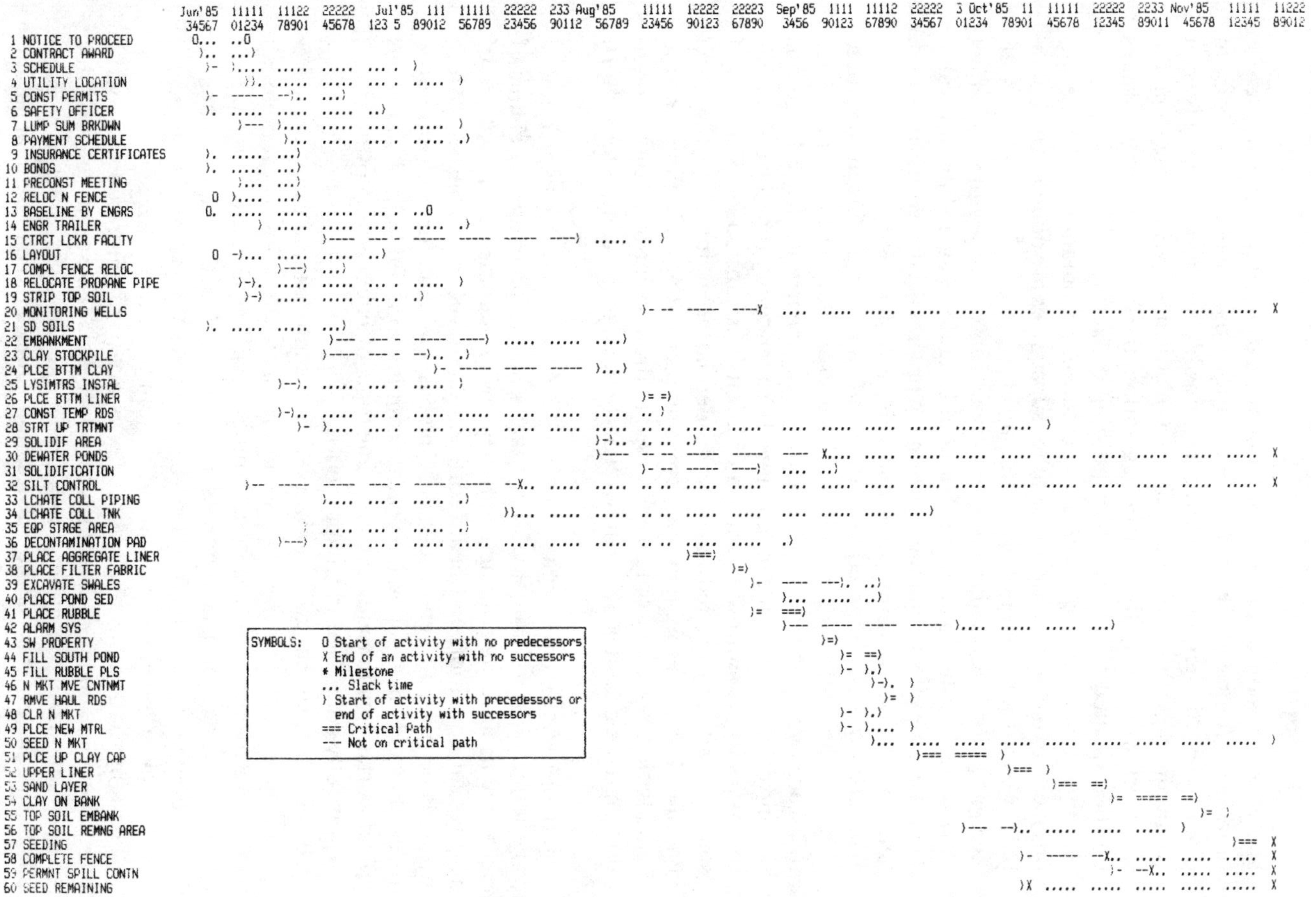

```
                          Jun'85 11111 11122 22222  Jul'85 111 11111 22222 233 Aug'85  11111 12222 22223 Sep'85 1111 11112 22222 3 Oct'85 11 11111 22222 2233 Nov'85 11111 11222
                          34567  01234 78901 45678  123 5 89012 56789 23456 90112 56789 23456 90123 67890 3456 90123 67890 34567 01234 78901 45678 12345 89011 45678 12345 89012
 1 NOTICE TO PROCEED      0...  ..0
 2 CONTRACT AWARD         )..   ...)
 3 SCHEDULE               )-  ).... ..... ..... ... .  )
 4 UTILITY LOCATION             )).  ..... ..... ... . ..... )
 5 CONST PERMITS          )-  ----- --).. ...)
 6 SAFETY OFFICER         ).  ..... ..... ..... ..)
 7 LUMP SUM BRKDWN            )--- )....  ..... ... . ..... )
 8 PAYMENT SCHEDULE                )....  ..... ... . ..... .)
 9 INSURANCE CERTIFICATES ).  ..... ...)
10 BONDS                  ).  ..... ...)
11 PRECONST MEETING             )....  ...)
12 RELOC N FENCE          0 )....  ...)
13 BASELINE BY ENGRS      0.  ..... ..... ..... ... . ..0
14 ENGR TRAILER                 )  ..... ..... ... . ..... .)
15 CTRCT LCKR FACLTY                  )---- --- - ----- ----- ----- ----) ..... .. )
16 LAYOUT                 0 -)... ..... ..... ..)
17 COMPL FENCE RELOC             )---)  ...)
18 RELOCATE PROPANE PIPE       )-). ..... ..... ... . ..... )
19 STRIP TOP SOIL              )-) ..... ..... ... . .)
20 MONITORING WELLS                                                   )- -- ----- ----X .... ..... ..... ..... ..... ..... ..... ..... ..... ..... ..... X
21 SD SOILS               ).  ..... ..... ...)
22 EMBANKMENT                         )--- --- - ----- ----) ..... ..... ....)
23 CLAY STOCKPILE                     )---- --- - --).. .)
24 PLCE BTTM CLAY                                  )- ----- ----- ----- )...)
25 LYSIMTRS INSTAL               )--). ..... ... . ..... )
26 PLCE BTTM LINER                                                    )= =)
27 CONST TEMP RDS                )-).. ..... ... . ..... ..... ..... ..... ..... .. )
28 STRT UP TRTMNT                  )- ).... ... . ..... ..... ..... ..... ..... .. .. ..... ..... .... ..... ..... ..... ..... ..... )
29 SOLIDIF AREA                                                  )-)... .. .. .)
30 DEWATER PONDS                                                 )---- -- -- ----- ----- ---- X.... ..... ..... ..... ..... ..... ..... ..... ..... ..... X
31 SOLIDIFICATION                                                     )- -- ----- ----) .... ..)
32 SILT CONTROL                )-- ----- ----- --- - ----- ----- --X.. ..... ..... .. .. ..... ..... .... ..... ..... ..... ..... ..... ..... ..... ..... ..... X
33 LCHATE COLL PIPING                 ).... ... . ..... .)
34 LCHATE COLL TNK                                  )).... ..... ..... .. .. ..... ..... .... ..... ..... ...)
35 EQP STRGE AREA                     ) ..... ... . ..... .)
36 DECONTAMINATION PAD           )---) ..... ... . ..... ..... ..... ..... ..... .. .. ..... ..... .)
37 PLACE AGGREGATE LINER                                                    )===)
38 PLACE FILTER FABRIC                                                            )=)
39 EXCAVATE SWALES                                                                  )- ---- ---). ..)
40 PLACE POND SED                                                                      ).... ..... ..)
41 PLACE RUBBLE                                                                     )= ===)
42 ALARM SYS                                                                           )--- ----- ----- ----- ).... ..... ..... ...)
43 SW PROPERTY                                                                              )=)
44 FILL SOUTH POND                                                                            )= ==)
45 FILL RUBBLE PLS                                                                            )- ).)
46 N MKT MVE CNTNMT                                                                              )-). )
47 RMVE HAUL RDS                                                                                   )= )
48 CLR N MKT                                                                                  )- ).)
49 PLCE NEW MTRL                                                                              )- ).... )
50 SEED N MKT                                                                                    )... ..... ..... ..... ..... ..... ..... ..... ..... )
51 PLCE UP CLAY CAP                                                                                  )=== ===== )
52 UPPER LINER                                                                                                 )=== )
53 SAND LAYER                                                                                                      )=== ==)
54 CLAY ON BANK                                                                                                        )= ===== ==)
55 TOP SOIL EMBANK                                                                                                               )= )
56 TOP SOIL REMNG AREA                                                                                 )--- --).. ..... ..... ..... )
57 SEEDING                                                                                                                          )=== X
58 COMPLETE FENCE                                                                                          )- ----- --X.. ..... ..... ..... X
59 PERMNT SPILL CONTN                                                                                              )- --X.. ..... ..... X
60 SEED REMAINING                                                                                          )X ..... ..... ..... ..... ..... X

SYMBOLS:  0 Start of activity with no predecessors
          X End of an activity with no successors
          * Milestone
          ... Slack time
          ) Start of activity with precedessors or
            end of activity with successors
          === Critical Path
          --- Not on critical path
```

Figure 15-2. Project schedule chart.

and water erosion can cause contamination to become concentrated or broadcast. To provide the contractor with the baseline data necessary to determine whether its work was responsible for modifying levels of contamination, the levels are measured in advance. (Where the remediation work will disturb the surface vegetation, the contractor compensates for the expected effect prior to disturbing the surface.)

- *Presence or absence of explosive conditions:* Contaminated materials often can explode if water, air, or other substances are added. Any potential for explosion is identified at the outset.
- *Average temperatures and temperature extremes:* Temperature can affect worker health, the volatilization of liquids, and the availability of water. Therefore, the effect of temperature on the progress of the work is reviewed, and any adverse conditions that may arise are identified.
- *Wind direction:* The direction of the wind can have significant ramifications for the success of a project. If, for example, the prevailing winds travel across neighboring residences, they might carry smells or dangerous organic compounds to the inhabitants. It is mandatory that the contractor find ways to prevent the migration of hazardous substances. However, if smells have been carried into the residential areas for some time, the baseline data can prove that they represent preexisting conditions, and the remediation work may only have served to draw attention to those conditions and to encourage some individuals to impugn the contractor's effort.

All right-to-know laws, regulations of OSHA, and labor laws are followed to the letter, both to protect workers' safety and to preempt any claims of negligence against the contractor after the work is complete. The status of the physical condition and health of each employee is established before anyone is permitted to enter the site. The baseline data for each employee are established in a general physical examination. Table 15-4 illustrates typical worker health requirements. (The topic of safety is not covered here, as it was fully addressed in Chapter 6.)

Certain physical limitations of workers may exclude them from consideration for activity on the site. At one hazardous waste site, for example, all workers had to carry a respirator at all times because the nature of the wastes was such that dangerous conditions could be produced abruptly; when an alarm sounded, respirators had to be put on. Anyone unable to use a respirator clearly had to be disqualified from working on the site.

Once remediation has begun, it is necessary to monitor continually the environmental parameters and the health of workers. Those data are recorded and maintained. All parameters monitored at the outset are checked periodically during the progress of the remediation, although the individual parameters are monitored at different frequencies.

TABLE 15-4. Employee physical requirements.

The Contractor shall utilize the services of an occupational physician to provide the minimum medical examinations and surveillance specified herein.

Medical surveillance protocol is the physician's responsibility but shall meet the requirements of OSHA standard 29 CFR Part 1910, Subpart A "Occupational Health and Environmental Control" for all personnel. In addition, the following protocol is suggested:

Baseline Examination: medical history; general physical:

a. Complete blood profile
b. Blood chemistry consisting of: chloride, CO_2, potassium, sodium, BUN, glucose, globulin, total protein, albumin, calcium, cholesterol, alkaline phosphatase, triglyceride, uric acid, creatinine, total bilirubin, phosphorous, lactic dehydrogenase, SGPT, hepatitis antigen, and 5′ nucleotidase.
c. Urine analysis.
d. Electrocardiogram.
e. Chest X-ray if not taken within the last year.
f. Pulmonary function.

All Contractor's on-site personnel involved in this project shall be provided with medical surveillance prior to onset of work, at the conclusion of project, and at any time there is suspected excessive exposure to toxic chemicals or physical agents.

The preservation of and access to all employee medical surveillance records shall be in conformance with OSHA Standard 29 CFR 1910.20.

The ability of on-site personnel to wear respiratory protection shall be guaranteed by the Contractor.

All health and safety protocols shall be in conformance with OSHA Standards 29 CFR 1910 and 29 CFR 1926.

Project Organization. The organization of the parties to a remediation project is a fundamental concern and is critical to the progress of the work. No site undergoing remediation is typical, and none proceeds according to the remediation plan without deviation. As the work progresses, additional and previously unknown hazards are exposed which were not addressed in the remediation plan. Under such circumstances, the plan of work has to be changed straightaway. Representatives of the parties of interest who have decision-making authority must be available and willing to make the necessary decisions expeditiously; and, so that they can reach a consensus, the lines of authority and communication must have been specified in advance.

Two types of organization are delineated and announced prior to the start of work: the internal organization of each party and the way the parties interact. Further, within each party, two components are defined:

- The degree of authority each individual has.
- The availability of each key decision-maker at all times.

As the work on the site progresses, it is important that supervision be provided by individuals with decision-making authority or that responsible individuals be available at all times. When potentially injurious but previously unsuspected conditions are discovered, the decision-makers meet and determine a new course of action. If accountable people cannot be reached for consultation, the environment, worker safety, and the success of the remediation itself could be jeopardized. To guarantee that decision-makers stay abreast of the remediation work and thus are able to act decisively when necessary, they are required to attend progress meetings at the site at least monthly, and they routinely receive copies of all correspondence related to the project.

Figure 15-3 shows types of organizations that could be structured. In every case, the organization is designed to maximize communication among decision-makers. In Figure 15-3(a), a traditional organization is displayed. The responsible party communicates with the regulatory agency and directs both the contractor and the engineer. The contractor and the engineer maintain an informal communication link, but they do not supervise each other's work except to the extent directed by the responsible party. When the project organization is turnkey, as displayed in Figure 15-3(b), the contractor supervises the engineer. The choice of organization depends upon the preference of the responsible party, which in every case interacts with the regulatory agency.

Adequate project organization is critical to facilitate communication and coordination; so after the project schedule and its particulars have been developed, a preconstruction meeting is conducted. This meeting, arranged and directed by the contractor, is held at the hazardous waste site, and the principal parties to the remediation effort attend: for the contractor, the construction superintendent; for the engineer, the project engineer—usually the person who supervised development of the remediation plan; for the U.S. EPA, the official who will monitor the work; and, for the responsible party, a representative of its interests. These representatives should have decision-making authority for their organizations in order to establish the basic arrangements for executing the work. Others interested in the remediation effort usually also attend, including concerned members of the public and news reporters. The purpose of the preconstruction meeting is to raise and answer all questions about the execution of the remediation plan, as well as to achieve a basic understanding of the goals of the remedial alternative. The organization of the parties to the project is delineated at this meeting, and the decision-making authority is clarified and acknowledged. The network

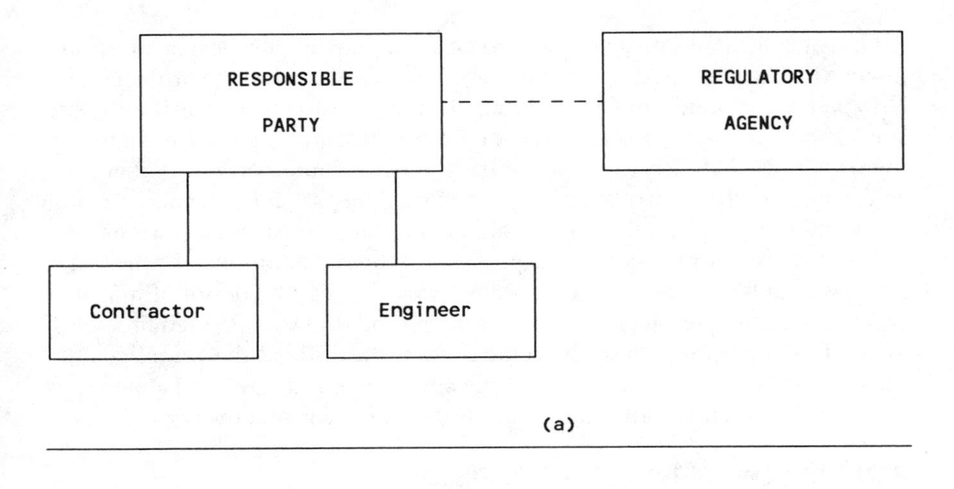

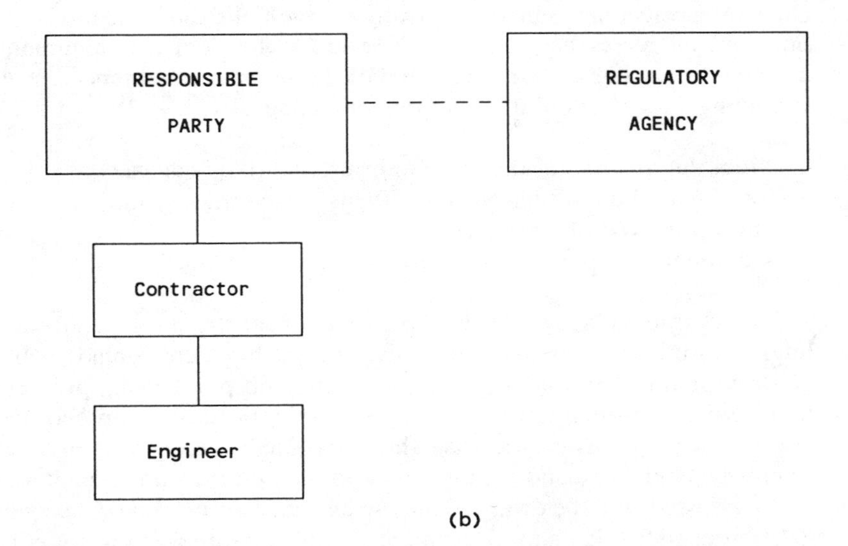

Figure 15-3. Prototype organizations for hazardous waste site remediation.

of information and communication specified at this meeting will be utilized throughout the project's implementation. At this meeting, too, the parties analyze the project requirements and examine the contractor's schedule for meeting those requirements. The agreements reached and the issues settled at this meeting are recorded carefully; the minutes of the meeting are transmitted to all parties for their review and concurrence or amendment.

On a traditional construction project, changes in the design often are made with the concurrence of only a few individuals within the project team. However, the ground rules are different for remediation of a hazardous waste site. If the contractor wishes to modify the remediation plan, the engineer, the owner, the U.S. EPA, and often the public become party to that action.

Throughout the term of the project, the principal parties confer periodically. These meetings are used to clarify all outstanding issues and questions, and they permit regular reviews of the progress of the remediation. The meetings are essential for the contractor because they provide the depth of information necessary to interpret the engineer's design and the U.S. EPA's requirements. When held near the site of the remediation, they afford the principals the opportunity to observe the work at the site. The minutes of these meetings also are documented, and they become part of the project record.

REMEDIATION ACTIVITIES IN THE FIELD

Once the project has been exhaustively planned, the schedule has been laid out, and all issues have been addressed by the contract documents or consent order, the actual remediation of the site may commence. From the beginning, it is essential that the project manager:

- Direct the progress of the work to insure that it advances correctly.
- Coordinate the multifarious individuals, corporations, and other organizations involved or interested in the work.
- Document all aspects of the work.

The contractor follows the basic practices of construction management in implementing the remedial alternative. These practices include schedule control, quality control, competitive pricing and purchasing, project cost monitoring, and manpower allocation. All administrative procedures are performed assiduously; otherwise, the years of painstaking work of the many scientists, engineers, and technicians who devised the remedial alternative may be wasted, and the owner of the site may face devastating consequences in heightened liability and government penalties. Moreover, the remediation of the site might fail, and the environmental damage would not be mitigated; so the site would require remediation again at some future time.

Besides implementing the remedial alternative properly, the contractor intends to make a profit on the work. There is not necessarily a conflict between the goals of scrupulous execution of the remedial alternative and profit making. A competently administered job should return a profit. On the other hand, a financial loss could cause the contractor to disregard the specifications of the project, and might jeopardize the project's success.

Monitoring Progress. Nearly every construction firm has a unique way of monitoring a project's performance, a method developed to accommodate its organizational needs. Therefore, no generalizations can be offered here. At a hazardous waste site, it is especially important to maintain firm control over the work, by any effective means. Such control is necessary because not only must the project proceed according to the accepted plan, but problems must be viewed with as much foresight as possible in order that positive solutions can be prepared. The monitoring of a hazardous waste site project includes quality control.

Throughout the course of a remediation project, it is the role of the manager to maintain high standards for the work. If quality is not controlled, tasks may have to be repeated, with both time and money lost. Remediation of hazardous waste sites incorporates assorted complex assignments, and quality control is specific to the site. Although no comprehensive list of quality control measures can be given for necessarily unique circumstances, the tests routinely performed are listed in Table 15-5. Whatever quality control system is developed, it should be sufficiently comprehensive to address every facet of the work on the hazardous waste site.

The quality control data are recorded on a specifically designed form; see Figure 15-4 for an example. This form then becomes the basis for the documentation of quality control efforts; they are an important component in the comprehensive documentation of the entire remediation project. In order to insure that the quality control component of tasks is not missed, it is highlighted on the project schedule, and persons responsible for the quality control are advised of it in advance to permit them to organize their data

TABLE 15-5. Routine quality control tests executed during remediation.

CONTAINMENT CELLS

- Gradation curves, compaction tests, optimum moisture analysis, in-place density tests.
- Synthetic liner manufacturing mill reports, seam tests performed in the field, leak tests.

STORAGE TANKS

- Concrete mix design test reports, batch plant test reports, field sample test reports.
- Pressure test on all influent and effluent lines, pressure test on tank, manufacturing mill certificates, coating soundings.
- Reinforcing steel mill certificates.

SITE CLEANUP

- Issuance and administration of manifests.
- Air quality and water quality monitoring.
- Soil analysis, surface wipe tests.

QUALITY CONTROL DATA

O = Owner E = Engineer C = Contractor A = Agency

Ref #	Description of Contract Requirement	Action	1st Review	2nd Review	3rd Review
00213	Notify owner intent to enter site 5 days	C			
	in advance.				
00702	Permits: Hauling and Borrow.	C	E	O	
00705	Commence 10 days after notice to pro-	C			
	ceed is issued.				
	Notify owner 4 days in advance.	C			
00714	Competent superintendent on-site.	C			
00720	Designation of Safety Officer.	C			
00748	Final payment affidavit.	C	E	O	
00725	Subcontractor Insurance Certificates.	C	E	O	
00735	Lump Sum Breakdown 10 days prior to	C	E	O	
	first payment.				
00752	Insurance Submissions.	C	E	O	
00756	Contract Documents.	E			
00757	Record Drawings.	C	E	O	
00811	Pre-Construction meeting.	O,E,C,A			
00814	Work schedule submitted within 10 days	C	E	O	A
	after contract award.				
00815	Contract work area security.	C			
00816	Lines, grades and elevations.				
00820	Contingency plan.	C	E	O	A
00822	Medical surveillance.	C	E	O	A
00823	Runoff control and treatment:				
	Control	C	E	O	
	Treatment	C	E	O	

Figure 15-4. Quality control data form.

collection in a timely manner. With such advance notice, too, no quality control items are omitted.

Progress Payments. One effective way for the owner of a site undergoing remediation to manage a project is to control the manner in which progress payments are made to the contractor. The payment method is determined in advance of any remediation work and should be equitable. A number of factors are recognized when the method is being negotiated:

- The owner may withhold a retainage to insure that the contractor will conclude the contract in a manner acceptable to all parties, especially the U.S. EPA.
- The owner makes certain that the contractor receives prompt payment for activities completed and that the money received represents fair value for the services performed. The contractor will have diverse financial demands to respond to and should not be expected to finance the work, unless that was a qualification in the prime contract.

The simplest arrangement of the payment schedule, when applicable, is compensation on a unit price basis. The contractor is paid for the number of units of work completed. For example, for every 1000 gal treated, the contractor would get \$100 (\$260/m^3); for installation of a clay liner, payment could be \$15 for every square foot (\$160/m^2); for restoration work, the rate might be \$2/ft^2 (\$21.50/m^2). Table 15-6 explains how costs for a portion of a remediation job were billed.

TABLE 15-6. Example: costs for tank cleaning and disposal.

Cost of Subcontractor A	\$ 271.00
Costs of Subcontractor B	
1. Excavation of Tank 27i	8,000.00
2. Remove contaminated soil (132 yd^3 × \$2.40/yd^3)	316.80
3. Transport and dispose soil (132 yd^3 × \$25/yd^3)	3,300.00
4. Concrete (18 yd^3 × \$90/yd^3)	1,620.00
5. Backfill material (188 yd^3 × \$8.50/yd^3)	1,598.00
6. Removal of tank in sidewalk; excavation and disposal of contaminated materials; backfilling and reinstallation of concrete sidewalk	2,799.82
Subtotal	\$17,905.62
10% Overhead and profit	1,790.56
TOTAL	\$19,696.18

Alternatively, if the project is arranged on a lump sum basis, whereby the contractor is paid a single price for the entire project regardless of the amount of work involved, a graduated schedule of payments can be established. With such an arrangement, the contractor is paid a proportion of the total price, less the retainage, at the end of each time period.

A third payment schedule could be a combination of lump sum and unit price. The contractor, in preparing the bid, sets a value on each of the tasks that comprise the project. As each task is completed according to the schedule, the contractor receives payment for that task, less the retainage. Both the engineer and the contractor independently estimate the progress payment schedule; they then negotiate to resolve any differences. Once accepted, it becomes established for the duration of the work. Determining the payment amounts is rather more complicated under this arrangement, but it is perhaps the most equitable method.

The responsible party, contractor, or engineer monitors the cash flow of a remediation project. A graph is drawn depicting the anticipated cash flow based upon budget estimates: money expended versus time. On that same graph, the actual cash flow is charted. See Figure 15-5. Having information about expenditures in visual form, as in Figure 15-5(a), permits the engineer to evaluate quickly the accuracy of the budget estimates and the actual progress of the work. The graph in Figure 15-5(b) indicates that significantly less money is being expended than was anticipated. Such a graph could be

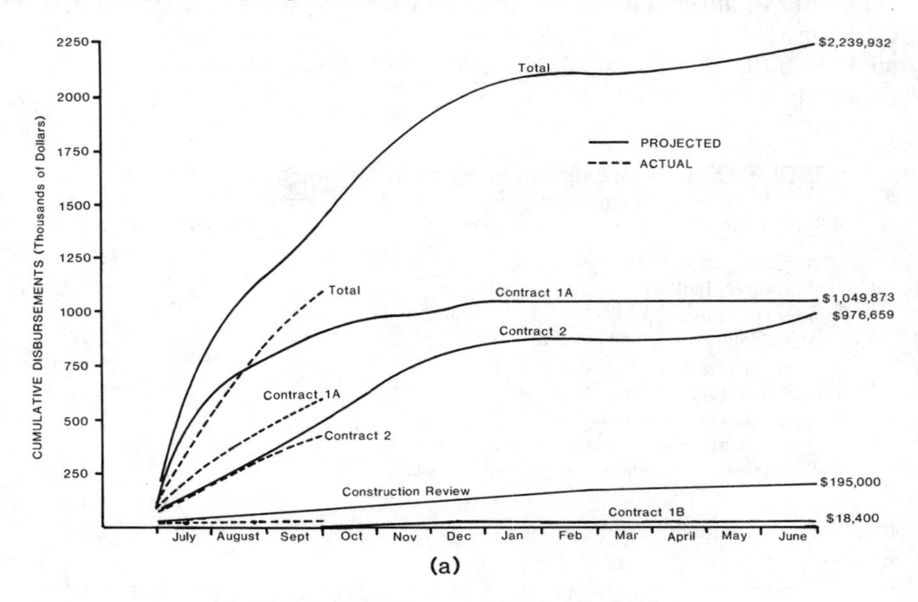

Figure 15-5. Cash flow curves for remediation project.

indicating that the contractor is behind schedule, and the responsible party would have to take steps to spur action.

Dialogue with Regulatory Agencies. Every aspect of the entire remediation —designation, initiation, planning, implementation, review—is executed according to government regulations. These regulations are presented in black and white in the code books, but shades of gray arise in the field when the regulations are applied. One function of the government official responsible for the remediation is to interpret specific requirements and to explain further the dictates of the regulations as applied to the specific job. Therefore, deliberate communication between, on the one hand, the contractor, the engineer, and the responsible party and, on the other, the U.S. EPA is maintained throughout the work effort. A single individual in the project organization, usually from the responsible party's camp, is authorized to maintain contact with the regulatory agency. That person's task is to work with the regulators to insure that no misunderstandings occur among the parties. The remediation effort would be scuttled if the various parties reached different conclusions concerning the same objective.

For example, on one project, the object was to remove a waste oil tank that contained some PCBs. The capacity of the tank was 150,000 gal (568 m^3); and it contained 15,000 gal (57 m^3) of oil, 25,000 gal (95 m^3) of water, and 47,000 gal (178 m^3) of sludge. Analysis of the contents of the tank indicated

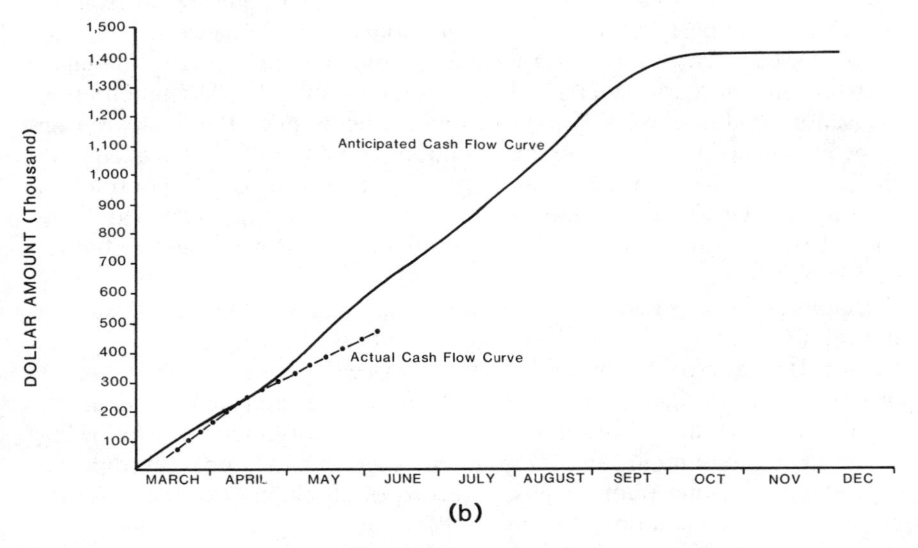

Figure 15-5. *(continued).*

that all levels contained PCBs in concentrations less than 20 ppm except for the bottom of the sludge; that layer contained PCBs with concentrations greater than 1000 ppm, the action limit according to federal regulations. A strict interpretation of the regulations would have dictated that all the contents of the tank be handled as if they contained PCBs, and the disposal would have cost in excess of $500,000. However, the fundamental environmental issue was proper disposal of the PCB-contaminated sludge layer. The other contents of the tank did not warrant the specialized handling appropriate to the contaminant. After negotiations extending over months, a solution acceptable to all parties was reached. The remediation of the contents of the tank was carried out at a substantially lower cost, yet the environmental problem was resolved satisfactorily. At times such as this, the engineer or contractor may require the technical services of a financial expert to assess the costs of remedial objectives. Information about the economics of the work is essential to insure that all money is expended in a prudent manner and that the effort addresses legitimate environmental concerns.

Documentation. Evidence of the remediation of a hazardous waste site is mandatory. Government agencies require substantive proof that the remediation was carried out in accord with the applicable control documents—contract, regulatory consent order, or approved cleanup plan. In addition, the corporation responsible for the site demands records describing the remediation effort in order to limit its future liability. All parties active on the site may be liable for the environmental hazards that could result from it; therefore, each party requires documentation specifying its work.

All of these records are necessary, and it is the task of the project manager to determine the amount of record keeping that is sufficient. Documentation costs money; although this is no place for false economies, there is no reason to be profligate. To determine the level of documentation that is adequate, the project manager considers the sensitivity of the project, the contractual relationship between the responsible party and the contractor, and the special need to develop cost information for future work and potential disputes in cost.

Documentation is necessary for several reasons. Its primary purpose is to permit investigators to reestablish the level of effort that was expended on the site. It also certifies how and when the work was performed, and whether or not the work was done in accord with the established remediation requirements. When a hazardous waste site is corrected, the following records, at a minimum, are maintained: project diaries or daily reports; reproducible 35-mm photographs; a record of all changes in the contract documents or remediation plan; as-built drawings, if applicable; and quality control reports.

A diary records daily events such as conversations, site visits, daily activities, and statistics on the labor and equipment resources expended. Daily reports are taken from the diaries and summarize these data. If more than one inspector is active on the remediation site, these data are taken from the diaries of each. The daily reports may be formally transmitted to the site's responsible party. Both the notes from a diary and the report derived from them are presented in Figure 15-6(a, b).

Prior to the start of the project, photographs are taken of the site to record its condition before remediation. During remediation, photographs are taken to document every phase and aspect of the work. Video recording equipment is especially useful in this regard. Additional data and documentation may be required if the project poses special concern over potential contractual claims, if cost information is especially important, or if the public insists upon it.

SUMMARY

Remediation of hazardous waste sites challenges engineers and contractors to implement the remedial alternative with discriminating and meticulous standards. The remedial alternative itself is the product of painstaking work on the part of specialized experts; it is a customized solution to a particular hazardous waste problem. In order for that alternative to be effective, it requires exact implementation.

The party responsible for the site engages a qualified contractor to implement the remedial alternative. The contractor works closely with the responsible party, with the technicians who developed the remedial alternative, with the regulatory agency, and with other parties of interest such as communities in the environs of the site. The contractor interprets the remedial alternative in the form of a remediation plan that can be implemented. All the details of the implementation are elucidated in the remediation plan, and that plan becomes part of a binding agreement, either the contract documents, the consent agreement, or the accepted cleanup plan.

In order to begin work on the hazardous waste site, the contractor is required to operationalize the remediation plan. Generally, the plan is expanded into a precise listing of tasks and responsibilities; next, a detailed project schedule is formulated with the critical path of tasks displayed. The project schedule becomes the manual for the remediation effort. The schedule is referenced continually, and it is updated whenever circumstances change. The schedule is the central guidance and management tool for the remediation effort because it contains not only the general requirements of government regulations but also the specific conditions of the job at hand. These stipulations are combined on the schedule with the exigencies of

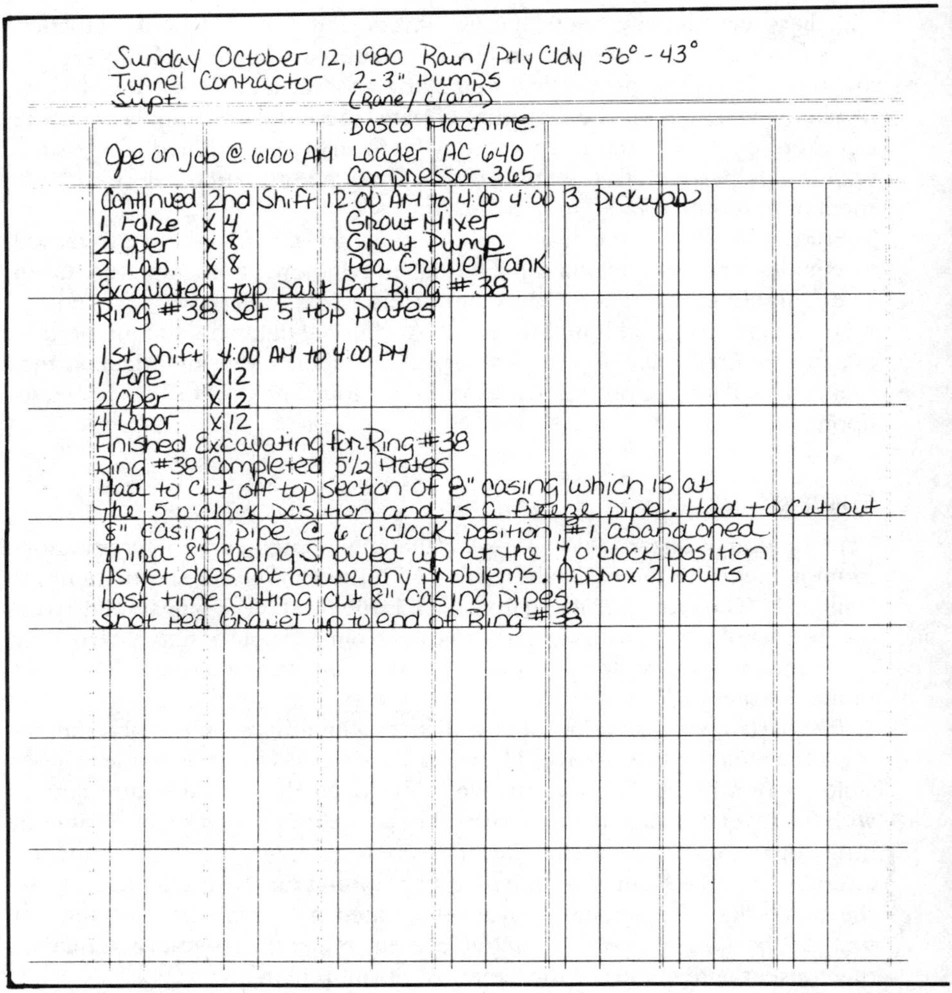

Sunday October 12, 1980 Rain / Ptly Cldy 56° - 43°
Tunnel Contractor 2-3" Pumps
Supt. (Rane / clam)
Dosco Machine
Joe on job @ 6:00 AM Loader AC 640
Compressor 365
Continued 2nd Shift 12:00 AM to 4:00 4:00 3 Pickups
1 Fore X 4 Grout Mixer
2 Oper X 8 Grout Pump
2 Lab. X 8 Pea Gravel Tank
Excavated top part for Ring #38
Ring #38 Set 5 top plates
1st Shift 4:00 AM to 4:00 PM
1 Fore X 12
2 Oper X 12
4 Labor X 12
Finished Excavating for Ring #38
Ring #38 Completed 5½ Plates
Had to cut off top section of 8" casing which is at
the 5 o'clock position and is a freeze pipe. Had to cut out
8" casing pipe @ 6 o'clock position, #1 abandoned
third 8" casing showed up at the 7 o'clock position
As yet does not cause any problems. Approx 2 hours
Lost time cutting cut 8" casing pipes,
Shot Pea Gravel up to end of Ring #38

Figure 15-6. Project diary.

time. The contractor is continually aware of the time required for each phase of the project because the contract designates the time frame within which the work must be completed, and the profitability of the job is tied to its timely completion.

As the remediation proceeds, there are many claims on the attention of the contractor. There are always hazards that must be dealt with or antici-

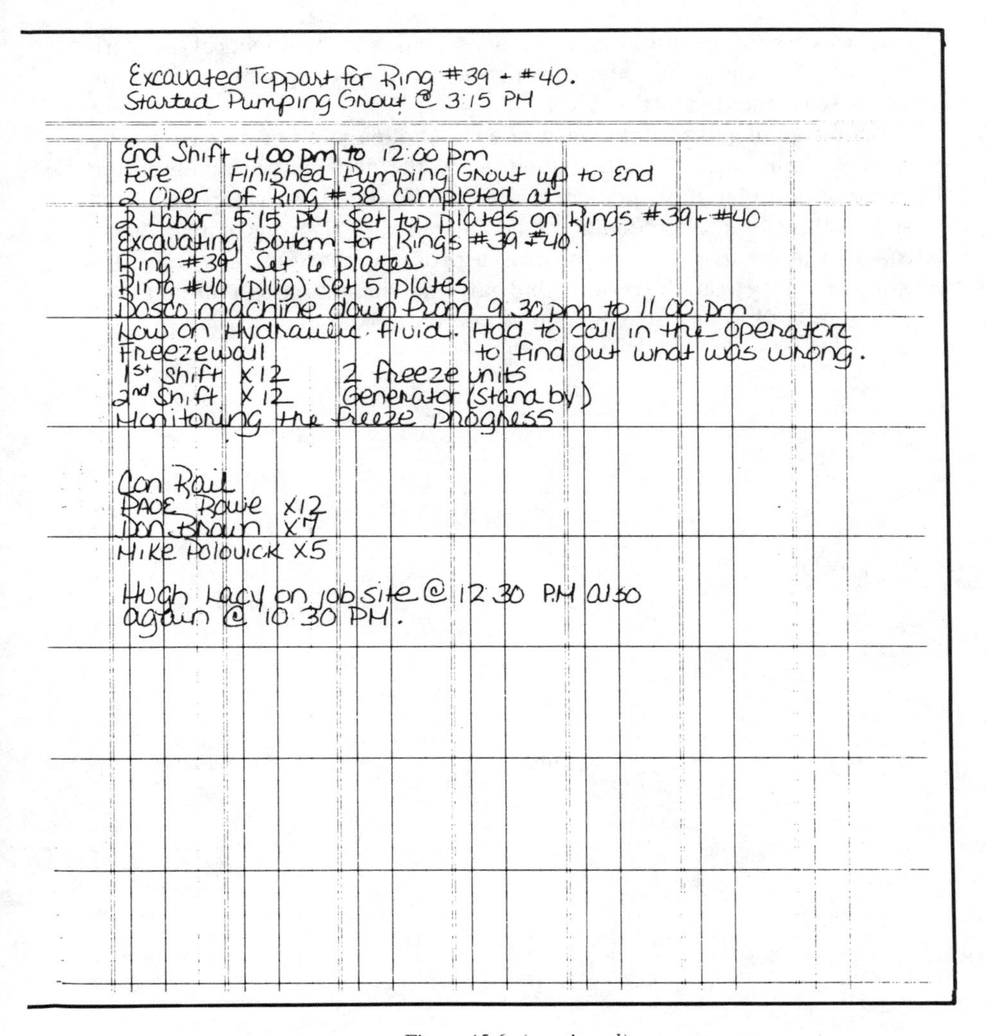

Excavated Toppart for Ring #39 + #40.
Started Pumping Grout @ 3:15 PM

End Shift 4 00 pm to 12:00 pm
Fore Finished Pumping Grout up to End
2 Oper of Ring #38 completed at
2 Labor 5:15 PM. Set top plates on Rings #39 + #40
Excavating bottom for Rings #39 #40.
Ring #39 Set 6 plates
Ring #40 (plug) Set 5 plates
Dosco machine down from 9 30 pm to 11 00 pm
Low on Hydraulic fluid. Had to call in the operator
Freezewall to find out what was wrong.
1st Shift X12 2 freeze units
2nd Shift X 12 Generator (stand by)
Monitoring the freeze progress

Con Rail
PAOE Rowe X12
Don Brown X7
Mike Polovick X5

Hugh Lacy on job site @ 12 30 P.M also
again @ 10 30 PM.

Figure 15-6. *(continued).*

pated; the health and safety of the workers and nearby populations is continually an important consideration. Government regulators are constantly reviewing the work. Throughout the course of the work, the contractor, including all employees and subcontractors, are bound by strict confidentiality limits on the information they may disclose. All aspects of the work must be documented to provide evidence of the work completed and the manner in which it was done. The public is especially sensitive to the effects of the work taking place. Constantly, too, there are financial pressures upon the contractor. All of these demands must be met.

Remediation of a hazardous waste site is a challenge. In order to meet the goals of the remedial alternative adequately, the contractor implementing the remediation plan must possess technical competence, awareness of the regulatory basis for the remediation, and an ability for rigorous management. Satisfactory remediation work returns a profit to the contractor; it also removes a liability from the responsible party, and it eliminates a threat to the environment and to any nearby human populations.

Index